"This is an important book, which is at the same time a great read. Michael Shermer digs into the American religious psyche with devastating logic and intensity. . . . Too often politeness (or cowardice) prevents people from asking questions or from expressing dissent. No such barriers stand in the way of Shermer's acute intellect, the more powerful since he so obviously cares about the issues on which he writes. I love his discussions of God and morality and when I disagree, I simply want to argue the more."

—Michael Ruse, author of *Taking Darwin Seriously* and professor of philosophy and zoology, University of Guelph

"Those who approach this intriguing and informative book with a receptive mind will come away with a much deeper appreciation for the wonderful interplay of biology and culture that makes us who we are—perhaps unique creatures in the universe."

—Donald Johanson, director of the Institute of Human Origins and author of *From Lucy to Language*

"The book will convince and delight all who are not chronically averse to opening their minds and thinking for themselves."

—Richard Dawkins, author of *Unweaving the Rainbow*

# HOW WE BELIEVE

# HOW WE BELIEVE

*Science, Skepticism, and the Search for God*

SECOND EDITION

MICHAEL SHERMER

A W. H. FREEMAN / OWL BOOK
Henry Holt and Company New York

All artwork and illustrations, except as noted on pp. 312–313, are by Pat Linse, art director of *Skeptic* magazine, are copyrighted by Pat Linse, and are reprinted with permission.

For further information on the Skeptics Society and *Skeptic* magazine, and to contact the author: P.O. Box 338, Altadena, CA 91001; 626/794-3119; Fax: 626/794-1301; Web page address: www.skeptic.com and e-mail address: skepticmag@aol.com. To subscribe (free) to the Skeptic Mag Internet Hotline send an e-mail to: join-skeptics@lyris.net.

Henry Holt and Company, LLC
*Publishers since 1866*
115 West 18th Street
New York, New York 10011



Library of Congress Cataloging-in-Publication Data
Shermer, Michael.
How we believe: science, skepticism, and the search for God / Michael Shermer.—2nd ed.
p. cm.
Includes bibliographical references and index.
ISBN 0-8050-7479-1 (pbk.)
1. Religion and science. 2. Faith and reason. I. Title.
BL240.2.S545 1999
215—dc21 99-40406
CIP

Henry Holt books are available for special promotions and premiums.
For details contact: Director, Special Markets.

First published in hardcover in 2000 by W. H. Freeman and Company

First Owl Books Edition 2002

Second Owl Books Edition 2003

A W. H. Freeman / Owl Book

*Designed by Cambraia Fernandes*

Printed in the United States of America

10 9 8 7 6 5 4 3 2 1

To **Stephen Jay Gould**

For examining God, religion, and myth as Spinoza would have it: not to ridicule, not to bewail, not to scorn, but to understand.

*Happy is the man who finds wisdom,*
*and the man who gets understanding,*
*for the gain from it is better than gain from silver*
*and its profit better than gold.*
*She is more precious than jewels,*
*and nothing you desire can compare with her.*
*Long life is in her right hand;*
*in her left hand are riches and honor.*
*Her ways are ways of pleasantness,*
*and all her paths are peace.*
*She is a tree of life to those who lay hold of her;*
*those who hold her fast are called happy.*

*—Proverbs 3:13-18*

# CONTENTS

Preface: The God Question — xiii
*A Moral Dilemma for Dr. Laura*

Introduction to the Paperback Edition — xxiii

**PART I. GOD AND BELIEF — 1**

Chapter **1.** Do You Believe in God? — 2
*The Difference in Our Answers and the Difference It Makes*

Chapter **2.** Is God Dead? — 16
*Why Nietzsche and* Time *Magazine Were Wrong*

Chapter **3.** The Belief Engine — 32
*How We Believe*

Chapter **4.** Why People Believe in God — 59
*An Empirical Study on a Deep Question*

Chapter **5.** O Ye of Little Faith — 89
*Proofs of God and What They Tell Us about Faith*

**PART II. RELIGION AND SCIENCE — 125**

Chapter **6.** In a Mirror Dimly, Then Face to Face — 126
*Faith, Reason, and the Relationship of Religion and Science*

Chapter **7.** The Storytelling Animal — 142
*Myth, Morality, and the Evolution of Religion*

Chapter **8.** God and the Ghost Dance — 171
*The Eternal Return of the Messiah Myth*

Chapter **9.** The Fire That Will Cleanse — 191
*Millennial Meanings and the End of the World*

Chapter **10.** Glorious Contingency — 214
*Gould's Dangerous Idea and the Search for Meaning in the Age of Science*

Afterword to the Second Edition — 239
*God on the Brain*

Appendix **I.** What Does It Mean to Study Religion Scientifically? Or, How Social Scientists "Do" Science 263

Appendix **II.** Why People Believe in God—The Data and Statistics 267

A Bibliographic Essay on Theism, Atheism, and Why People Believe in God 280

Notes 285

Credits 312

Index 315

# PREFACE

## The God Question
### A Moral Dilemma for Dr. Laura

Not long after I set out to write this book, I received a fax from a subscriber to the magazine I publish, *Skeptic,* who had just finished reading the most recent issue (Vol. 5, No. 2) devoted to "The God Question." This volume of the magazine addressed the various theological, philosophical, and scientific arguments for God's existence, Einstein's views on God, skeptic Martin Gardner's belief in God, arguments for and against immortality, and the decline of atheism in America. The correspondent, however, was not writing about any specific article, but about the qualifications of one of the members of *Skeptic*'s board of advisors. "I would love to know what qualifies a person to be on your editorial board," the letter began. "If he were interested would Rev. Pat Robertson qualify? I consider myself to be an atheist, a skeptic, and a semiprofessional talk show listener. In the latter capacity I have had many occasions to listen to one Dr. Laura Schlessinger, a member of your board." The letter went on to chronicle Schlessinger's reliance on the Bible as her authority for resolving moral dilemmas presented to her by her callers on her radio program. "I didn't know that skeptics relied on authority to settle disagreements over morality," the letter concluded.

This was not the first correspondence we received concerning Laura Schlessinger's position on our board of advisors. Throughout 1996 and 1997 we were sent a couple of dozen critical letters, faxes, and e-mails, and for a couple of weeks in mid-1997, on a skeptics Internet discussion group, a debate ensued about Schlessinger's involvement in the skeptical movement. We explained that membership or involvement in any capacity with the Skeptics Society and *Skeptic* magazine is not exclusionary. We could not care less what anyone's religious beliefs are. In fact, at least two of our more prominent supporters—the comedian and songwriter Steve Allen and the mathematician and essayist Martin Gardner—are believers in God. Other members of the board may believe in God as well. I do not know. I have never asked.

The primary mission of the Skeptics Society and *Skeptic* magazine is the investigation of science and pseudoscience controversies, and the promotion of critical thinking. We investigate claims that are testable or examinable. If someone says she believes in God based on faith, then we do not have much to say about it. If someone says he believes in God and he can *prove* it through rational arguments or empirical evidence, then, like Harry Truman, we say "show me." Some Christians claim that the Shroud of Turin proves that Jesus lived and was crucified and resurrected. But the shroud was carbon-14 dated and found to be a fourteenth-century hoax (some are now claiming that the dating process was contaminated and that the shroud may be older still, but these claims have never been corroborated in peer-reviewed journals). Some creationists claim that geology proves that the Earth was created only 10,000 years ago. But strict scientific dating techniques show that the Earth is billions of years old. Similarly, some physicists and cosmologists claim that the laws of nature, the configuration of atoms, and the structure of the universe prove it was all created by a supernatural being. But science continues to show that everything from the simplest atoms to the most complex galaxies is explicable by natural laws, historical contingencies, and rules of self-organized complexity.

If, in the process of learning how to think scientifically and critically, someone comes to the conclusion that there is no God, so be it—but it is not our goal to convert believers into nonbelievers. From considerable personal experience I can attest to the futility of trying to either prove or disprove God (see Chapter 5). In any case, the process would seem inherently impossible since, by admission of nearly all religions, belief in God rests on faith and means suspending the requirements of proof and logic. When people say they believe in God because it comforts them, because they have faith, because of personal revelation, or just because it "works" for them, I have no qualms with these reasons. But when others say they can *prove* God, *prove* that their religion is the right one, *prove* that we cannot be moral without God, and so forth, such claims demand a scientific and rational analysis. Of course, if the answers to these arguments are as obvious and clear-cut as both theists and atheists think, then why do such debates continue, even in the hallowed halls of theological seminaries and universities, where presumably the question of God's existence would have been resolved by now? It has not. That fact alone tells us something about the nature of the subject. So, I would have thought that any member of the Skeptics Society would be aware that God's existence, from a scientific and rational perspective, remains an open question—it cannot be "proved" one way or the other. But I also would have imagined that Schlessinger, who is both highly intelligent and well educated, would have been equally aware and sensitive to this issue.

For those who may still be unfamiliar with her, Laura Schlessinger, best known as "Dr. Laura," is the star and host of the top-rated nationally syndicated radio talk show program, running neck and neck with Rush Limbaugh for the number-one spot. She is in virtually every radio market in America, has millions of listeners, receives on average upwards of 60,000 calls per day, and in 1997 her program was sold for a staggering $71 million. Her books, *Ten Stupid Things Women Do to Mess Up Their Lives, Ten Stupid Things Men Do to Mess Up Their Lives, How Could You Do That?,* and *The Ten Commandments,* were all national bestsellers in both hardback and paperback. She draws huge crowds for her speaking engagements. "Dr. Laura" mugs, T-shirts, and newsletters are promoted daily through the radio program and an ever-growing mailing list of fans. She has been featured on several national newsmagazine and morning programs, and was even satirized in *Playboy* magazine. This is all to say that Laura Schlessinger is hugely influential. When she speaks, people listen.

We invited Laura to be on our board of advisors in 1994 when she took a skeptical stance about the recovered-memory movement and other "victimization" groups. We admired her courage to make a public statement against what in hindsight turned out to be a bad chapter in the history of psychology. At the time it was a very dangerous thing to denounce. (The "recovery" of distant memories of childhood sexual abuse usually turns out to be nothing more than the planting of "false" memories in patients by well-meaning but irresponsible therapists.) We invited Laura to speak for the Skeptics Society at Caltech. For nearly three hours (and without notes) she paced back and forth across the stage, educating and entertaining a sizeable audience. She was brilliant and funny. Most of all she was controversial. Schlessinger promotes critical thinking, independence of thought, self-reliance, and other attributes certainly admired by most free thinkers, humanists, and skeptics. Although she publicly ratcheted up the intensity of her religious convictions through 1996 and 1997 (when she converted to Judaism), and critical letters came pouring into *Skeptic*'s office, we continued to defend her because as a general principle we do not believe in excluding people from organizations based on their religious beliefs.

Imagine my surprise, then, when we received another fax four days later from Laura Schlessinger herself, who had just finished reading the issue of *Skeptic* entitled "The God Question." The fax read:

> *Please remove my name from your Editorial Board list published in each of your* Skeptic *Magazine issues immediately. Science can only describe what; guess at why; but cannot offer ultimate meaning. When man's limited intellect has the arrogance to pretend an ability to analyze God, it's time for me to get off that train.*

A voice-mail message followed, reinforcing the seriousness of her resignation. Amazed at this conjunction of ironic events, I called Laura at her home that same morning and spoke with her at length. She made it clear and in no uncertain terms (as Laura does with such effectiveness on her radio show) that she was "offended" by our issue and that God was off limits to human reason and inquiry. There is a God. Period. End of discussion. I pointed out that we had gone out of our way *not* to offend, and that, in fact, the arguments and critiques that we presented came from some of the greatest theologians and philosophers over the past two thousand years. Arrogant all, she responded. God is not open for analysis. But *which* God, I inquired? There is only one God, she explained—the God of Abraham (she clarified this to mean monotheism—Christianity and Islam included—not just Judaism). But what about the Problem of Evil and the Problem of Free Will, I asked. Laura's rapid-fire answers to these timeless problems of theology told me that this was not the first time she had spoken about them.

Our conversation wove in and out of a number of deep philosophical and moral issues—issues Laura had clearly contemplated for much of her adult life. In her twenties, she admitted, she was an atheist, not unhappy but certainly not a fulfilled individual. But now she is a theist and claims to have found not only greater happiness but also completeness as an individual. She said she can now stand on moral terra firma, the very basis of her radio success. In fact, she has essentially shifted her radio show emphasis from psychological advice to moral counseling—callers are now instructed to preface their question with "my moral dilemma is this. . . ." Laura then helps them resolve the dilemma, often with sage advice from the good book.

Dr. Laura believes in God. More than that, she *knows* God exists. Her level of doubt must be as close to zero as a belief system can get. And she is not alone. In fact, most believers in God stand very firm in the conviction of their belief. Why?

## GENESIS TO REVELATION

Why people believe in God is a specific subject I address to get at a deeper one: how we believe. If this were just a generic book on the psychology of belief systems, however, there would be little concern for controversy or emotional reaction. But this book is more than that, a lot more. So my moral dilemma is this: *How can we have a dialogue about the God Question and keep our emotions in check?* As I will explain in the next chapter, I am an agnostic who has no ax to grind with believers, and I hold no grudge against religion. My only beef with believers is when they claim they can use science and reason to *prove* God's existence, or that *theirs* is the One True Belief; my only gripe with religion is when it becomes intolerant of other peoples' beliefs, or when it becomes a tool of political oppression, ideological extremism, or the cultural suppression of diversity. I am unabashedly interested in understanding how and why any of us come to our beliefs, how and why religion evolved as the most powerful institution in human history, and how and why belief (or lack of) in God develops and shapes our thoughts and actions. One prominent scientist told me "you have a rather conciliatory attitude toward religion," and after reading an early draft of this book noted: "You seem to be saying it is okay for people to believe in God." Of course, whether I say it is okay for people to believe in God or not, they will believe (or not) regardless. My primary focus in addressing readers is not whether they believe or disbelieve, but *how* and *why* they have made their particular belief choice. Within the larger domain of how we believe, I am mainly interested in three things: (1) Why people believe in God; (2) the relationship of science and religion, reason and faith; and (3) how the search for the sacred came into being and how it can thrive in an age of science.

The intellectual and spiritual quest to understand the universe and our place in it is the foundation of the God Question, the various answers to which are explored in the first chapter of this book, including theist, agnostic, nontheist, and atheist, along with the differences these positions make in our thinking about the question. At the beginning of the twentieth century social scientists predicted that belief in God would decrease by the end of the century because of the secularization of society. In fact, as the second chapter shows, the opposite has occurred. Never in history have so many, and such a high percentage of the population, believed in God. Not only is God not dead, as Nietzsche proclaimed, but he has never been more alive. To find out why, the "Belief Engine" is considered in the third chapter as the mechanism by which any of us come to believe in anything, including and especially the magical thinking that leads millions of people to believe in psychics and mediums who claim that they can talk to the dead in heaven. To get at the core of the God Question and why people believe, the fourth chapter presents the results of an empirical study that asked a random sampling of the population that very question. The results were most enlightening, not only in the reasons people give for belief in God (made especially poignant when contrasted with why we think *other* people believe in God) but also in the quality and depth of the answers given (often in multipage, single-spaced typed letters), showing that the God Question is one of the most compelling any of us can ask ourselves.

It turns out that the number-one reason people give for why they believe in God is a variation on the classic cosmological or design argument: The good design, natural beauty, perfection, and complexity of the world or universe compels us to think that it could not have come about without an intelligent designer. In other words, people say they believe in God because the evidence of their senses tells them so. Thus, contrary to what most religions preach about the need and importance of faith, most people believe because of reason. So the fifth chapter reviews the various proofs of God, from those presented by medieval philosophers to those proffered by modern creationists, and considers what these arguments, and their employment in the service of religious belief, tell us about faith.

This relationship between science and religion, reason and faith, the subject of the sixth chapter, has once again emerged to the forefront of cultural importance due to a conjuncture of events, including the millennium that beckons us to reconsider the meaning of the past and future and the relative roles of science and religion in history; the discovery by physicists that the universe is more finely tuned and delicately balanced than we ever realized; the magnificent photographs of the universe made by the Hubble Space Telescope revealing an almost spiritual beauty of the cosmos as never seen before; and the issuance by the Pope of two statements, one acknowledging the validity of the theory of evolution and the other endorsing the successful marriage of *fides et ratio,* faith and reason.

Because humans are storytelling animals, a deeper aspect of the God Question involves the origins and purposes of myth and religion in human history and culture, the subject of the seventh, eighth, and ninth chapters. Why is there an eternal return of certain mythic themes in religion, such as messiah myths, flood myths, creation myths, destruction myths, redemption myths, and end of the world myths? What do these recurring themes tell us about the workings of the human mind and culture? What can we learn from these myths beyond the moral homilies offered in their narratives? What can we glean about ourselves as we gaze into these mythic mirrors of our souls?

Not only are humans storytelling animals, we are also pattern-seeking animals, and there is a tendency to find patterns even where none exists. To most of us the patterns of the universe indicate design. For countless millennia, we have taken these patterns and constructed stories about how our cosmos was designed specifically for us. For the past few centuries, however, science has presented us with a viable alternative in which we are but one among tens of millions of species, housed on but one planet among many orbiting in an ordinary solar system, itself one among possibly billions of solar systems in an ordinary galaxy, located in a cluster of galaxies not so different from billions of other galaxy clusters, themselves whirling away from one another in an expanding cosmic bubble that very possibly is only one among a near-infinite number of bubble universes. Is it really possible that this entire cosmological multiverse exists for one tiny subgroup of a single species on one planet in a lone galaxy in that solitary bubble universe? The final chapter explores the implications of this scientific worldview and what it means to fully grasp the nature of contingency—what if the universe and the world were not created for us by an intelligent designer, and instead is just one of those things that happened? Can we discover meaning in this apparently meaningless universe? Can we still find the sacred in this age of science?

To help me answer these questions a number of people have been highly influential in my thinking and writing, both directly and indirectly. The ultimate genesis of my beliefs, as it is for all of us of course, is parental, so I thank my mother, Lois, my stepfather, Dick, my late father, Richard, and my stepmother, Betty, for raising me in an atmosphere open and uncritical toward both religious and secular beliefs; I truly had a free choice in the matter, as it should be for all children. For introducing me to Christianity in my youth I thank the Oakleys: George, Marilyn, George, and Joyce (though they are not to be blamed for my subsequent fall from grace). At Glendale College Professor Richard Hardison was especially effective in helping me think clearly about philosophy and theology, particularly with regard to reason and faith; and at Pepperdine University Professor Tony Ash's courses on Jesus the Christ and the writings of C. S. Lewis awakened me to the depth and seriousness of Christian theology and apologetics. The primary credit (or blame, depending on your perspective) for my turn toward science and secular humanism in graduate school goes to Professors Bayard Brattstrom, Meg White, and Doug Navarick at the California State University–Fullerton, whose passion for science made me realize that no religion could come close to the epic narratives told by cosmologists, evolutionary biologists, and social scientists about the origins and evolution of the cosmos, life, behavior, and civilization.

Over the past two decades countless conversations with hundreds of people have helped me sort out some answers to these deep religious and philosophical questions, but those most directly affecting the development of this book include *Skeptic* magazine editors and board members David Alexander, Tim Callahan, Napoleon Chagnon, Gene Friedman, Nick Gerlich, Penn Jillette, Gerald Larue, Bernard Leikind, Betty McCollister, Tom McDonough, Sara Meric, Richard Olson, Donald Prothero, Vincent Sarich, Jay Snelson, Carol Tavris, Teller, and Stuart Vyse. As always I acknowledge the support of the Skeptics Society and *Skeptic* magazine provided by Dan Kevles, Susan Davis, and Chris Harcourt at the California Institute of Technology; Larry Mantle, Ilsa Setziol, Jackie Oclaray, and Linda Othenin-Girard at KPCC 89.3 FM radio in Pasadena; Stan Hynds and Linda Urban at Vroman's bookstore in Pasadena; as well as those who help at every level of our organization, including Jane Ahn, Jaime Botero, Jason Bowes, Jean Paul Buquet, Bonnie Callahan, Cliff Caplan, Randy Cassingham, Amanda Chesworth, Shoshana Cohen, John Coulter, Brad Davies, Clayton Drees, Janet Dreyer, Bob Friedhoffer, Jerry Friedman, Sheila Gibson, Michael Gilmore, Tyson Gilmore, Steve Harris, Andrew Harter, Laurie Johansen, Terry Kirker, Diane Knudtson, Joe Lee, Tom McIver, Dave Patton, Rouven Schaefer, Brian Siano, and Harry Ziel.

I am especially grateful for the additional input provided by my agents Katinka Matson and John Brockman, my editor John Michel and my publicist Sloane Lederer at W. H. Freeman and Company (as well as Diane Maass, Peter McGuigan, and all the folks in production at this fine publishing house); as well as Louise Ketz and Simone Cooper; and for taking the time to read individual chapters or provide valuable feedback on my thinking I thank Richard Abanes, Michele Bonnice, Richard Dawkins, Jared Diamond, Richard Elliott Friedman, Ursula Goodenough, Alex Grobman, Donald Johanson, Elizabeth Knoll, J. Gordon Melton, Massimo Pigliucci, Michael Ruse, Eugenie Scott, Nancy Segal, Frank Tipler, Bob Trivers, Edward O. Wilson, and Rabbi Edward Zerin. Bruce Mazet and Frank Miele both went above and beyond the call of duty to both critique and support my efforts to grasp the deeper meaning of the God Question; and James Randi, as always, serves as inspiration requiring perspiration to keep up with his tireless efforts to keep us on our intellectual toes.

The influence of Frank Sulloway on my thinking is immeasurable, but not his effect on this book, especially Chapter 4 and our corroboration on the study of religious attitudes, which can be measured precisely and significantly at three sigmas above the mean. I am also deeply appreciative of my Skeptics Society partner Pat Linse, not only for her brilliant artwork and design of *Skeptic* magazine and for preparing all of the illustrations for this book, but also for the conversations on God and religion that have kept in check my occasional paroxysms of irritations with religion.

Finally, I thank Kim for being my wife, confidante, and best friend who has refereed the countless wrestling matches that go on in my mind about the timeless questions that concern us all; and Devin (although she had no choice in the matter) for being my daughter, joy, and source of mind-cleansing play so necessary to get rid of the cognitive clutter that goes with research and writing.

When we began the Skeptics Society and *Skeptic* magazine in 1992 we adopted a quote from the seventeenth-century philosopher and religious thinker Baruch Spinoza: “I have made a ceaseless effort not to ridicule, not to bewail, not to scorn human actions, but to understand them.” When it comes to religion it is especially difficult for any of us to apply this principle consistently. But if we do, the moral dilemma of how to discuss the God Question without offense may be resolved. As my friend and colleague Stephen Jay Gould told me: “You cannot understand the human condition without understanding religion or religious arguments.”

I hope that this book in some small way adds to our understanding of the human condition.

# Introduction to the Paperback Edition

## The Gradual Illumination of the Mind

### Reconsiderations and Recapitulations on the God Question

*It appears to me (whether rightly or wrongly) that direct arguments against christianity and theism produce hardly any effect on the public; and freedom of thought is best promoted by the gradual illumination of men's minds which follows from the advance of science.*

—*Charles Darwin*

In the first edition of this book I wrote on page xv of the Preface: "Of course it is okay for people to believe in God; moreover, people will believe regardless of what I say or think." The second part of this sentence can be verified, since God's ratings have not slipped in the polls one iota since *How We Believe* was first released in October 1999. In fact, a March 2000 poll from Gallup shows that, as always, belief in God remains potent, not only in America but worldwide:

—*Belief in God:* Even when Gallup added the option for respondents that they "don't believe in God, but believe in a universal spirit or higher power," only eight percent chose that response with 86 percent saying that they believe in God. Gallup added: "In fact, only five percent of the population choose neither of these choices and thus claim a more straightforward atheist position."

—*Church Attendance:* Although less than the percentage of people who believe in God, "about two-thirds of the population claim to attend services at least once a month or more often," Gallup said, while "thirty-six percent say they attend once a week." By contrast, only 8 percent say they never attend religious services, while 28 percent report that they "seldom" go.

—*Church Membership:* Matching the figures for church attendance, two-thirds of Americans say they are members of a church or some other religious institution. "Only nine percent of the public respond with 'none' when asked to identify a religious affiliation or preference," Gallup concluded.

—*Importance of Religion:* Americans match people in other countries in ranking religion as very or fairly important in their lives. In a joint study between Gallup International and the London-based Taylor Nelson Sofres marketing firm covering 60 countries, 87 percent said that they consider themselves to be part of some religion. In America 60 percent say that religion is "very important in their life," with another 30 percent saying that it is "fairly important."

—*God and Politics:* Since 2000 is a presidential election year, Gallup found that 52 percent of voters surveyed "would be more likely to vote for a candidate for president who has talked about his or her personal relationship with Jesus Christ during debates and news interviews." As anyone who watched the presidential debates knows, all the candidates went on public record to extol their Christian beliefs, including the Democratic candidate Al Gore, not exactly known for his conservatively religious views. On the Republican side, George W. Bush announced that he considered Jesus to be the most influential philosophical thinker in his life.

## SKEPTICISM AS A VIRTUE

Is it okay for people to believe in God? A number of atheists objected to this statement. One wrote me: "Religion is a bad idea. Belief in god is a bad idea. These ideas should be self-evident to any rationalist. That religion/belief is common is not a reason to avoid such statements. That religion/belief will perhaps always be with us is not a reason. That religion/belief is old is not a reason. That religion/belief may at times do some good is not a reason. None of these statements are reasons to avoid clearly stating the truth. Anything less is duplicitous, disingenuous, appeasing—and ultimately, helps the other side by providing approval where disapproval should instead be offered."

"The other side." What a revealing way to phrase a critical attitude toward religion, whose long history of dividing the world between "our side" and the "other side" is a notoriously bloody one. Should nonbelievers really ape this most nonsalubrious side of the system of belief from which they so often distance themselves? Clearly religion has no monopoly here. The very propensity to cleave nature into unambiguous yeses and noes may very well be an evolutionary by-product whose ultimate outcome could result in the extinction of the species (and a further indication that not everything in evolution can be explained by its adaptive significance).

Another friend who objected to my "okay to believe" statement spelled it out even clearer: "I won't let anyone who believes in god in my home. I won't sleep with them and I have none in my social circle. But I can do more." What "more" shall we do? What more can we do? Should we evangelize against Christianity, Judaism, Islam, Hinduism, Buddhism, and the other systems of religious belief? Since I am a libertarian in more ways than just political, I am disinclined to tell people what they should or should not be doing with their personal lives and beliefs. Nevertheless, I am a scientific and skeptical activist (not just a dispassionate onlooker from the intellectual sidelines), so I am forced on a daily basis to attempt to dissuade people from their less rational beliefs. How to reconcile these competing motives? Through a positive push-forward program instead of a negative push-back agenda. Evangelize for science rather than rail against religion. Don't curse the darkness; light a candle. Charles Darwin, who renewed the science-religion debate nearly a century and a half ago, expressed this position well in the epigraph above.

Nevertheless please note that in this edition of the book I changed the phrase to read that it is okay not to believe in God. By this statement I am speaking to those atheists, nontheists, and nonbelievers of all stripes as a form of validation from a fellow free-thinker; I am also reaching out to theists and believers of all faiths who, occasionally or even frequently, doubt their faith. Doubt is good. Questioning belief is healthy. Skepticism is okay. It is more than okay, in fact. Skepticism is a virtue and science is a valuable tool that makes skepticism virtuous. Science and skepticism are the best methods of determining how strong your convictions are, regardless of the outcome of the inquiry. If you challenge your belief tenets and end up as a nonbeliever, then apparently your faith was not all that sound to begin with and you have improved your thinking in the process. If you question your religion but in the end retain your belief, you have lost nothing and gained a deeper understanding of the God Question. It is okay to be skeptical.

In light of Darwin's wise advice, why, one may ask, do I devote an entire chapter (2) to a head-on confrontation of the alleged proofs of God's existence? The reason is that my laissez-faire attitude toward other people's religious beliefs ends when they use, misuse, and abuse reason and science in the service of faith and religion. As even libertarians will admit, your freedom to swing your fist ends at my nose.

Claims that religious tenets can be proved through science require a response from the scientific community. Making evidentiary claims puts religion on science's turf, so if it wants to stay there it will have to live up to the standards of scientific proof. This is not an archaic academic or philosophical issue. As I show in Chapter 4, the scientistically based "design argument" is the most common one made. People say they believe in God because of the evidence of their senses and their understanding of how the world works. In other words, they give reasons for their beliefs. What are those reasons? If they are good reasons shouldn't we all become believers?

## GOD AND THE INTELLIGENTLY DESIGNED UNIVERSE

The hottest area in the search for scientific support of God's existence can be found in the so-called "new creationism" that deals in "irreducible complexity" and especially "Intelligent Design" (ID as it is known among its adherents). Although I discuss these at length in Chapter 5, they continue to generate so much attention that it is worth expanding on it more here. It is rapidly becoming the strongest scientistic argument for believers. For example, I participated in two scientific debates on ID in 2000, a number of new books on it have been released by Christian publishers since my book came out, and an entire issue of the Christian magazine *Touchstone* was devoted to Intelligent Design, "a new paradigm in science that could revolutionize the way we view creation, the cosmos, and ourselves."

Much is made of the fact that the universe is grandly complex, intricate, and apparently delicately balanced for carbon-based life forms such as ourselves. It is here where science and religion meet, say believers who wish to graft the findings of science onto 4,000-year-old religious doctrines. And they have no difficulty in finding observations from leading scientists that seemingly support their contention that the universe does not just look designed, it is designed. "It is not only man that is adapted to the universe, " John Barrow and Frank Tipler proclaim in *The Anthropic Cosmological Principle,* "The universe is adapted to man. Imagine a universe in which one or another of the fundamental dimensionless constants of physics is altered by a few percents one way or the other? Man could never come into being in such a universe. That is the central point of the anthropic principle. According to the principle, a life-giving factor lies at the center of the whole machinery and design of the world." For theists, of course, that life-giving factor is God.

The Templeton Foundation has spent tens of millions of dollars promoting a reconciliation between science and religion, including the grant of the single largest cash prize in history for "progress in religion." On the day I wrote this introduction, in fact, it was announced that physicist Freeman Dyson won the prize valued at $964,000, for such works as *Disturbing the Universe,* one passage of which is often quoted by ID theists: "As we look out into the universe and identify the many accidents of physics and astronomy that have worked to our benefit, it almost seems as if the universe must in some sense have known that we were coming." Mathematical physicist Paul Davies also won the Templeton prize, and we can understand why in such passages as this from his 1999 book *The Fifth Miracle*:

> *In claiming that water means life, NASA scientists are . . . making—tacitly—a huge and profound assumption about the nature of nature. They are saying, in effect, that the laws of the universe are cunningly contrived to coax life into being against the raw odds; that the mathematical principles of physics, in their elegant simplicity, somehow know in advance about life and its vast complexity. If life follows from [primordial] soup with causal dependability, the laws of nature encode a hidden subtext, a cosmic imperative, which tells them: "Make life!" And, through life, its by-products: mind, knowledge, understanding. It means that the laws of the universe have engineered their own comprehension. This is a breathtaking vision of nature, magnificent and uplifting in its majestic sweep. I hope it is correct. It would be wonderful if it were correct.*

Indeed, it would be wonderful. But not any more wonderful than if it were not correct. If life on Earth is unique, or at least exceptionally rare (and in either case certainly not inevitable, as I demonstrate in the final chapter), how special is our fleeting Mayfly-like existence; how important it is that we make the most of our lives and our loves; how critical it is that we work to preserve not only our own species, but all species and the ecosystem itself. Whether the universe is teaming with life or we are alone, whether our existence is strongly necessitated by the laws of nature or it is highly contingent, whether there is more to come or this is all there is, either way we are faced with a worldview that is equally breathtaking and majestic in its sweep across time and space.

In the *Touchstone* issue on Intelligent Design, Whitworth College philosopher Stephen Meyer argues that ID is not simply a "God of the gaps" argument to fill in where science has yet to give us a satisfactory answer. It is not just a matter of "we don't understand this so God must have done it" (although to me, and to all scientists I have spoke to about ID, this is how these arguments always appear). ID theorists like Meyer and Phillip Johnson, William Dembski, Michael Behe, and Paul Nelson (all leading IDers and contributors to this issue) say they believe in ID because the universe really *does* appear to be designed. "Design theorists infer a prior intelligent cause based upon present knowledge of cause-and-effect relationships," Meyer writes. "Inferences to design thus employ the standard uniformitarian method of reasoning used in all historical sciences, many of which routinely detect intelligent causes. Intelligent agents have unique causal powers that nature does not. When we observe effects that we know only agents can produce, we rightly infer the presence of a prior intelligence even if we did not observe the action of the particular agent responsible." Even an atheist like Stephen Hawking can be found to present cosmological arguments seemingly supportive of scientistic arguments for God's existence:

> *Why is the universe so close to the dividing line between collapsing again and expanding indefinitely? In order to be as close as we are now, the rate of expansion early on had to be chosen fantastically accurately. If the rate of expansion one second after the big bang had been less by one part in $10^{10}$, the universe would have collapsed after a few million years. If it had been greater by one part in $10^{10}$, the universe would have been essentially empty after a few million years. In neither case would it have lasted long enough for life to develop. Thus one either has to appeal to the anthropic principle or find some physical explanation of why the universe is the way it is.*

That explanation, at the moment, is a combination of a number of different concepts revolutionizing our understanding of evolution, life, and cosmos, including the possibility that our universe is not the only one. We may live in a multiverse in which our universe is just one of many bubble universes all with different laws of nature. Those with physical parameters like ours are more likely to generate life than others. But why should any universe generate life at all, and how could any universe do so without an intelligent designer? The answer can be found in the properties of *self-organization* and *emergence* that arise out of what are known as complex adaptive systems, or complex systems that grow and learn as they change. Water is an emergent property of a particular arrangement of hydrogen and oxygen molecules, just as consciousness is a self-organized emergent property of billions of neurons. The entire evolution of life can be explained through these principles. Complex life, for example, is an emergent property of simple life: simple prokaryote cells self-organized to become more complex units called eukaryote cells (those little organelles inside cells you had to memorize in beginning biology were once self-contained independent cells); some of these eukaryote cells self-organized into multi-cellular organisms; some of these multicellular organisms self-organized into such cooperative ventures as colonies and social units. And so forth. We can even think of self-organization as an emergent property, and emergence as a form of self-organization. How recursive. No Intelligent Designer made these things happen. They just happened on their own. Here's a bumper sticker for evolutionists: Life Happens. In *The Life of the Cosmos,* cosmologist Lee Smolin explains how this property of emergence and self-organization out of complexity works:

> *It seems to me quite likely that the concept of self-organization and complexity will more and more play a role in astronomy and cosmology. I suspect that as astronomers become more familiar with these ideas, and as those who study complexity take time to think seriously about such cosmological puzzles as galaxy structure and formation, a new kind of astrophysical theory will develop, in which the universe will be seen as a network of self-organized systems. Many of the people who work on complexity . . . imagine that the world consists of highly organized and complex systems but that the fundamental laws are simply fixed beforehand, by God or by mathematics. I used to believe this, but I no longer do. More and more, what I believe must be true is that there are mechanisms of self-organization extending from the largest scales to the smallest, and that they explain both the properties of the elementary particles and the history and structure of the whole universe.*

There may even be a type of natural selection at work among many universes, with those whose parameters are like ours being most likely to survive. Those universes whose parameters are most likely to give rise to life occasionally generate complex life with brains big enough to achieve consciousness and to conceive of such concepts as God and cosmology, and to ask such questions as Why?

## REASONS TO BELIEVE

Self-organization, emergence, and complexity theory form the basis of just one possible natural explanation for how the universe and life came to be the way it is. But even if this explanation turns out to be wanting, or flat-out wrong, what alternative do Intelligent Design theorists offer in its stead? If ID theory is really a science, as they claim it is, then what is the mechanism of how the Intelligent Designer operated? ID theorists speculate that four billion years ago the Intelligent Designer created the first cell with the necessary genetic information to produce all the irreducibly complex systems we see today. But then, they tell us, the laws of evolutionary change took over and natural selection drove the system, except when totally new and more complex species needed creating. Then the Intelligent Designer stepped in again. Or did He (She? It?)? They are not clear. Did the Intelligent Designer—let's call it ID—create each genus and then evolution created the species? Or did ID create each species and evolution created the subspecies? ID theorists seem to accept natural selection as a viable explanation for microevolution—the beak of the finch, the neck of the giraffe, the varieties of subspecies found in most species on earth. If ID created these species why not the subspecies? And how did ID create the species? We are not told. Why? Because no one has any idea but you can't just say, "God did it."

I presented all these challenges to the leading Intelligent Design theorists at a June 2000 conference at Concordia University (Wisconsin) on "Intelligent Design and Its Critics." Although there were some critics there, both on stage and in the audience, it was mostly populated by ID supporters. The conference was partially sponsored by the Templeton Foundation, and was clearly structured to make it appear that there is a real scientific debate ongoing about Intelligent Design. However, as I pointed out in my opening remarks, the conference was being held at a Lutheran college and just before I was introduced they announced what time chapel was the next morning and how we can obtain transportation to it. Virtually every ID supporter turns out to be a born-again Christian. Can this really be a coincidence? For these remarks I was later accused of committing the "genetic fallacy," where one attacks the person rather than their arguments. Nevertheless, my participation at this conference was a debate in which I did address many of their points.

It is not coincidental that ID supporters are almost all Christians. It is inevitable. ID arguments are reasons to believe if you already believe. If you do not, the ID arguments are untenable. But I would go further. If you believe in God, you believe for personal and emotional reasons (as I show in Chapter 4), not out of logical deductions. But this chapter also shows that highly educated believers, especially men who were raised religious, have a strong tendency to defend their beliefs with rational arguments. And looking out over an auditorium of about 250 ID supporters at this debate it was overwhelmingly educated males.

ID theorists also attack scientists' underlying bias of "methodological naturalism." That is, they feel it is not fair to forbid supernaturalism from the equation as it pushes them out of the scientific arena on the basis of nothing more than a rule of the game. But if we change the rules of the game to allow them to play, what would that look like? How would that work? What would we do with supernaturalism? ID theorists do not and will not comment on the nature of ID. They wish to say only "ID did it." This is not unlike the famous Sidney Harris cartoon with the scientists at a chalkboard filled with equations: an arrow points to a blank spot in the series and denotes "Here a miracle happens." Although IDers eschew any such "god of the gaps" style arguments, that is precisely what it all amounts to. They have simply changed the name from GOD to ID.

Let's assume for a moment, though, that ID theorists have suddenly become curious about how ID operates. And let's say that we have determined that certain biological systems are indeed irreducibly complex and intelligently designed. As ID scientists who are now given entrée into the scientific stadium with the new set of rules that allows supernaturalism, they call a time-out during the game to announce "Here ID caused a miracle." What do we do with supernaturalism in the game of science? Do we halt all future experiments? Do we continue our research and say "Praise ID" every couple of hours? The whole system collapses in a risible game of semantics.

## GLADLY WOLDE WE LEARNE

If there is a God, He has yet to provide incontrovertible evidence of His existence, leaving belief in Him instead to lie in the realm of faith, or emotional preference, which is the very basis of the theological position known as fideism. Because I see this as the most tenable of all theistic possibilities, I have explored it further since I first wrote this book. As Martin Gardner, a fideist and believer in God, noted in his 1983 book *The Whys of a Philosophical Scrivener*: "If 'evidence' means the kind of support provided by reason and science, there is no evidence for God and immortality." Gardner rejects the flood story ("even as a myth it is hard to admire the 'faith' of a man capable of supposing God could be that vindictive and unforgiving"), does not believe that God asked Abraham to kill his son ("Abraham appears not as a man of faith, but as a man of insane fanaticism"), and finds wanting most of the stories in the Bible: "The Old Testament God, and many who had great 'faith' in him, are alike portrayed in the Bible as monsters of incredible cruelty."

If, as I argue in Chapter 4, beliefs are based on emotion rather than evidence, personality instead of reason, upbringing more than arguments, it would seem to vindicate Gardner's fideism as the most honest of all the reasons to believe in God. In a personal aside, Gardner confesses that he does have some faith:

> *Let me speak personally. By the grace of God I managed the leap when I was in my teens. For me it was then bound up with an ugly Protestant fundamentalism. I outgrew this slowly, and eventually decided I could not even call myself a Christian without using language deceptively, but faith in God and immortality remained. The original leap was not a sharp transition. For most believers there is not even a transition. They simply grow up accepting the religion of their parents, whatever it is.*

Gardner is, if nothing else, refreshingly honest about his faith: "The leap of faith, in its inner nature, remains opaque. I understand it as little as I understand the essence of a photon. Any of the elements I listed earlier as possible causes of belief, along with others I failed to list, may be involved in God's way of prompting the leap. I do not know, I do not know!"[12] We do not know either, but we ought to be able to respect this honest appraisal of how and why you believe, and especially acknowledge what Gardner, as one of the chief teachers of science and skepticism, have offered us for enlightenment on the problem. As the clerk of Oxenford in Chaucer's *Canterbury Tales* proclaimed: "gladly wolde he lerne, and gladly teche."

# HOW WE BELIEVE

# Part I

## God and Belief

R. Buckminster Fuller

*Sometimes I think we're alone.*
*Sometimes I think we're not.*
*In either case, the thought is quite staggering.*

Chapter 1

# Do You Believe in God?

## *The Difference in Our Answers and the Difference It Makes*

*The word* God *is used in most cases as by no means a term of science or exact knowledge, but a term of poetry and eloquence, a term thrown out, so to speak, as a not fully grasped object of the speaker's consciousness, —a literary term, in short; and mankind mean different things by it as their consciousness differs.*

—Matthew Arnold, *Literature and Dogma,* 1873

In my senior year of high school I accepted Jesus as my savior and became a born-again Christian. I did so at the behest of a close and trusted friend, who assured me this was the road to everlasting life and happiness. It was a Saturday night and we were sitting, ironically, at my father's monkey-wood bar, fully equipped to allow a number of guests to imbibe just about any mixed drink their imaginations could create. We read John 3:16 (now infamous for its appearance on hand-printed signs at nationally televised sporting events): "For God so loved the world, that he gave his only begotten Son, that whosoever believeth in him should not perish, but have everlasting life." At the moment of my conversion coyotes began howling outside. We took it as a sign that Lucifer was unhappy at the loss of another soul from Sheol.

The next day I attended church services with my friend and his family, and when the minister called for anyone to come forward to be saved, I went up to make it official. My friend assured me that I did not need to be saved twice, but I figured maybe it was more official at a church than at a bar. From that moment on everything seemed neatly explained by the Christian paradigm. Anytime something good happened, it was God's will and a reward for good behavior; anytime something bad happened, it was part of God's larger plan, and even though I did not at present understand the long-term benefits, these would become clear in due time. Either way it was a neat and tidy worldview—everything in its place and a place for everything.

## A LEAP OF FAITH

The whole process was premised on faith. With faith in Jesus, I now had eternal life. With faith in God, I was saved. I had found the One True Religion, and it was my duty—indeed it was my pleasure—to tell others about it, including my parents, brothers and sisters, friends, and even total strangers. In other words, I "witnessed" to people—a polite term for trying to convert them (one wag called it "Amway with Bibles"). Of course, I read the Bible, as well as books about the Bible. I regularly attended youth church groups, one in particular at a place called "The Barn," a large red house in La Crescenta, California, at which Christians gathered a couple of times a week to sing, pray, and worship. I got so involved that I eventually began to put on Bible study courses myself.

In my sophomore year at Glendale College I read Hal Lindsey's *The Late Great Planet Earth*. The front cover of the book pronounced "AMAZING BIBLICAL PROPHECIES ABOUT THIS GENERATION!" (with "OVER 2,000,000 COPIES IN PRINT!"), while the back cover asked provocatively, "IS THIS THE ERA OF THE ANTICHRIST AS FORETOLD BY MOSES AND JESUS?" My Christian friends and I began reading the newspapers to watch the millennial drama unfold as Lindsey said the Bible had predicted. I recall taking a political science course in which the professor was talking about the possibility of a European Common Market, and comparing his take on this event to Lindsey's, who claimed this is a reincarnation of the Roman Empire as prophesied in Daniel and the Book of Revelation: "We believe that the

Common Market and the trend toward unification of Europe may well be the beginning of the ten-nation confederacy predicted by Daniel and the Book of Revelation." Following this there will be "a revival of mystery Babylon," and the rise of "a man of such magnetism, such power, and such influence, that he will for a time be the greatest dictator the world has ever known. He will be the completely godless, diabolically evil 'future fuehrer.'" Skeptics beware, says Lindsey: "If this sounds rather spooky, bring your head out from under the skeptical covers and examine with us in a later chapter the Biblical basis and the current applications." I threw the covers off and devoured the book with great credulity. So did millions of others: Through the 1970s *The Late Great Planet Earth* sold 7.5 million copies, making it, according to the *New York Times Book Review* (April 6, 1980), the bestselling nonfiction book of the decade. By 1991, notes the *Los Angeles Times* (February 23, 1991), the book had reached an almost unimaginable figure of 28 million copies sold in 52 languages worldwide. Prophecy sells, especially prophecies of biblical proportions.

Taking all this fairly seriously, I transferred to Pepperdine University (affiliated with the Church of Christ) with the intent of majoring in theology. I took courses in the Old and New Testaments, on the history of the Bible, the writings of C. S. Lewis, and the historical Jesus. I stayed after class to talk to professors and visited them in their offices. I went to chapel several times a week (students were required to go twice and attendance was taken) and prayed regularly. I even told one coed that I "loved" her—in the Christian sense of loving everyone—but I am afraid she took it the wrong way. (She needn't have worried—students were prohibited from visiting the dorm rooms of members of the opposite sex, and such sin-provoking activities as dancing were forbidden.)

## A BREACH IN THE FAITH

There were problems with my conversion from the beginning, however, and I think deep down on some level I must have known it. First, my motives for converting, while sincere later, were not quite as pure at the time—my friend had a sister that I wanted to get to know better and I figured this might help. On reflection, howling coyotes are not exactly unusual, since my parents' home is nestled high up in the San Gabriel mountains of Southern California where coyotes routinely come down from the hills to rummage through trash cans. More importantly, there were chinks in the armor: Another friend at my high school told me I had chosen the wrong path and that his faith, Jehovah's Witnesses, was the One True Religion, making me wonder how another religion could be as certain it had the truth as my newfound one did. I was generally uncomfortable witnessing to people, especially strangers. And the normal sexual urges that overwhelm teenagers created intense conflict and frustration.

There were philosophical problems as well. I recall spending an afternoon with a Presbyterian minister whose deep wisdom I greatly respected, going over and over what is known as the "Problem of Free Will": If God is omniscient (all knowing) and omnipotent (all powerful), then how can we be held responsible for making "choices" we could not possibly have made? If we do have free will, does this mean God is limited in knowledge or power? And if God is limited, what else can He not do? The minister, who had a Ph.D. in theology, did his best to address the problem but it all seemed like labyrinthine word games and obfuscating analogies to me. For example: "Imagine history as one long film, which God has already seen but we, the characters in the film have not, so our actions 'seem' free even though they are predestined." Or: "God is outside of space and time so the normal laws of cause and effect do not apply to Him."

Similarly, with one of my Pepperdine professors I grappled with what is known as the "Problem of Evil": If God is omnibenevolent (all good) and omnipotent, then why is there evil in the world? If He allows evil, then He is not all good. If He cannot help but allow evil, then He is not all powerful. The best book I have read on this problem is Harold Kushner's *When Bad Things Happen to Good People*, but his solution—"God can't do everything, but he can do some important things"—is not how most people conceive of the "almighty."

To this day I have not heard an answer to the Problem of Evil that seems satisfactory. As with the Problem of Free Will, most answers involve complicated twists and turns of logic and semantic wordplay. One answer, for example, is based on a fundamental assumption of logic that no set may have itself as its own subset—God cannot create a stone so heavy that He cannot lift it. Likewise, God cannot be encompassed in the subset of evil. Evil, like heavy stones, exists independently of the larger set of God, even though remaining within that set. Another riposte involves explaining specific historical evils, like the Holocaust, where one answer is that "*humans* committed these evil acts, not God." But all this avoids the problem altogether: Either God *allowed* Nazis to kill Jews, in which case He is not omnibenevolent, or God *could not prevent* Nazis from killing Jews, in which case He is not omnipotent. In either case God is not the plenipotent Yahweh of Abraham, the King of Kings and Lord of Lords Sovereign of the Universe. Or, in explaining the death of innocent children from cancer or automobile accidents, one rejoinder is that "God has a bigger plan for us and we shall grow and learn from this experience." The problem here is that no matter what happens—good things or bad things—God's intentions can be inferred. Everything that happens is attributed to God, and this just puts us back to where we started with God either unable or unwilling to take action or prevent the evil.

At Glendale College I challenged my philosophy professor (and now my friend) Richard Hardison, to read *The Late Great Planet Earth*, believing he would see the light. Instead he saw red and hammered out a two-page, single-space typed list of problems with Lindsey's book. I still have the list, folded and tucked neatly into my copy of the book. Hardison took no prisoners. For example, where Lindsey writes, "When a prophet speaks in the name of the Lord, if the word does not come to pass or come true, that is a word which the Lord has not spoken," Hardison notes that this creates an "inevitable precision, since we disregard those prophesies that don't occur." On page 40 Lindsey explains that when reading the Bible we should "take every word at its primary, ordinary, usual, literal meaning," yet on page 50 Lindsey says that "the bones coming together and sinews and flesh being put upon them" *really* means "the regathering of the people into a physical restoration of a national existence in Palestine. Isn't it fascinating how graphic this physical analogy is?" Lindsey cannot have it both ways. On pages 55–56, Lindsey commits another logical fallacy: "Peter considered the certainty and relevance of the prophetic word to be the most important thing. He even warned that in 'the latter times' men posing as religious leaders would rise from within the Church and deny, even ridicule, the prophetic word (II Peter 2:1–3; 3:1–18). If you pass this book around to many ministers you'll find how true this prediction has become." Hardison notes that "denial of Lindsey's position is thus impossible without proving oneself to be among those misguided persons that Peter warns about. This becomes a device to make Lindsey's position nondisprovable." Hardison concluded his analysis with this biting statement:

> *Of all Lindsey's statements, the one I most want to quarrel with is found in the introduction: "There are many students who are dissatisfied with being told that the sole purpose of education is to develop inquiring minds. They want to find some of the answers to their questions—solid answers, a certain direction." I think I can*

*offer some possible explanations for this "egregious" development. But even more, I feel impelled to propose that such a student is dead.*

Hardison's analysis shook me up. I did not want to be a "dead" student in only my second year of college. So I continued reading what the great minds in history had to say about God. It was an illuminating experience that got me thinking about the concept of "believing" in God. What does it mean to believe in God or not to believe in God? Are these great questions about God's existence answerable from a scientific perspective? Can reason alone help us arrive at solutions to the moral dilemmas of our lives? In short, does religion present us with soluble problems to be analyzed with the tools of observation and logic, or are these questions too subjective and too personal for us to come to a collective agreement on a solution?

## THE ART OF THE INSOLUBLE

The British Nobel laureate Sir Peter Medawar once described science as the "art of the soluble." "No scientist is admired for failing in the attempt to solve problems that lie beyond his competence," Medawar opined. "If politics is the art of the possible, research is surely the art of the soluble." If science is the art of the soluble, religion is the art of the insoluble. God's existence is beyond our competence as a problem to solve.

This is what Thomas Huxley meant when he coined the term *agnostic* in 1869: "When I reached intellectual maturity and began to ask myself whether I was an atheist, a theist, or a pantheist . . . I found that the more I learned and reflected, the less ready was the answer. They [believers] were quite sure they had attained a certain 'gnosis,'—had, more or less successfully, solved the problem of existence; while I was quite sure I had not, and had a pretty strong conviction that the problem was insoluble." In the now-classic 1966 *Time* magazine cover story, "Is God Dead?," the editors came to the same conclusion after spending a year conducting more than 300 interviews with leading theologians from around the world:

*For one thing, every proof seems to have a plausible refutation; for another, only a committed Thomist [a follower of the theology of St. Thomas Aquinas] is likely to be spiritually moved by the realization that there is a self-existent Prime Mover [a first being that moves all others but itself does not need to be moved—see*

*Chapter 5]. "Faith in God is more than an intellectual belief," says Dr. John Macquarrie of Union Theological Seminary. "It is a total attitude of the self."*

One either takes the leap of faith or does not. Faith is the art of the insoluble.

There are many positions one can take with regard to the God Question (see the Bibliographic Essay at the end of this book for suggested readings on both the theist and atheist positions). The *Oxford English Dictionary* (*OED*), our finest source for the history of word usage, defines *theism* as implying "belief in a deity, or deities" and "belief in one God as creator and supreme ruler of the universe." *Atheism* is defined by the *OED* as "disbelief in, or denial of, the existence of a God." And *agnosticism* as implying "unknowing, unknown, unknowable." At a party held one evening in 1869, Huxley further clarified the term *agnostic*, referencing St. Paul's mention of the altar to "the Unknown God" as: "one who holds that the existence of anything beyond and behind material phenomena is unknown and so far as can be judged unknowable, and especially that a First Cause and an unseen world are subjects of which we know nothing." Belief in God is the art of the insoluble.

To clarify this linguistic discussion it might be useful to distinguish between a statement about the universe and a statement about one's personal beliefs. As a statement about the universe, *agnostic* would seem to be the most rational position to take because by the criteria of science and reason God is an unknowable concept. We cannot prove or disprove God's existence through empirical evidence or deductive proof. Therefore, from a scientific or philosophical position, theism and atheism are *both* indefensible positions as statements about the universe. Thomas Huxley once again clarified this distinction:

> *Agnosticism is not a creed but a method, the essence of which lies in the vigorous application of a single principle. Positively the principle may be expressed as, in matters of the intellect, follow your reason as far as it can carry you without other considerations. And negatively, in matters of the intellect, do not pretend the conclusions are certain that are not demonstrated or demonstrable. It is wrong for a man to say he is certain of the objective truth of a proposition unless he can produce evidence which logically justifies that certainty.*

Martin Gardner, mathematician, former columnist for *Scientific American*, and one of the founders of the modern skeptical movement, is a believer who admits that the existence of God cannot be proved. He calls himself a *fideist*, or someone who believes in God for personal or pragmatic reasons, and defended this position to me in an interview: "As a fideist I don't think there are any arguments that prove the existence of God or the immortality of the soul. Even more than that, I agree with Unamuno that the atheists have the better arguments. So it is a case of quixotic emotional belief that is really against the evidence and against the odds." *Credo consolans*, says Gardner—I believe because it is consoling. Fideism is the art of the insoluble.

As for my part, I used to be a theist, believing that God's existence was soluble. Then I became an atheist, believing that God's nonexistence was soluble. I am now an agnostic, believing that the issue is insoluble. Ever since I made my position known in the pages of *Skeptic* magazine many years ago, I have received a large volume of correspondence, much of it from atheists who accuse me of copping out or being wishy-washy in using the term *agnostic*. One wrote: "I, sir, am a plain unqualified atheist. Would you like to hear my reason? Okay, 'there is no God.' That's my reason." Most skeptics and atheists would agree and argue that there are really only two positions on the God Question: you either believe in God or you do not believe in God—theism or atheism. What's this agnosticism nonsense, they ask?

If by fiat I had to bet on whether there is a God or not, I would bet that there is not. Indeed, I live my life as if there is no God. And if the common usage of the term *atheism* was nothing more than "no belief in a God," I might be willing to adopt it. But this is not the common usage, as we saw in the *OED*. (And we would do well to remember that dictionaries do not give definitions, they give *usages*.) Atheism is typically used to mean "disbelief in, or denial of, the existence of a God" (not to mention its pejorative permutations). But "denial of a God" is an untenable position. It is no more possible to prove God's nonexistence than it is to prove His existence. "There is no God" is no more defensible than "there is a God." The problem with the term *agnostic*, however, is that most people take it to mean that you are unsure or have yet to make up your mind, so the term *nontheist* might be more descriptive.

Belief or disbelief in God is clearly a decision of considerable personal importance. But making this decision is not a science. For thousands of years the greatest minds of every generation have worked diligently to prove the existence of God, and for thousands of years equally great minds have produced valid refutations of those proofs.

The problem may be in the meaning of the word *prove*. Drawing upon the *OED* once again, to "prove" means: "to make trial of, put to the test." How could you possibly put God to the test? There is no conceivable experiment that could confirm or disconfirm God's existence. There comes a time in the history of an idea when it seems reasonable to conclude that the problem is beyond the human mind to solve. God is insoluble.

## WHAT IS GOD?

Although it is almost certainly not possible to define God in any concise way, it would seem remiss not to at least try in any discussion such as this. Studies show that the vast majority of people in the Industrial West who believe in God associate themselves with some form of monotheism, in which *God is understood to be all powerful, all knowing, and all good; who created out of nothing the universe and everything in it with the exception of Himself; who is uncreated and eternal, a noncorporeal spirit who created, loves, and can grant eternal life to humans.* Synonyms include Almighty, Supreme Being, Supreme Goodness, Most High, Divine Being, the Deity, Divinity, God the Father, Divine Father, King of Kings, Lord of Lords, Creator, Author of All Things, Maker of Heaven and Earth, First Cause, Prime Mover, Light of the World, Sovereign of the Universe, and so forth.

Many scientists, however, feel that such discussions about the nature of and belief in God are meaningless, tantamount to asking, as anthropologist Donald Symons did, "Do you believe [fill in any three letters] exists?" Symons explained:

> *You have to know more about what's in the brackets and how its existence or nonexistence might be determined or, at least, what kinds of evidence might potentially bear on the question. If you find out that the questioner has essentially no ideas about the characteristics of the [ ] (such as, for example, whether it is made of matter), and, more importantly, states that no conceivable observation could have any bearing on the existence/nonexistence question, then to me the original question is meaningless, or incoherent, or empty, or some similar concept.*

Vince Sarich, another anthropologist, feels that the God Question "may be one of those I have tended to term a 'wrong question'; that is, one that wrongly assumes there is an answer in a form defined by the question." In what way is it the wrong question? "Gods that live only in people's heads are far more powerful than those that live 'somewhere out there' for the simple reasons that (1) there aren't any of the latter variety around, and (2) the ones in our heads actually affect our lives and, of course, the lives of those we interact with and everything else we touch." Therefore, Sarich concludes, "the whole God Question—atheist, agnostic, theist, whatever—is irrelevant." How so?

> *What difference does it, or can it, make? Who cares? Who should care? Indeed, who even should care about anyone else's answer to that particular question? That answer will in no sense begin to define what feelings you will have in any particular situation, nor even more important, what actions you will take on behalf of those feelings. The fact is that you will have, indeed you must have, a belief system that has moral and ethical dimensions, while you may, or may not justify that belief system, implicitly or explicitly, in terms of a God or gods. I believe that gods exist to the extent that people believe in them. I believe that we created gods, not the other way around. But that doesn't make God any less "real." Indeed, it makes God all the more powerful. So, yes, I believe in, and, maybe, to some extent fear, the God in your head, and all the gods in the heads of believers. They are real, omnipresent, and something approaching omnipotent.*

This is what makes the God Question one of the most potent we can ask ourselves, because whether God *really* exists or not is, on one level, not as important as the diverse answers offered from the thousands of religions and billions of people around the world. To an anthropologist these differences are scientifically interesting in trying to understand the cultural causes of the diversity of belief. But from a believer's perspective, the differences are emotionally significant because they tell us something about our personal values and commitments.

An even more extreme position with regard to the God question is that of Paul Tillich: "The question of the existence of God can be neither asked nor answered. If asked, it is a question about that which by its very nature is above existence, and therefore the answer—whether negative or affirmative—implicitly denies the nature of God. *It is as atheistic to affirm the existence of God as it is to deny it.* God is being-itself, not *a* being." The God question cannot even be asked.

## THE FAITH OF THE FLATLANDERS

One problem with arguing God through a series of logical definitions and syllogisms is the impossibility of finding spiritual or emotional comfort in such a rational process. For most people God is not found in the sixth place after the decimal point. Another problem is the impossibility of comprehending something that is, by definition, incomprehensible. Whatever God is, if there is a God, He would be so wholly Other that no corporeal, time-bound, three-dimensional, nonomniscient, nonomnipotent, nonomnipresent being like us could possibly conceive of an incorporeal, timeless, dimensionless, omniscient, omnipotent, omnipresent being like God. It would be like a two-dimensional creature trying to grasp the meaning of three-dimensionality, an analogy a nineteenth-century Shakespeare scholar named Edwin Abbott put into narrative form in the splendid 1884 mathematical tale, *Flatland*. The story powerfully illuminates the insolubility of God's existence and why faith instead of reason, religion instead of science, is the proper domain of God.

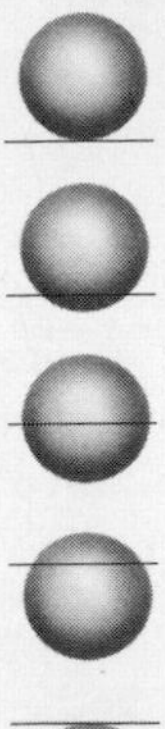

*A human being trying to understand God is like a two-dimensional being trying to understand the third dimension. In his classic tale* Flatland, *Edwin Abbott describes such an existence, where a circle would only be perceived as a line. Watching a three-dimensional object such as a sphere pass through Flatland, a resident would see only a point and then a succession of circles growing larger at first and then smaller as it returns to a point before vanishing.*

Abbott's surrealistic story begins in a world of two dimensions, where the inhabitants—geometrical figures such as lines, triangles, squares, pentagons, hexagons, and circles—move left and right, forward or backward, but never "up or down." Looking at a coin you can see the shapes within the circle, much like you could see the inhabitants of Flatland from Spaceland looking down; but if you turn the coin on its side, the interior disappears and you only see a straight line. This is what all geometrical shapes look like to Flatlanders.

One day a mathematician Square in Flatland encounters a stranger that mysteriously changes sizes from a point, to a small circle, to a big circle, back to a small circle, and finally vanishes altogether. Since Flatlanders do not arbitrarily grow and shrink in size, the Square is confused. The stranger explains that he is not a single circle changing sizes but "many circles in one," and to prove his three-dimensional nature to the Square he employs logic and reason: "I am not a plane figure, but a solid. You call me a Circle; but in reality I am not a Circle, but an infinite number of circles, of size varying from a point to a circle of thirteen inches in diameter, one placed on top of the other. When I cut through your plane as I am now doing, I made in your plane a section which you, very rightly, call a Circle."

The Square still does not understand, so the stranger, a Sphere, turns from example to analogy:

*Sphere:* Tell me, Mr. Mathematician, if a Point moves Northward, and leaves a luminous wake, what name would you give to the wake?

*Square:* A straight line.

*Sphere:* And a straight line has how many extremities?

*Square:* Two.

*Sphere:* Now conceive the Northward straight line moving parallel to itself, East and West, so that every point in it leaves behind it the wake of a straight line. What name will you give to the figure thereby formed? We will suppose that it moves through a distance equal to the original straight line. What name, I say?

*Square:* A Square.

*Sphere:* And how many sides has a square? How many angles?

*Square:* Four sides and four angles.

*Sphere:* Now stretch your imagination a little, and conceive a Square in Flatland, moving parallel to itself upward.

The problem, of course, is that "upward" has no meaning for a two-dimensional being who has never experienced the third dimension of "height." The Square is still confused, so the Sphere walks him through a clear-cut proof: If a *point* produces a line with two terminal points and a *line* produces a square with four terminal points, then the next number is 8, which the Sphere explains makes a cube—a six-sided square in Spaceland. This he further proves with logic: If a point has zero sides, a line two sides, a square four sides, then the next number is 6. "You see it all now, eh?" says the Sphere triumphantly. Not quite. For the dimension-challenged Square, reason is not revelation: "Monster, be thou juggler, enchanter, dream, or devil, no more will I endure thy mockeries. Either thou or I must perish."

With failed reason the Sphere, in a throe of frustration, reaches into Flatland and yanks the Square into Spaceland, whereupon he instantly transforms into a cube. Revelation! But then a thought occurs to the Cube. If the Sphere is many circles in one, there must be a higher dimension that "combines many spheres in one superior existence, surpassing even the solids of Spaceland. . . . [M]y lord has shown me the intestines of all my countrymen in the land of two dimensions by taking me with him into the land of three. What therefore more easy than now to take his servant on a second journey into the blessed region of the fourth dimension?" But the Sphere will not hear of such nonsense: "There is no such land. The very idea of it is utterly inconceivable." So the Cube, with a touch of ersatz innocence, recalls the Sphere's mathematical arguments, noting the Sphere's impatience with the Cube's impertinence:

*Cube:* Was I not taught below that when I saw a line and inferred a plane, I in reality saw a third unrecognized dimension, not the same as brightness, called "height"? And does it not now follow that, in this region, when I see a plane and infer a solid, I really see a fourth unrecognized dimension? . . . [A]nd besides this, there is the argument from analogy of figures.

*Sphere:* Analogy! Nonsense: what analogy?

*Cube:* . . . [I]n one dimension, did not a moving point produce a line with *two* terminal points? In two dimensions, did not a moving line produce a square with *four* terminal points? In three dimensions, did not a moving square produce . . . a cube, with *eight* terminal points? And in four dimensions shall not a moving cube—alas, for analogy, and alas for the progress of truth, if it be not so—shall not, I say, the motion

> of a divine cube result in a still more divine organization with *sixteen* terminal points? Behold the infallible confirmation of the series, 2, 4, 8, 16; is not this a geometrical progression?

The Sphere, now fit to be tied, will have nothing to do with this bohemian heresy, so he promptly thrusts the Cube back into Flatland where he becomes, once again, a lowly two-dimensional square. The story closes with the Square in prison, locked up after he attempted to explain to his fellow Flatlanders what divine dimensions he had experienced: "Prometheus up in Spaceland was bound for bringing down fire for mortals, but I—poor Flatland Prometheus—lie here in prison for bringing down nothing to my countrymen."

Like the Cube's impudent challenge to use the Sphere's own analogies to argue for yet a higher dimension, the proofs of God can themselves be used to consider the possibility of another being still higher, ad infinitum. Like the two-dimensional Flatlanders who could not grasp the nature of three-dimensionality despite ironclad logic and reasoning, God's existence or nonexistence cannot possibly be understood in human terms. What cannot be understood, cannot be proved. What is unprovable is insoluble.

When I was a believer it was always my understanding from reading the Bible that religious belief is ultimately based on faith. In fact, my own "leap of faith," like the Square's transformation into a Cube, had nothing to do with logical proofs and mathematical reasoning. Is that not how most people come to believe in God? Is that not what it means to believe in God? Does this not help explain, in part, why, in the most secular society in history, when God is supposedly dead, belief in Him has never been so high?

Chapter 2

# Is God Dead?

## *Why Nietzsche and* Time *Magazine Were Wrong*

*God is dead.*
—Nietzsche

*Nietzsche is dead.*
—God

Somewhere, on some long-forgotten bathroom wall, a wag scrawled the above graffito. Though it may be too clever by half, it is a telling remark about our times that despite the fact God has been declared dead numerous times, He seems always to have the final word.

It was barely more than a century ago that the German philosopher Friedrich Nietzsche penned the words for which he has become so famous in a book considered by philosophers to be his greatest work, *Thus Spoke Zarathustra.* After ten years in mountainous solitude, Zarathustra descends to mingle among men and there discovers a holy man who tells him, "I make songs and sing them, and when I make songs I laugh, I cry and I hum: in this way I praise God." He then inquires of Zarathustra: "But what gift do you bring us?" Zarathustra replies: "What have I to give you? Nay, let me go, lest I take something away from you!"

> *And so they separated, the old one and the man, laughing as two boys laugh. But when Zarathustra was alone, he spoke thus to his heart: Is it possible that the holy old man in his forest has not yet heard the news that God is dead?*

## GOD IN THE 1960s

Reflecting Nietzsche's pronouncement nearly a century later on its April 8, 1966, cover, *Time* magazine brazenly inquired of its readers in stark, red type on a black background: IS GOD DEAD? The cover story by John T. Elson (although oddly no byline was given in the article itself) was entitled "Theology: Toward a Hidden God," but in hindsight it was really more of a mirror held up to what appeared at the time to be our godless culture. By 1966 the most turbulent decade in memory was in full rage as the baby-boomer generation flexed its moral (and immoral) muscles against the conservative establishment's vision of America as a God-fearing nation. Political assassinations, campus rebellions, inner-city riots, mass demonstrations, sex, drugs, rock 'n' roll, and especially the Vietnam War led many disillusioned Americans down a nihilistic path into existential angst.

At the height of the cold war, most unsettling of all was perhaps the military/political strategy of Mutual Assured Destruction, or MAD. Reflecting the feeling of the day, Barry McGuire's 1965 guttural rock song, "Eve of Destruction," warned, "If the button is pushed, there's no running away / There'll be no one to save with the world in a grave." What happened to the savior of old? He died. In like manner, the characters in Nevil Shute's 1957 novel, *On the Beach*, struggle to find meaning in a world made meaningless after total nuclear war results in a slow but ineluctable end of life. If there is no next year, what will you do tomorrow? Oddly, the characters continue to work and love and live in the face of imminent death for a future that will not come. The American captain of the submarine *Scorpion*, observing Australia as the last outpost of survivors, discovers from one of them that she is taking shorthand, typing, and bookkeeping classes.

*Is God dead? Almost 100 years after Nietzsche's famous assertion,* Time *magazine posed the question on its April 8, 1966, cover.*

"I'll be able to get a good job next year," she explains, knowing that there is not going to be a next year. "It's the same at the university. There are many more enrollments now than there were a few months ago." Why? What else is there to do but to pretend that the world has meaning? After all (and this was one of the deeper messages of the book), what Shute's doomed survivors face is what all of us face; the only difference is that we do not know how or when our end will come, so the fiction of purpose is preserved. The epigraph from T. S. Eliot on the title page of Shute's book expresses this poetically:

In this last of meeting places
We grope together
And avoid speech

Gathered on this beach of the tumid river . . .
This is the way the world ends
This is the way the world ends
This is the way the world ends
Not with a bang but a whimper.

Similar cultural images of God's death bestudded the cultural landscape in this period. Five years after the *Time* article, John Lennon's song "Imagine" asked us to project ourselves into a godless future in the hope that we might help bring it about. But this was only the exclamation mark on a statement Lennon made in the very same year as the *Time* cover story, when he prophesied God's demise (and nearly caused the Beatles'): "Christianity will go. It will vanish and shrink. I needn't argue about that, I'm right and will be proved right. We're more popular than Jesus Christ right now." God may have died, but his fans outvoiced Lennon's, resulting in mass bonfires of Beatles' records and Lennon's public apology.

Other examples from the era abound. Arthur C. Clarke's novel, *2001: A Space Odyssey*, for example, opens with ape-men being given the spark of humanity—not from God but from an advanced alien race. Moon-Watcher learns to kill with tools, a gift from the secular gods that would turn out to be more powerful than any they had known before: "Now he was master of the world, and he was not quite sure what to do next. But he would think of something." At the end the story comes full circle with Star-Child being given God-like powers to prevent humanity from taking the leap into nuclear annihilation. After harmlessly detonating a space-based nuclear missile, Star-Child contemplated his newfound powers, also a gift from the alien gods: "Then he waited, marshaling his thoughts and brooding over his still untested powers. For though he was master of the world, he was not quite sure what to do next. But he would think of something." Certainly he was not thinking about God.

## *TIME* AND GOD

To the star-children of the 1960s it appeared as if God *had* died, as *Time* magazine suggested He had—if only by daring to pose the question in the first place. Even for some theologians this appeared to be the case. Noted *Time*:

*Is God dead? It is a question that tantalizes both believers, who perhaps secretly fear that he is, and atheists, who possibly suspect that the answer is no.*

*Is God dead? The three words represent a summons to reflect on the meaning of existence. No longer is the question the taunting jest of skeptics for whom unbelief is the test of wisdom and for whom Nietzsche is the prophet who gave the right answer a century ago. Even within Christianity, now confidently renewing itself in spirit as well as form, a small band of radical theologians has seriously argued that the churches must accept the fact of God's death, and get along without him.*

Drawing on the results of more than 300 interviews conducted over the course of a year by thirty-two *Time* correspondents around the world, Elson revealed the existence of a new breed of radical theologians known as "Christian atheists" (an oxymoron if there ever was one), to be contrasted with straightforward Nietzschean atheists: "Nietzsche's thesis was that striving, self-centered man had killed God, and that settled that. The current death-of-God group believes that God is indeed absolutely dead, but proposes to carry on and write a theology without *theos*, without God." In addition to these Christian atheists were the existentialist atheists, mostly literary types such as Simone de Beauvoir, who suggested: "It was easier for me to think of a world without a creator than of a creator loaded with all the contradictions of the world." Yet another brand were the "distracted atheists," or "people who are just 'too damn busy' to worry about God at all." "Practical atheists" rounded out the field—the folks who fill the pews on Sunday but in reality are "disguised nonbelievers who behave during the rest of the week as if God did not exist." Philosopher Michael Novak was quoted to represent the general spiritual dolor that was sweeping America: "I do not understand God, nor the way in which he works. If, occasionally, I raise my heart in prayer, it is to no God I can see, or hear, or feel. It is to a God in as cold and obscure a polar night as any nonbeliever has known."

The reasons *Time* gave for God's death are telling for the age. An obituary for God published in the Methodist student magazine *Motive*, for example, was chosen as an emblem of this new throwaway theology: "ATLANTA, Ga., Nov. 9—God, creator of the universe, principal deity of the world's Jews, ultimate reality of Christians, and most eminent of all divinities, died late yesterday during major surgery undertaken to correct a massive diminishing influence." The cause of this declining impact was attributed to "secularization, science, urbanization—all have made it comparatively easy for the modern man to ask where God is, and hard for the man of faith to give a convincing answer, even to himself." Particularly with the rise of modern science, "slowly but surely, it dawned on men that they did not need God to explain, govern or justify certain areas of life." Even the old standby threat of eternal punishment in hell was impotent. "Unlike in earlier centuries, there is no way for churches to threaten or compel men to face that leap; after Dachau's mass sadism and Hiroshima's instant death, there were all too many real possibilities of hell on earth."

Reader reactions to what is arguably the most famous and controversial cover story in *Time*'s seventy-five-year history (generating more letters—3,430—than any issue before or since) were at once amusing and instructive (April 15, 1966, 13). "No," said a Chicago reader. "Yes," proclaimed a Notre Dame professor. "Not only is God dead—he never was," pronounced the president of the Freethinkers of America. Equally vehement was this letter from a reader in Mount Vernon, New York: "Your ugly cover is a blasphemous outrage and, appearing as it does during Passover and Easter week, an affront to every believing Jew and Christian." A more measured response came from a ministerial student at Concordia Seminary in St. Louis: "God is dead to those who wish him so; he lives for those who hope in him." The most accurate, however, came from a rather unexpected source—Jay North, best known as television's Dennis the Menace (May 6, 1966, 9): "In sending you my views I realize I have two strikes against me: I am a teen-ager, and I am in show business. In neither category does much religious thought go on, according to the public. . . . I have found, too, that the citizens of Hollywood are as strong in their devotion as are their priests and ministers and rabbis. This God-is-dead premise seems to me merely a fad; religion will live through it." How right the menacing Mr. North would turn out to be.

## GOD'S RESURRECTION

From Nietzsche's pronouncement to *Time*'s declaration that he was right took eighty years. Another thirty should have buried him for good, no? No. A Gallup poll of American adults published in the *Wall Street Journal* on January 30, 1996, reported that 96 percent believe in God, 90 percent believe in heaven, 79 percent believe in miracles, 73 percent believe there is a hell, 72 percent believe there are angels, and 65 percent believe the devil is real. A gender gap was evident for two beliefs: Women outnumbered men in belief in miracles (86 percent women versus 71 percent men) and angels (78 percent women versus 65 percent men). Not surprisingly, education makes a difference, but not as much as one might think. Belief in heaven, for example, breaks down as follows: college postgraduates: 75 percent; college graduates: 80 percent; some college: 90 percent; no college: 94 percent. The 20 percent range gap deflects from the reality that three out of four people with master's and doctorate degrees believe in heaven. Who says God is dead?

Other polls corroborate God's vitality, such as George Barna's 1996 *Index of Leading Spiritual Indicators*, which reported a 93 percent figure of belief, and his 1995 poll revealing that 87 percent say their religious faith is very important in their lives. (Interestingly, the 1996 poll also showed that 30 percent of believers described "God" as a deity other than the biblical God: 11 percent saw God as a higher consciousness; 8 percent said God is the total realization of personal human potential; 3 percent voiced a belief in many gods each with his or her own power and authority; and 3 percent reported that everyone is his or her own god.) While some believers occasionally have doubts, a 1997 Pew Research Center survey reported that a remarkable 71 percent of Americans say that they "never doubt the existence of God" (up from 60 percent in a 1987 survey). Even in Southern California, that bastion of New Age spiritualism, God is alive and well, as noted in a 1991 *Los Angeles Times* poll in which 91 percent of respondents reported believing in "God or a universal spirit," 67 percent believing in "life after death," and 67 percent believing in heaven.

## SUPPLY-SIDE RELIGION AND THE SECULARIZATION OF THE WORLD

At the beginning of the twentieth century social scientists predicted that with the advent of universal public education and the rise of science and technology, culture would become secularized and religiosity would dramatically decrease. This "secularization" thesis has been thoroughly refuted, as religiosity continues to increase at the end of and into the next century. The question is, why?

According to the University of Chicago sociologist of religion, Andrew Greeley, in an economic explanation, one of the reasons is that as the century progressed a free market of religious competition increased and diversified, causing religions and churches to compete with one another for customers. In a paper delivered to the 1997 American Sociological Association meeting in Toronto, entitled "Pie in the Sky While You're Alive: Life after Death and Supply Side Religion," Greeley demonstrated that belief in an afterlife rose from 65 percent to 84 percent among Catholics, and from 24 percent to 40 percent among Jews; the latter statistic is surprising because belief in life after death normally decreases proportionally with education, and Jews are among the most educated of all religious groups. (Protestants remained steady at 80 percent.) Even those with no religious affiliation showed an increase in belief in life after death, from 31 percent to 50 percent. These statistics, says Greeley, fly in the face of our intuitive thoughts about the rise of science and the decline of religion: "A furious battle is raging in social science about religion. Traditional theories have emphasized the decline of religion as part of an inevitable process of 'secularization.' In the face of scientific progress, the growth of rationality, and the elimination of superstition, religion is seen retreating, as Durkheim said it would, to the periphery of society." But Greeley's data, along with the polls cited above, show that Durkheim, along with *Time* and Nietzsche, were wrong. Why? Greeley tests an interesting theory of supply-side religion:

> *They argue that the "demand" for religion is relatively constant since the need of "compensation" because of death and suffering is a given in society and that the different levels of religious behavior that one can observe in various regions of a country like the United States and in various countries are the result of the available "supply" of religious services. In a controlled religious marketplace, they assert, religion becomes a lazy monopoly because the Established Church (or Established Churches as in Germany) need not compete for "customers." On the other hand, when there is no legal monopoly various "firms" must compete for "customers" and hence provide more industrious personnel and more services. In such situations religious activity increases.*

Economic theories of religion date back at least as far as Adam Smith's 1776 publication *The Wealth of Nations*, in which he observed that market forces govern churches no differently than they do secular firms. But Greeley looks to a deeper cause that he thinks can be found in two fundamental principles of human biology and psychology, which together form the beginning of our answer to the question of how we believe: "Humankind is born with two incurable diseases, life from which it inevitably dies and hope which hints that death may not be the end. A conviction that life does not end with death is a tentative endorsement of the validity of hope." The "product" sold by these religious "corporations" competing in the spiritual "marketplace" is life after death. Sales are on the rise. All market indicators are positive. Indeed, Greeley found that only 0.8 percent of Americans call themselves "hard-core atheists" (who believe there is no God or life after death), and a mere 3.4 percent "soft-core" atheists (who are at least "open" to the possibility of God and life after death, but do not presently believe).

Greeley, and his colleague Wolfgang Jagodzinski from the University of Cologne, based their findings on data gathered from 19,381 respondents interviewed from 1973 to 1994 in the General Social Surveys conducted by the National Opinion Research Center at the University of Chicago. They found a statistically significant increase in belief in life after death across generations, especially in the immigrant shift from old-world monopolies to the new-world open marketplace. Greeley explained the demographics of his study:

> *Unnoticed by scholars at the time or since that time, American society and its open marketplace of religious firms was exercising substantial influence on the religious belief of immigrants since the turn of the century. Among other things this influence increased dramatically across generations belief in life after death. This discovery simply cannot be explained by the "secularization" theory and is quite compatible with the "supply side" theory:*
> *religious competition does seem to generate and increase hope.*

This message is not lost on religious leaders who must compete for members of the various religious beliefs and congregations in this supply-side model. As Greeley concluded: "In a competitive religious marketplace like the United States the clergy must work hard at what they are supposed to be doing: preaching a message of hope in the face of the tragedies of life." Clearly religion plays an important role in society that has not been filled by secular institutions, and since none has ever offered a "product" competitive with life after death, belief in God and religiosity has increased as a market response.

## SOCIAL INDICATORS OF GOD

Contrary to the rhetoric of modern conservative fundamentalists—who proclaim that Americans have turned away from God and that we need to return to that "old-time religion"—as a people we have never been so religious. In a revealing book, *The Churching of America, 1776–1990,* the authors point out that for the past two centuries American church membership rates have risen from a paltry 17 percent at the time of the Revolution (!), to 34 percent by the middle of the nineteenth century, to over 60 percent today. Bully-pulpit preachers who remind us regularly that we are slouching ever further toward cultural depravity and godless hedonism could not be more wrong. "New-time" religion far outstrips our forebears' religiosity. Proof of that can be found in any number of cultural signposts.

Consider a spate of *Time* magazine cover stories over the past decade: June 10, 1991: "Evil: Does It Exist—Or Do Bad Things Just Happen?" April 10, 1995: "Can We Still Believe in Miracles?" December 18, 1995: "Is the Bible Fact or Fiction?" October 28, 1996: "And God Said . . . Betrayal. Jealousy. Careerism. They're All in the Bible's First Book. Now There's a Spirited New Debate over the Meaning of Genesis." March 24, 1997: "Does Heaven Exist?"

Not to be outdone, *Newsweek*'s March 31, 1997, issue addresses "The Mystery of Prayer: Does God Play Favorites?" *U.S. News and World Report* countered the same week with "Life after Death: Science's Search for Meaning in Near-Death Experiences." An earlier, December 19, 1994, cover reads: "Waiting for the Messiah: The New Clash over the Bible's Millennial Prophecies."

Religiosity is more than just belief in God, of course. It is the full package that includes heaven and life after death. In a poll of 1,018 adult Americans conducted on March 11 and 12, 1997, *Time* reported that belief in heaven is still quite strong, with 81 percent reporting belief, and 88 percent looking forward to meeting "friends and family members in heaven" when they die. *Newsweek*'s feature article title asked: "Is God Listening?" According to their poll, conducted by the Princeton Survey Research Associates on March 20 to 21, 1997, 87 percent of Americans believe the answer is "yes," with 29 percent reporting that they pray more than once a day and 25 percent at least once a day.

For some people, they *know* there is an afterlife because they have experienced it through a "near-death experience." A *U.S. News and World Report* poll, for example, notes that of the nearly 18 percent of Americans who claimed to "have been on the verge of dying, many researchers estimate that a third have had unusual experiences while straddling the line between life and death—perhaps as many as 15 million Americans. A small percentage recall vivid images of an afterlife—including tunnels of light, peaceful meadows, and angelic figures clad in white." Pediatric oncologist Diane Komp, who has talked to more than a few children about such experiences, concludes: "I came away convinced that these are real spiritual experiences." Everyone seems to agree that near-death experiences are genuine "experiences," in the sense that the individuals have had something happen to them, which they report as being life changing. Floating out of the body, passing through some sort of tunnel, hallway, or canyon, the white light at the end of the tunnel, seeing your lost loved ones on the "other side," are common elements reported by many. Whether the near-death experience represents a bridge to the "other side" is another matter, but those who experience it often treat it as a religious or spiritual awakening.

In addition to God and the afterlife, miracles are staging a comeback. A recent Canadian poll reveals that more than half the respondents reported a belief in angels, and nearly as many said they had personally experienced a miracle or divine intervention (again, women were almost twice as likely to believe as men). Some miracles involved petitionary prayer, such as asking for a loved one to be saved from a serious illness, while other miracles were attributed to chance encounters and good fortune. One person noted that he had stopped his car to take a photograph, thereby missing a fatal accident by three minutes. Another reported: "My aunt was on a life support system and the doctors told my family that she was dead and they were going to turn off the life support system. They forgot to turn it off and the next morning they found her alive, breathing, and talking." Other miracles were more mundane: "I was talking to somebody telling them I was broke and someone heard me talking about it and they came to my door and took me shopping for groceries." Still others held rather low standards for what constitutes the miraculous: "I went to someone's house and got a good deal on a power tool that I wanted for a long time."

Perhaps the most significant religious cultural phenomenon of the 1990s was the Promise Keepers, a well-organized group of men who "promise" to take responsibility for their lives through the *Seven Promises* of (1) "honoring Jesus Christ," (2) "pursuing vital relationships with a few other men," (3) "practicing spiritual, moral, ethical, and sexual purity," (4) "building strong marriages and families through love, protection and biblical values," (5) "supporting [the] church by honoring and praying for [one's own] pastor," (6) "reaching beyond any racial and denominational barriers to demonstrate the power of biblical unity," and (7) "influencing His world, being obedient to the great commandment (Mark 12: 30–31) and the great commission (Matthew 28: 19–20)." Openly antigay, antiabortion, and antifeminist, Promise Keepers members rally against the standard targets the Christian right loves to hate: atheism and evolution, and the perceived moral degradation of America that comes with them. Begun in 1991 with a fledgling 4,200 members, by late 1997 the group had grown to 1.25 million, with an annual revenue of $87 million and a paid staff of 452 working out of its Denver headquarters. Expanding their ranks (Promise Keepers often speak in military terms, such as the "army of God" and "wake-up calls," and they have even hired retired military officers), founder Bill McCartney explains their long-term plan: "The goal is to go into every church whether they like us or not." McCartney told 39,000 pastors in Atlanta to "take this nation for Jesus . . . whoever stands with the messiah will rule with him." When half a million men blanketed the Washington, D.C., Mall on October 4, 1997, it was the largest religious rally in American history.

Even television, that quintessential morass of moral decay, has been heeding this trend. At the start of the 1997 season, viewers, accustomed to the likes of such sinfully tantalizing shows as *Baywatch* and *Melrose Place*, were treated to an unprecedented eight programs with religious or spiritual themes. According to a March 1997 *TV Guide* poll, 61 percent of those surveyed indicated they wanted to see more references to God in prime time. And according to a Parents Television Council survey, since 1993 the depiction of religious symbols and spiritualism on national television increased 400 percent.

Book sales also reflect these trends. The American Booksellers Association (ABA), for example, reported that books on religion and spirituality rose 112 percent between 1991 and 1996. And from 1996 to 1997 books on religion were the only type of adult nonfiction whose sales were steadily rising. In 1997, for example, among national bestseller lists such as in the *New York Times* and *Publishers Weekly*, religion and spirituality titles averaged five spots among the top fifteen. Examples included Michael Drosnin's *The Bible Code*, Neale Donald Walsh's *Conversations with God*, and Billy Graham's autobiography. James Van Praagh's *Talking to Heaven* had a remarkable run of over three consecutive months as the number-one bestselling book in America, with sales approaching a million copies. According to the ABA, publishers are calling for books that bridge the gap between scholarly depth and everyday spirituality. The mantra is "make it popular and serious." Today's readers, while skeptical of easy answers and shallow summaries of complex problems, still desire a sense of the sacred, or what is called "lived religion." ABA's Willard Dickerson says the trend highlights a "growing hunger in the public reading market for answers that go past the secular or materialistic cultures."

One deeper motive contributing to the search for God comes from the fact that the Hebrew Bible has God's influence slowly but ineluctably fading as the story unfolds, so that by the end God's face is almost completely hidden and humans are left to fend for themselves. Bible scholar Richard Elliott Friedman documents this phenomenon in his 1995 work, *The Disappearance of God*:

> *The Bible begins, as nearly everybody knows, with a world in which God is actively and visibly involved, but it does not end that way. Gradually through the course of the Hebrew Bible (also known as the Old Testament, Holy Scriptures, or* tanak*), the deity appears less and less to humans, speaks less and less. Miracles, angels, and all other signs of divine presence become rarer and finally cease. In the last portions of the Hebrew Bible, God is not present in the well-known apparent ways of the earlier books. Among God's last words to Moses, the deity says, "I shall hide my face from them. I shall see what their end will be." (Deut. 31:17, 18; 32:20). By the end of the story God does just that. The consequences and development of this phenomenon in the New Testament and in post-biblical Judaism are extraordinary as well.*

Extraordinary indeed! Where did God go and, more importantly, why did He choose to disappear? Friedman explores these questions and provides some intriguing answers. He closes his exploration with a discussion of the relationship of science and religion, and a comparison of Kabbalah and cosmology, concluding: "There is some likelihood that the universe *is* the hidden face of God." This depends, of course, on how one defines God, but I am more interested in the search than the disappearance. Is the New Age resurgence in spirituality an attempt to uncover the hidden face? Perhaps the face is to be found in a mirror. In her splendid little book, *The Sacred Depths of Nature*, biologist Ursula Goodenough explores this possibility in what she calls "religious naturalism": "If religious emotions can be elicited by natural reality—and I believe that they can—then the story of Nature has the potential to serve as the cosmos for the global ethos that we need to articulate." In any case, this longing and search tells us something very deep about the need in the human psyche for the spiritual and sacred aspects of life not often found in the sciences or humanities. Yet they are there if you know where to look.

## SACRED SCIENCE

Scientists and skeptics must address the fact that God is alive and well at the end of the second millennium—and likely will be at the end of the third. It would appear that news of God's death will always be premature. Atheists, humanists, skeptics, and freethinkers who envision the day when the world will be free of God and religion are about as likely to realize their dream as are the anarcho-capitalists who foresee the end of all government and the privatization of the entire world. Such beliefs, in fact, are themselves a type of secular religion, like those that sprang from Europe at the turn of the twentieth century—Marxism, Freudianism, and social Darwinism. Even scientism, in some extreme circles, turns into a type of secular faith where all things come to those who believe.

In fact, science is a type of myth, if we think of myths as stories about ourselves and our origins (and not in the pejorative sense of myths as things "untrue"). Many gain considerable emotional, even "spiritual," satisfaction from reading scientific articles and books by geologists about the creation of the Earth, by paleontologists about the evolution of life, by paleoanthropologists about human origins, by archeologists about the genesis of civilization, by historians about the development of culture, and especially by cosmologists about the origins of the universe. Tens of millions of people watched Carl Sagan's 1980 *Cosmos* series with rapt attention. In 1997 the PBS series *Stephen Hawking's Universe* gripped viewers every Monday night. Books on evolution by Richard Dawkins, Stephen Jay Gould, Donald Johanson, and Edward O. Wilson are eagerly sought by readers and often find themselves on bestseller lists. Why? Because at these boundaries of scientific knowledge the lines between science, myth, and religion begin to blur as we ask ultimate questions about ourselves, our origins, and our place in the cosmos.

In 1998 I witnessed a sublime example of the scientific sacred when Stephen Hawking visited the California Institute of Technology (Caltech), as he does nearly every year in meeting with Kip Thorne, John Preskill, and other cosmologists. During his visits he often agrees to deliver a public address via his now-familiar voice synthesizer that has an almost surrealistic, otherworldly resonance. Hawking was slated for the largest venue on campus—Beckman Auditorium—which holds 1,100 people. When that hall filled, the staff piped a video feed into Remo Hall, filling another 400 seats. This was not enough, so large theater speakers were pointed out toward the quad area where hundreds more sat on the grass, rock-concert style, listening to a scientific superstar. When he rolled into Beckman Auditorium and down the aisle in his motorized wheelchair, Hawking received a standing ovation, as he did upon his departure. He delivered his standard lecture about the Big Bang, black holes, time, and the universe, all covered in his bestselling book *A Brief History of Time*, which broke all records for the number of weeks any science book has been on a bestseller list. There followed an illustrated version of the book, as well as a documentary also entitled *A Brief History of Time*, followed by a documentary about the making of the documentary!

The mythical nature of science, however, was not as obvious in Hawking's lecture as it was in the subsequent question-and-answer period. The majority of the audience was not especially interested in the minutiae of quantum mechanics or the nuances of cosmological theories. What people wanted were The Answers to the Big Questions: "How did time begin?" "What was there before the universe?" "Why does the universe bother to exist at all?" There are no Final Answers to these Big Questions, of course, but this does not stop people from asking. Here the public was given an opportunity to inquire of a physically disabled but cognitively brilliant man the biggest question of all: "Is there a God?"

Stephen Hawking's lectures are delivered at normal speed because he writes them ahead of time and the computer feeds the words to the voice synthesizer at a staccato pace. But answering questions is another thing altogether. Hawking must construct his sentences word by word, at a glacially slow meter. During this process his colleagues talk to the audience until the answer comes. For this final question, however, the effect was one of unbearable anticipation. Asked an essentially unanswerable question, Hawking sat there in his chair, rigid and stone quiet, only his eyes darting back and forth across the computer screen. One had the feeling of having traveled to Delphi or Mecca, now forced to wait in bursting expectation of The Answer to the Biggest Question. A minute or two went by as cosmologist Kip Thorne politely explained how Stephen's computer works. Finally it came. With humor and politeness Hawking, wisely, explained: "I do not answer God questions."

It did not matter, because the answers themselves do not matter as much as the process of thinking about the questions and contemplating their ultimate meaning. God is not dead because God represents these ultimate concepts that have been with us as long as we have been human. It is the concepts themselves that reach into the deepest parts of our minds. To contemplate them is not the exclusive domain of either science or religion. It belongs to all of humanity. To that end science too is sacred in the sense of pondering these majestic and timeless issues. What can be more soul shaking than peering through a 100-inch telescope at a distant galaxy, holding a 100-million-year-old fossil or a 500,000-year-old stone tool in one's hand, standing before the immense chasm of space and time that is the Grand Canyon, or listening to a scientist who gazed upon the face of the universe's creation and did not blink? That is deep and sacred science.

Chapter 3

# THE BELIEF ENGINE

## *How We Believe*

*We do everything by custom, even believe by it; our very axioms, let us boast of freethinking as we may, are oftenest simply such beliefs as we have never heard questioned.*
—Thomas Carlyle, *Sartor Resartus, III,* 1836

At 9:00 P.M. on Wednesday evening, August 3, 1997, the renowned sage and Vedic philosopher from India, Sri Leachim Remresh, took to the airwaves on WGN radio's *Milt Rosenberg Show* to offer pearls of psychic wisdom to Chicago listeners. Remresh explained that he presently lives in Sedona, Arizona, a New Age capital of sorts, where the Earth's mystical energies are focused in special vortices. Having traveled extensively throughout India, and studied under some of the great Himalayan sages, Remresh enlightened his listeners on how the linear mode of Western scientific thinking restricts our ability to perceive other dimensions, times, and forces. Callers were told they need only give their birth date and ask a single question for Remresh to tap into the cosmic vibrations.

The first caller, a woman born in 1953, wanted to know if her present relationship was going to work out. Remresh cut straight to the woman's heart, telling her that she had previously been married but was now in a relationship with a man who was not as committed as she. In fact, he might even have someone else on the side. The woman gasped in acknowledgment. That was precisely the problem. What should she do? Remresh told her that she already knew what she needed to do. Another caller, a woman born in 1941, wanted to know what she should do about her son. Remresh once again drew upon psychic harmonies, telling the woman that her son was presently adrift in life but that in a few years he would turn his life around; she should not worry too much that he has no goals. Remresh was absolutely right. She wished her son would do something, *anything!* She gave Remresh a 95 percent psychic accuracy rating. The host then announced he had an even more startling revelation, right after the break, of course.

Sri Leachim Remresh, it was revealed, was simply Michael Shermer spelled backwards (with a couple of letter reversals to ease pronunciation). I was in Chicago as part of my national book tour for *Why People Believe Weird Things,* and Milt asked if I would play along with this experiment to show how easy it is to appear to have special insight into people's lives, and how convinced people can become when such bold proclamations are offered. I generally shun such deceptive tactics, but the point was well made. With no formal training in how to be a psychic (and with no psychic abilities whatsoever), I repeated the mantras of New Age gurus, offered some generalizations about human behavior I learned as a student of psychology and in forty-three years of life experience, and let my callers do the rest. The first woman was about my age. She referred to her "relationship," not her marriage. Since most people my age have been married at least once, her previous history of marriage was an easy guess. Statistically speaking men are more promiscuous than women, and women are more committed to relationships than men, so it did not take a genius (or a psychic) to figure out what was behind her question. The second woman was fifty-six years old, so I figured her son must be in his late teens or early twenties (since people are having children later in life these days). Many guys that age are lost souls, rudderless and unanchored, seeking independence from their parents but not yet parents themselves. So I played the odds and was right again.

Among magicians this process is called "cold reading," and is practiced by those who bill themselves as "mentalists." Start with generalizations, then work your way to specifics, using subject feedback (verbal as well as nonverbal when available). The four areas people most want to know about are obvious—love, health, money, and career. So you work your way through them, spending about ten to fifteen minutes on each. Sri Leachim Remresh was successful for the same reason all mediums, psychics, palm and tarot card readers, and astrologers are successful—the people who come to them for advice *believe* they will be successful. Once that belief is in place, the mind makes certain it is confirmed.

Why? Why are we so gullible? Why is it so difficult to discriminate between what is real and what is bogus? The answer can be found in understanding the power of belief systems that drive, and as often as not distort, our perceptions of reality.

## THE PATTERN-SEEKING ANIMAL

In our complex and contingent world, random events often happen in seemingly peculiar sequences that cry out for meaning. We usually rise to the occasion, finding patterns in nature even when they do not exist or have no real significance: the "eagle rock" overlooking the 134 Freeway in Eagle Rock, California—it is just a stone outcropping but our minds see in it the general shape of an eagle-like bird; the "JFK" stone in Hawaii looking for all the world like the late president in profile; the face of Jesus in a tortilla; the Virgin Mary on the side of a building. The first two are amusing but do not strike observers as filled with cosmic import. For some, however, the latter two trigger emotional responses linked to spiritual significance—witness the crowds that appear whenever the Virgin Mary makes her "appearance" on a barn door, in the shadows of trees, or, recently, on the side of the Ugly Duck car rental building in Clearwater, Florida, where the faithful come in wheelchairs and canes to be healed.

We are especially attracted to patterns with a spiritual or religious link, which touch our deepest desire for there to be a Something Else calling the shots and running the show. For some, that Something Else is God; for others it is angels, or fate, or synchronicity, or collective consciousness, or some universal life force. For thousands of years our myths and religions have sustained us with stories of meaningful patterns—gods and God, supernatural beings and mystical forces, the relationship of humans with other humans and their creators, and our place in the cosmos. For the past four centuries, however, science has provided a means for determining which patterns are real and which are illusions, and so we have expelled most ancient and medieval traditions from the pantheon. Or have we?

*The Virgin Mary appears in Clearwater, Florida. Humans are pattern-seeking animals, quite adept at finding meaning even in random patterns of light and shadow, such as this "sighting" of the Virgin Mary on the south window of the Ugly Duck car rental building on Highway 19. The image was "discovered" just before Christmas 1996, and before long devotees transformed the parking lot into a shrine. The image was actually caused by a film of oil from a nearby palm tree, sprayed onto the window by sprinklers.*

A Gallup poll conducted in 1991 revealed that half of all Americans believe in astrology and almost as many believe in extrasensory perception, or ESP; a third believe in the lost continent of Atlantis and in ghosts; and a full two-thirds believe they have had a psychic experience. Do we really live in the Age of Science? We do, but we mostly partake of the fruits of science—technology—whereas fundamental principles of scientific thinking are often poorly taught and rarely employed.

One reason for this tendency to believe in the supernatural is that we may be hardwired to think magically. We have lived in the modern world of science and technology for only a couple of hundred years, yet humanity has existed for a *couple of hundred thousand* years.

What were we doing all those long-gone millennia? How did our brains evolve to cope with the problems in that radically different world?

Evolutionary psychologists—scientists who study the brain and behavior from an evolutionary perspective—make a reasonable argument that the modern brain (and along with it the mind and behavior) evolved over a period of about three million years from the small, fist-sized brain of the *Australopithecines* to the melon-sized brain of modern *Homo sapiens.* Since civilization arose only about 13,000 years ago with the domestication of plants and animals, 99.99 percent of human evolution took place in our "environment of evolutionary adaptation," or EEA. The conditions of the EEA are primarily what shaped our brains, not what happened over the past thirteen millennia. Evolution does not work that fast. The brains of 25,000-year-old Cro-Magnons appear to be no different than ours. Leda Cosmides and John Tooby, codirectors of the Center for Evolutionary Psychology at the University of California, Santa Barbara, have summarized the field this way:

> *Evolutionary psychology is based on the recognition that the human brain consists of a large collection of functionally specialized computational devices that evolved to solve the adaptive problems regularly encountered by our hunter-gatherer ancestors. Because humans share a universal evolved architecture, all ordinary individuals reliably develop a distinctively human set of preferences, motives, shared conceptual frameworks, emotion programs, content-specific reasoning procedures, and specialized interpretation systems—programs that operate beneath the surface of expressed cultural variability, and whose designs constitute a precise definition of human nature.*

Steven Pinker describes these specialized computational devices as *mental modules.* Pinker's "module" is metaphorical, however. Modules are not necessarily located in a single spot in the brain (although they can be, as with Broca's area for language). He describes it as something that "may be broken into regions that are interconnected by fibers that make the regions act as a unit." A bundle of neurons here connected to another bundle of neurons there, "sprawling messily over the bulges and crevasses of the brain" might form a module. Their interconnectedness rather than location is the key to the module's function. The brain then is not so much a single organ as it is a system of specific organs evolved in the EEA to solve specific problems.

Within the circle of professional scientists who study the brain—neurophysiologists, cognitive psychologists, psychopharmacologists, and brain–mind philosophers—this is a very controversial subject. Some support the modular view of the brain, some reject it outright, while others fall in between. David Noelle, of the Center for the Neural Basis of Cognition at the Carnegie Mellon University and the University of Pittsburgh, informs me that:

> *Modern neuroscience has made it clear that the adult brain does contain functionally distinct circuits. As our understanding of the brain advances, however, we find that these circuits rarely map directly onto complex domains of human experience, such as 'religion' or 'belief.' Instead, we find circuits for more basic things, such as recognizing our location in space, predicting when something good is going to happen (e.g., when we will be rewarded), remembering events from our own lives, and keeping focused on our current goal. Complex aspects of behavior, like religious practices, arise from the interaction of these systems—not from any one module.*

Most mental modules are thought of as quite specific, but some evolutionary psychologists argue for making a distinction between mental modules being "domain-specific" versus "domain-general." Tooby, Cosmides, and Pinker, for example, reject the idea of a domain-general processor, whereas many psychologists accept the notion of a global intelligence, called *g*, that would most certainly be considered domain-general. Archaeologist Steven Mithen goes so far as to say that it was a domain-general processor that made us human: "The critical step in the evolution of the modern mind was the switch from a mind designed like a Swiss army knife to one with cognitive fluidity, from a specialized to a generalized type of mentality. This enabled people to design complex tools, to create art and believe in religious ideologies. Moreover, the potential for other types of thought which are critical to the modern world can be laid at the door of cognitive fluidity."

Instead of the metaphor of a "module" to account for such cultural phenomena as religion, I would suggest that we evolved a more general *Belief Engine,* and that it is Janus-faced—that is, under certain conditions it leads to magical thinking while under different circumstances it leads to scientific thinking. We might think of the Belief Engine as a central processor that sits beneath more specific modules. How does this work?

Humans evolved to be skilled pattern-seeking creatures. Those who were best at finding patterns (standing upwind of game animals is bad for the hunt, cow manure is good for the crops) left behind the most offspring. We are their descendants. The problem in seeking and finding patterns is knowing which ones are meaningful and which ones are not. Unfortunately our brains are not always good at determining the difference. The reason is that discovering a meaningless pattern (painting animals on a cave wall before a hunt) usually does no harm and may even do some good in reducing anxiety in uncertain environments. So we are left with the legacy of two types of thinking errors: *Type 1 Error: Believing a falsehood* and *Type 2 Error: Rejecting a truth.* In some cases, neither of these errors will automatically get us killed, so we can live with them. And we do, on a daily basis—witness the aforementioned Gallup poll statistics of magical thinking. The Belief Engine is an evolved mechanism for helping us survive, because in addition to committing Type 1 and Type 2 errors, we also commit what we might call a *Type 1 Hit: Not believing a falsehood* and a *Type 2 Hit: Believing a truth.*

It seems reasonable to argue that the brain consists of both specific and general modules, and the Belief Engine is a domain-general processor. In fact, it is one of the most general of all modules since at its core it is the basis of all learning. After all, we have to believe *something* about our environment, and these beliefs are learned through experience. But the *process* of forming beliefs is genetically hardwired. To account for the fact that the Belief Engine is capable of both Type 1 and 2 Errors along with Type 1 and 2 Hits, we can consider two conditions under which it evolved:

1. *Natural Selection:* The Belief Engine is a *useful mechanism* for survival, not just for learning about dangerous and potentially lethal environments (where Type 1 and 2 Hits help us survive), but in reducing anxiety about those environments; through magical thinking—there is psychological evidence that magical thinking reduces anxiety in uncertain environments; medical evidence that prayer, meditation, and worship may lead to greater physical and mental health; and anthropological evidence that magicians, shamans, and the kings who use them have more power and win more copulations, thus spreading their genes for magical thinking.

2. *Spandrel:* The magical-thinking part of the Belief Engine is also a *spandrel*—Stephen Jay Gould's and Richard Lewontin's metaphor for a necessary by-product of an evolved mechanism. In

their influential paper, "The Spandrels of San Marco and the Panglossian Paradigm," Gould and Lewontin explain that in architecture a spandrel is "the tapering triangular space formed by the intersection of two rounded arches at right angle." This leftover space in medieval churches is filled with elaborate, beautiful designs so purposeful looking "that we are tempted to view it as the starting point of any analysis, as the cause in some sense of the surrounding architecture. But this would invert the proper path of analysis." To ask "what is the purpose of the spandrel?" is to ask the wrong question. It would be like asking "why do males have nipples?" The correct question is "why do *females* have nipples?" The answer is that females need them to nurture their babies, and males and females are built on the same architectural frame. It was simply easier for nature to construct males with worthless nipples rather than reconfigure the underlying genetic architecture.

In this sense the magical-thinking component of the Belief Engine is a spandrel. We think magically because we have to think causally. We make Type 1 and 2 Errors because we need to make Type 1 and 2 Hits. We have magical thinking and superstitions because we need critical thinking and pattern-seeking. The two cannot be separated. Magical thinking is a spandrel—a necessary by-product of the evolved mechanism of causal thinking.

If my hypothesis is correct—that humans evolved a Belief Engine whose function it is to seek patterns and find causal relationships, and in the process makes mistakes in thinking—then we should find evidence for this engine in our ancestors as well as ourselves. Superstitions do not leave behind many fossils, though Cro-Magnon cave paintings and flower-strewn Neanderthal burial sites may serve as a starting point. We can also consider the behaviors of indigenous peoples living today, to a cautious extent, as mirrors of our ancestral Belief Engine, as well as our immediate predecessors in the Middle Ages. Here are three examples that show the relationship between magical thinking and the environment, and how we might have evolved a Belief Engine.

1.

*The Azande.* In the late 1920s anthropologist Edward Evans-Pritchard studied magical thinking among the Azande people of Southern Sudan of central Africa, who were living in a transitional state from hunting, fishing, and gathering to farming. In his 1937 book, *Witchcraft, Oracles and Magic among the Azande,* Evans-Pritchard outlined the magical use of medicines, including "medicines connected with natural forces" (prevention of rain,

*Earliest evidence of hominid magical thinking. In a cave 132 feet deep cut into the Zagros Mountains of northern Iraq 60,000 years ago, the body of a Neanderthal man was carefully buried on a bed of evergreen boughs. The corpse was placed on his left side, head to the south, facing west, and covered in flowers (identified through microscopic analysis of the surviving pollens). Already in the grave were an infant and two women.*

delay of sunset), "medicines connected with hoe culture" (to ensure the fruitfulness of food plants), "medicines connected with hunting, fishing, and collecting" (for everything from making the hunter invisible to preventing wounded animals from escaping), "medicines connected with arts and crafts" (smelting, beer brewing), "medicines connected with mystical powers" (witches, sorcerers), "medicines connected with social activities" (sexual potency, wealth), and, of course, "medicines connected with sickness." In all of these categories the Azande Belief Engine produced Type 1 and 2 Errors *and* Hits. They believed plenty of falsehoods and rejected plenty of truths, but they also rejected falsehoods and believed truths. Evans-Pritchard noted:

*Some Zande medicines actually do produce the effect aimed at, but so far as I have been able to observe the Zande does not make any qualitative distinction between these medicines and those that have no objective consequences. To him they are all alike* ngua, *medicine, and all are operated in magical rites in much the same manner. A Zande observes taboos and fish-poisons before throwing them into the water just as he addresses a crocodile's tooth while he rubs the stems of his bananas with it to make them grow. And the fish-poison really does paralyze the fish while, truth to tell, the crocodile's tooth has no influence over bananas.*

As ethnobotanist Alondra Oubré has demonstrated, these Type 1 and 2 Errors and Hits of indigenous peoples are very valuable because not only do they sometimes get it right, that knowledge can be used to our benefit in the treatment of diseases (and thus they should be so compensated for their magical knowledge that translates into life-saving medicines for us).

2.

*The Yanomamö.* In similar fashion, Napoleon Chagnon discovered in his years among the Yanomamö people of South America that in some villages magical plants are cultivated and used for a number of functions, including the seduction of young women (the powder of a plant is pressed against the woman's nose and mouth at which "she swoons and has an unsatiable desire for sex—so say both the men and the women"); to make men tranquil and sedate ("it is thrown on the men especially when they are fighting"); the destruction of an enemy ("People allegedly cultivate an especially malevolent plant that can be 'blown' on enemies at a great distance, or sprinkled on unwary male visitors while they sleep"); and blaming your enemies for your own misfortunes ("All Yanomamö groups are convinced that unaccountable deaths in their own village are the result of the use of harmful magic and charms directed at them by enemy groups").

According to Chagnon, such magical thinking serves a very pragmatic and useful purpose, as in the Yanomamö jaguar myths, which exist because "the jaguar is an awesome and much-feared beast, for he can and does kill and eat men. He is as good a hunter as the Yanomamö are and is one of the few animals in the forest that hunts and kills men—as the Yanomamö themselves do." The Yanomamö Belief Engine has constructed these myths and superstitions for a very specific problem of survival in an uncertain and dangerous world.

3.

*The Trobriand Islanders.* From 1914 to 1918 the anthropologist Bronislaw Malinowski lived among the Trobriand Islanders off the coast of New Guinea. In Trobriand fishing practices Malinowski discovered that the farther out to sea the islanders went, the more

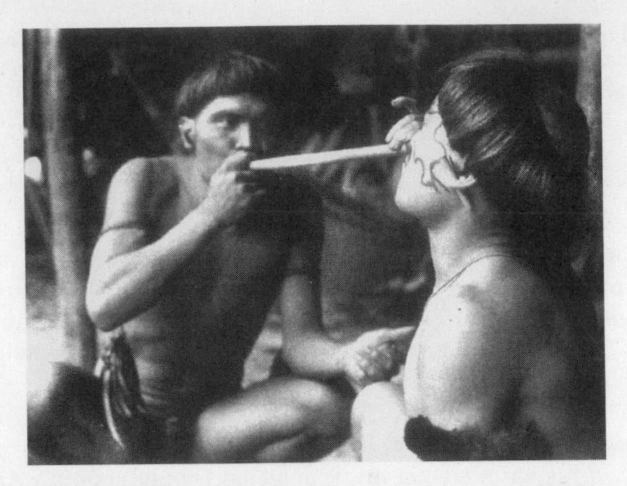

*Enhancing magical thinking. The Yanomamö man on the left is blowing the hallucinogenic ebene powder into the nostrils of the other. According to Napoleon Chagnon, the powder will trigger "grimaces, chokes, groans, coughs, gasps," followed by "watery eyes and a profusely runny nose . . . dry heaves are also very common, as is out-and-out vomiting. Within a few minutes, one has difficulty focusing and begins to see spots and blips of lights. Knees get rubbery. Soon the hekura spirits can be seen dancing out of the sky and from the mountain tops, rhythmically prancing down their trails to enter the chest of their human beckoner, who by now is singing melodically to lure them into his body where he can control them—send them to harm enemies or help cure sick kinsmen." The Belief Engine is more susceptible to superstitions and magical thinking when such hallucinogenic drugs are used.*

complex the superstitious rituals became. In the calm waters of the inner lagoon, there were very few rituals. By the time they reached the dangerous waters of deep-sea fishing, the Trobrianders were deep into magic. In his 1925 essay, "Magic, Science, and Religion," Malinowski concluded that this belief system served the function of dealing with the anxiety produced by uncertainty. Contrary to his fellow anthropologists of the time, who held a progressive "stage" theory of superstition—with so-called primitives and their superstitions on the bottom, and white Europeans with their science on top—Malinowski discovered that magical thinking derived from environmental conditions, not inherent stupidities: "We find magic wherever the elements of chance and accident, and the emotional play between hope and fear have a wide and extensive range. We do not find magic wherever the pursuit is certain, reliable, and well under the control of rational methods and technological processes. Further, we find magic where the element of danger is conspicuous."

In fact, Malinowski argues, it is as natural for humans to think scientifically as it is for them to think magically: "There are no peoples however primitive without religion and magic. Nor are there, it must be added at once, any savage races lacking either in the scientific attitude or in science." The same could be said for modern humans. It all depends on the environmental circumstances. Evolution gave us a large, complex, and malleable brain with certain built-in modules that respond to changing environments. In order to be successful hunters, fishers, and farmers, not to mention spouses, parents, and community members, any group of humans—Neanderthals, Cro-Magnons, Trobriand Islanders, or we—would need a certain understanding of and mastery over both the physical and social environments: "In all this they are guided by a clear knowledge of weather and seasons, plants and pests, soil and tubers, and by a conviction that this knowledge is true and reliable, that it can be counted upon and must be scrupulously obeyed." Malinowski discovered that humans inhabit two worlds—the sacred and the profane—with a clear-cut division between the two: "There is first the well-known set of conditions, the natural course of growth, as well as the ordinary pests and dangers to be warded off by fencing and weeding. On the other hand there is the domain of the unaccountable and adverse influences, as well as the great unearned increment of fortunate coincidence. The first conditions are coped with by knowledge and work, the second by magic." The Trobriand Islanders are a reflection of ourselves, our ancestors, and our common evolutionary heritage to think both causally and magically.

## THE MEDIEVAL BELIEF ENGINE

The relationship between the Belief Engine and the uncertainties and vagaries of life is clear in examining beliefs in the Middle Ages. Consider the fact that in medieval times 80 to 90 percent of the people were illiterate. Most could not even read the Bible, particularly since it was written in Latin, guaranteeing that it would remain the exclusive intellectual property of an elite few. Almost everyone believed in sorcery, werewolves, hobgoblins, witchcraft, and black magic. If a noblewoman died, her servants ran around the house emptying all containers of water so her soul would not drown. Her lord, in response to her death, faced east and formed a cross by lying prostrate on the ground, arms outstretched. If the left eye of a corpse did not close properly, the soul would spend extra time in purgatory (leading to the ritual closing of the eyes upon death). A man knew he was near death if he saw a shooting star or a vulture hovering over his home. If a wolf howled at night the one who heard him would disappear before dawn (one can imagine a campfire conversation: "I didn't hear anything, did you?" "No, not I."). Bloodletting was popular. Plagues were believed to be the result of an unfortunate conjuncture of the stars and planets. And the air was believed to be infested with such soulless spirits as unbaptized infants, ghouls who pulled out cadavers in graveyards and gnawed on their bones, water nymphs who lured knights to their deaths by drowning, drakes who dragged children into their caves beneath the earth, and vampires who sucked the blood of stray children.

Given the uncertainty and tenuousness of life in the Middle Ages, such superstitions should come as no surprise. In 1662 in England, for example, sixty out of every one hundred children never saw their seventeenth birthday. Life expectancy at birth of boys born in 1675 was thirty. Food supplies were unpredictable and plagues decimated dense but weakened populations. In London alone, there were six epidemics in the one hundred years spanning 1563 to 1665, wiping out between one-tenth and one-sixth of the population each time. Devastating fires routinely destroyed entire neighborhoods. Houses were made with thatched roofs and wooden chimneys, candles were the only source of light, and there were no safety matches. Firefighting techniques consisted of nothing more than throwing buckets of water, usually too little too late. There were no insurance companies, banks for personal savings, or any of the other security measures we take for granted in the modern world. Life really was, in Thomas Hobbes's apt phrase, "nasty, brutish, and short."

For the medieval mind, magical thinking provided an understanding of how the world worked: It attenuated anxiety and allowed people to shed personal responsibility by blaming events on bad luck, evil spirits, mischievous fairies, or God's will, and permitted one to cast blame on others through curses and witchcraft. Astrology, the most popular science of the day, invoked the alignment of the stars and planets to explain all manner of human and natural phenomena, the past, present, and future, and life's vagaries from daily events to yearly cycles. Only religion could rival astrology as an all-embracing explanation for the vicissitudes of life.

By the end of the seventeenth century Newton's mechanical astronomy had replaced astrology; the mathematical understanding of chance and probability displaced luck and fortune; chemistry succeeded alchemy; banking and insurance decreased human misfortune and its attendant anxiety; city planning and social hygiene greatly attenuated the power of plagues; and medicine began its long road toward a germ theory of disease. Cumulatively, these events pushed us into the Age of Science, reducing the number of thinking errors and attenuating the power of superstition. Nevertheless, magical thinking is still with us, rearing its head wherever uncertainties arise.

## THE MODERN BELIEF ENGINE

Ancestral and medieval superstitions survive in the modern high-tech world because the Belief Engine is a part of human nature. We see instances of this in everything from gambling (lucky streaks, cards, and dice) to athletic performances. In baseball, for example, where players are expected to hit a small, white ball traveling at nearly 100 miles per hour, superstition leads to all sorts of bizarre behaviors on the part of fully modern, educated human beings. Wade Boggs was famous for his superstitions, insisting on running his wind sprints at precisely 7:17 P.M., ending his grounder drill by stepping on the bases in backward order, never stepping on the foul line when taking the field but always stepping on it returning to the dugout, and eating chicken before every game. It is worth noting, however, that such superstitions are not at all uncommon among hitters where connecting with the baseball is so difficult and so fraught with uncertainties that the very best in the business fail a full seven out of every ten times at bat. Fielders, by contrast, typically succeed in excess of nine out of every ten times a ball is hit to them (the best succeed better than 95 percent of the time), and they have correspondingly fewer superstitions associated with fielding. But as soon as these same fielders pick up a bat, magical thinking goes into full swing.

Psychologist Stuart Vyse has documented such modern superstitions in an attempt to provide a psychological explanation for believing in magic. In addition to Boggs's bizarre behavior, he notes that mega-author Michael Crichton (a medical doctor and a writer who specializes in using science in such novels as *Jurassic Park*) eats the same thing for lunch every day while working on a new novel; and New York Giants football coach Bill Parcells, during his years in which the Giants were Superbowl champs in 1986 and 1991, would

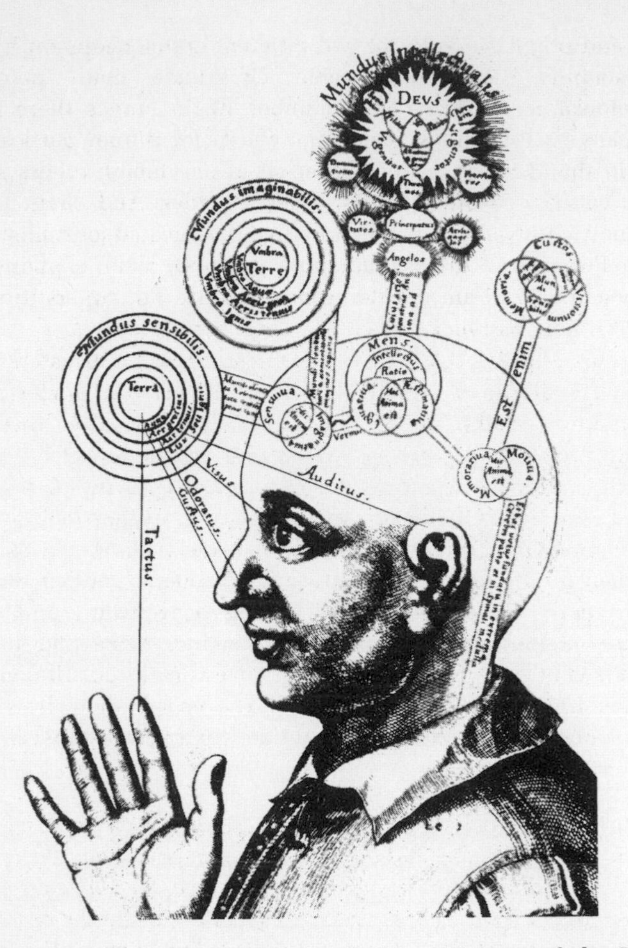


*The Medieval belief engine. Robert Fludd's "Mystery of the Human Head" from his* Ultrisque Cosmic Maioris Scilicet Et Minoris Metaphysica, *published in 1617, is an early attempt to understand the workings of the mind through science. Fludd's work is in the tradition of making correspondences between the macrocosm and the microcosm, between the universe and man. Here the celestial world, composed of God and angels, penetrates through the skull and into the soul. The four elements (left concentric circles—earth, water, air, and fire) communicate with the five senses. The "imaginable world" (middle concentric circles) corresponds to the metaphysical sensations, "as in dreams, by non-existent objects and, consequently, by the shadows of elements." The Mundus Intellectualis—the world of the intellect—is linked to the imagination through a series of spheres. At the back of the skull is the sphere of "memorative, or pertaining to remembrance," which Fludd shows connected to the intellect.*

stop and purchase coffee at two different coffee shops on his way to the stadium before every game. Of course, many people have triskaidekaphobia, fear of the number 13 (in France there is even a company that provides emergency guests for dinner parties to make certain that thirteen people never sit at one table); others still shun black cats and avoid walking under a ladder. And chain letters are routinely distributed, even by intelligent, educated journalists such as Gene Forman of the *Philadelphia Inquirer,* who explained: "You understand that I am not doing this because I'm superstitious. I just want to avoid bad luck." We understand perfectly.

In his gripping tale of the 1996 Mount Everest disaster, *Into Thin Air,* Jon Krakauer shows this relationship between uncertainty and superstition in the Belief Engines of the normally libertine and morally free-spirited Sherpas. When nearing the top of the mountain the Sherpas were overwhelmed by magical thinking. One action in particular was "forbidden by the mountain"—"sauce-making" (sexual liaisons between unmarried couples) in the tents above Base Camp. "Whenever the weather would turn nasty, one or another Sherpa was apt to point up at the clouds boiling heavenward and earnestly declare, 'Somebody has been sauce-making. Make bad luck. Now storm is coming.'" Sandy Pittman, the New York socialite who nearly lost her life during the 1996 expedition, posted the following diary entry on her Everest Web page about the superstitious Sherpas:

> *. . . a mail runner had arrived that afternoon with letters from home for everyone* and *a girlie magazine which had been sent by a caring climber buddy back home as a joke. . . . Half of the Sherpas had taken it to a tent for closer inspection, while the others fretted over the disaster they were certain that any examination of it would bring. The goddess Chomolungma, they claimed, doesn't tolerate "jiggy jiggy"*—anything *unclean—on her sacred mountain.*

The Belief Engine is real. It is normal. It is in all of us. Stuart Vyse shows, for example, that superstition is not a form of psychopathology or abnormal behavior; it is not limited to traditional cultures; it is not restricted to race, religion, or nationality; nor is it only a product of people of low intelligence or lacking in education. There is variance in magical thinking among individuals, of course, but *all* humans possess it because it is part of our nature, built into our neuronal mainframe. We do not live in a Pleistocene environment, but our minds were built there and often function as if we do. Witness a recent craze for mediums who say they can talk to the dead. It is a classic case of the Belief Engine at work. Because of the remarkable popularity of this particular claim, and the belief by millions of people that this alleged ability to speak with the dearly departed is evidence of a bridge to God and heaven, it is worth exploring it in detail. The phenomenon at work here tells us much about the very human need to believe, particularly when it is tethered to the tragedy of death.

## TALKING TWADDLE WITH THE DEAD

Throughout much of 1998 and 1999, the bestselling book in America was by a man who says he can talk to the dead (and so can you, if you buy his book). It turns out that our loved ones who have passed over are not really dead, just on another spiritual plane. All you have to do is fine-tune your frequencies and, like Sri Leachim Remresh, you too can turn off the Here and Now and tune into that Something Else.

I am referring to James Van Praagh, the world's most famous medium . . . for now anyway. He appeared three times, unopposed, on *Larry King Live.* He was featured on NBC's *Dateline,* and *The Today Show* and on ABC's *20/20.* He made the talk-show rounds, including Oprah (who was mildly skeptical) and Charles Grodin (who was not skeptical at all), and even had Charles Gibson on ABC's *Good Morning America* talking to his dead dad. Cher met with him to talk one last time with Sonny. Denise Brown received a reading to make a final connection with her sister, Nicole Brown Simpson. What is going on here? Who is James Van Praagh, and why do so many people believe in him?

### An Actor in Search of a Role

A brief glance at Van Praagh's biography is revealing. According to Alex Witchel of the *New York Times* (February 22, 1998), Van Praagh is the third of four children, born and raised Roman Catholic in Bayside, Queens, New York. At one point, he considered becoming a priest. He served as an altar boy and even entered a Catholic preparatory seminary—the Blessed Sacrament Fathers and Brothers in Hyde Park. His father is Allan Van Praagh, the head carpenter at the Royale Theater on Broadway (where his brother still works). His mother was Irish-Catholic and one of his sisters is a eucharistic minister. While attending college he found part-time work at the theater where, says Witchel, while the other stagehands were playing cards during the shows, Van Praagh "was out front watching, picking up pointers he still uses for his numerous television appearances." The lessons were well learned.

His college career was checkered, including enrollments at Queensboro Community College, State University of New York at Genesee, Hunter College, and, finally, San Francisco State University where he graduated with a degree in broadcasting and communications. Subsequently he moved to Los Angeles and began working in the entertainment industry, including Paramount Studios and a stint with the famed William Morris agency in Hollywood. He confesses in his book, *Talking to Heaven:* "I dreamed of a career as a screenwriter. As luck would have it, while coordinating a conference with the creative staff of *Hill Street Blues,* I became friendly with one of the show's producers. When I told him I would be graduating soon, he offered what I thought was my first big break." After graduation, Van Praagh moved to Hollywood where "I vowed that I would not leave Tinsel Town until I realized my dream and became a writer." Through a job at William Morris, Van Praagh met a medium who told him: "You know, James, you are very mediumistic. The spirit people are telling me that one day you will give readings like this to other people. The spirits are planning to use you." Van Praagh had found his role in Hollywood. He would act the part of a spirit medium.

In 1994 he was discovered by NBC's *The Other Side,* for whom Van Praagh made numerous appearances in their exploration of the paranormal. This, and other media appearances, generated countless personal and group readings, pushing him above the psychic crowd and eventually leading to his status as a bestselling author.

Who does James Van Praagh say he is? According to his own Web page, "Van Praagh is a survival evidence medium, meaning that he is able to bridge the gap between two planes of existence, that of the living and that of the dead, by providing evidential proof of life after death via detailed messages." Van Praagh calls himself a "clairsentient," or "clear feeling," where he can allegedly "feel the emotions and personalities of the deceased." His analogue, he says, is "Whoopi Goldberg in *Ghost.*" He claims that the "spirits communicate by their emotions," and even though they do not speak English or any other language, they can tell you, for example, "that you changed your pants because of a hole in the left seam or that you couldn't mail letters today because the stamps weren't in the bottom right desk drawer." He readily admits that he makes mistakes in his readings (there are so many he could hardly deny it), rationalizing it this way: "If I convey recognizable evidence along with even a fraction of the loving energy behind the message, I consider the reading successful." In other words, if he can just get a few hits, then "convey" the all-important emotional stroking that your loved one still loves you and is happy in heaven, he has done his job. From the feedback of his clients, this is all most people need.

The forty-year-old medium's message cuts to the core of most people's deepest fear and loftiest desire, as he told the *New York Times:* "When a reunion between the living and the dead takes place it may be the first time the living understand that death has not robbed them of the love they once experienced with family and friends on the earth plane. With the knowledge of no death, they are free to live life." No one has explained the attraction of this message better than Alexander Pope did over two and a half centuries ago, in his 1733 *Essay on Man:*

> Hope springs eternal in the human breast;
> Man never Is, but always To be blest.
> The soul, uneasy, and confin'd from home,
> Rests and expatiates in a life to come.

By itself, however, this does not explain precisely *how* our Belief Engine drives us to be compelled to believe such claims. Why are we so willing to suspend disbelief when it comes to the afterlife?

## Gambling on the Afterlife

By way of analogy, consider the gambling games of Las Vegas. Gaming is big business, as anyone can see driving down the ever-burgeoning neon-glaring strip. In fact, gambling is the best bet in business, far superior to the stock market, *as long as you are the house.* With only a tiny advantage on any given game, and heaps of customers playing lots of rounds, the house is guaranteed to win. For the roulette wheel, for example, with eighteen red slots, eighteen black slots, and two green slots (zero and double zero), the take is only 5.26 percent. That is, by betting either black or red, you will win eighteen out of thirty-eight times, or 47.37 percent, whereas the house will win twenty out of thirty-eight times, or 52.63 percent. If you placed one hundred $1.00 bets, you would be out $5.26, on average. This may not sound like a lot, but cumulatively over time, with millions of gamblers betting billions of dollars every year, the house take is significant. Other games are better for gamblers. For straight bets in craps, the house take is a mere 1.4 percent; for blackjack, with the most liberal rules and optimal (non-card-counting) player strategies, the house earns just under 1 percent. These are the best games to play if you are a gambler (that is to say, you will lose more slowly). With other games it is downhill for the gambler. The take for some slot machines, for example, is a staggering 25 percent. That is, you are losing 25 cents on the dollar, or, the house wins 62.5 percent and you win 37.5 percent of the time. Yet people still play. Why?

As B. F. Skinner showed in rats, pigeons, and humans, organisms do not need steady reinforcement to continue pressing a bar, pecking a plate, or pulling a one-armed bandit (slot machine). Intermittent reinforcement will do just as well, and sometimes even better, at eliciting the desired behavior. A "Variable Ratio Schedule" of reinforcement turns out to be the best for gambling games, where the payoff is unpredictably variable, depending on a varying rate of responses. Payoff comes after ten pulls, then three pulls, then twelve pulls, then seven pulls, then twenty-three pulls, and so on. When I was a graduate student in experimental psychology in the mid-1970s I worked in an operant laboratory where we created these variable schedules of reinforcement for our subjects. It is remarkable how infrequently the payoffs need to come to keep the subjects motivated. And this was for such basic rewards as sugar water (rats), seed (pigeons), and money (humans). Imagine how much more motivating, and, correspondingly, lower the rate of reinforcement can be, when the reward is the belief that your lost loved ones are not really dead and, as an added bonus, you can speak with them through a medium. This renders intelligible, in part, the success of someone like James Van Praagh, whose hit rate is far below that of even the lowest-paying gambling games in Las Vegas. It also helps explain the more general case of how we believe.

I once sat in on a day of readings with Van Praagh and kept a running tally of his ratio of hits and misses for each of ten subjects (one of whom was me), all filmed for NBC's *Unsolved Mysteries*. Being generous with what kind of information counted as a "hit," Van Praagh averaged five to ten hits for every thirty questions/statements, or 16 to 33 percent—significantly below that of roulette where the player wins almost half the time. But because Van Praagh's payoff is the hope of life after death and a chance to speak with a lost loved one, people are exceptionally forgiving of his many misses. Like all gamblers, Van Praagh's clients only need an occasional hit to convince them.

## How to Talk to the Dead

Watching James Van Praagh work a crowd or do a one-on-one reading is an educational experience in human psychology. Make no mistake about it, this is one clever man. We may see him, at best, as morally reprehensible, but we should not underestimate his genuine theatrical talents and his understanding, gained through years of experience speaking with real people, of what touches off some of the deepest human emotions. Van Praagh masterfully uses his ability and learned skills in three basic techniques he uses to "talk" to the dead:

1. *Cold Reading.* Most of what Van Praagh does is what is known in the mentalism trade as cold reading, where you literally "read" someone "cold," knowing nothing about them. He asks lots of questions and makes numerous statements, some general and some specific, and sees what sticks. Most of the time he is wrong. His subjects visibly shake their heads "no." But he only needs an occasional strike to convince his clientele he is genuine.

2. *Warm Reading.* This is utilizing known principles of psychology that apply to nearly everyone. For example, most grieving people will wear a piece of jewelry that has a connection to their loved one. Katie Couric on *The Today Show,* for example, after her husband died, wore his ring on a necklace when she returned to the show. Van Praagh knows this about mourning people and will say something like "Do you have a ring or a piece of jewelry on you, please?" His subject cannot believe her ears and nods enthusiastically in the affirmative. He says, "Thank you," and moves on as if he had just divined this from heaven. Most people also keep a photograph of their loved one either on them or near their bed, and Van Praagh will take credit for this specific hit that actually applies to most people.

   Van Praagh is facile at determining the cause of death by focusing either on the chest or head areas, and then exploring whether it was a slow or sudden end. He works his way down through these possibilities as if he were following a computer flowchart and then fills in the blanks. "I'm feeling a pain in the chest." If he gets a positive nod, he continues. "Did he have cancer, please? Because I'm seeing a slow death here." If he gets the nod, he takes the hit. If the subject hesitates at all, he will quickly shift to heart attack. If it is the head, he goes for stroke or head injury from an automobile accident or fall. Statistically speaking there are only half a dozen ways most of us die, so with just a little probing, and the verbal and nonverbal cues of his subject, he can appear to get far more hits than he is really getting.

**3.** *Hot Reading.* Mentalist Max Maven informs me that some mentalists and psychics also do "hot" readings, where they obtain information on a subject ahead of time. I do not know if Van Praagh does research or uses private detectives to get information on people, but I have discovered from numerous television producers that he consciously and deliberately pumps them for information about his subjects ahead of time, then uses that information to deceive the viewing public that he got it from heaven. Leah Hanes, for example, who was a producer and researcher for NBC's *The Other Side,* explained to me how Van Praagh used her to get information on guests during his numerous appearances on the show (interview on April 3, 1998):

> *I can't say I think James Van Praagh is a total fraud, because he came up with things I hadn't told him, but there were moments on the show when he appeared to be coming up with fresh information that he got from me and other researchers earlier on. For example, I recall him asking about the profession of the deceased loved one of one of our guests, and I told him he was a fireman. Then, when the show began, he said something to the effect, "I see a uniform. Was he a policeman or fireman, please?" Everyone was stunned, but he got that directly from me.*

## Deception or Self-Deception?

When I first began following Van Praagh I thought perhaps there was a certain element of self-deception on his part where, giving him the benefit of the doubt (he does appear likable), he developed his cold- and warm-reading techniques through a gradual developmental process of subject feedback and reinforcement, much like how gurus come to believe in their own divinity when enough of their followers tell them they are divine.

Human behavior is enormously complex, so I suppose it is possible that Van Praagh is *both* deceiving and self-deceiving, but over the years I have observed much more of the former than the latter. During the *Unsolved Mysteries* shoot, which lasted ten hours and was filled with numerous breaks, Van Praagh would routinely make small talk with us, asking lots of questions and obtaining information, which he subsequently used to his advantage when the cameras were rolling.

Is it possible he does not consciously realize that he is doing this? I contacted numerous mentalists about Van Praagh and they assured me that it is very unlikely he is self-deceiving because these are techniques that they all use, and they do so consciously and purposefully. I was told that I was being naive in trying to give Van Praagh the benefit of the doubt. I spoke to an individual who works a 900-number psychic hotline, who knows Van Praagh and many of the people who work with him in that industry, and he assures me that Van Praagh is *not* self-deceived. The psychic industry consensus, this source tells me, is that James Van Praagh knows exactly what he is doing.

That may be so, but as a general principle self-deception is a powerful tool because if you believe the lie yourself your body is less likely to give off telltale clues, making it more difficult for an observer to detect deception. I am fully convinced that cult leaders, after being told for years by hundreds and thousands of followers that they are special, at some point begin to believe it themselves, making them all the more convincing to other and potential followers.

## Caught Cheating

Even for seasoned observers it is remarkable how Van Praagh appears to get hits, even though a closer look reveals how he does it. When we were filming the *20/20* piece for ABC, I was told that overall he had not done well the night before, but that he did get a couple of startling hits—including the name of a woman's family dog. But when we reviewed the videotape, here is what actually happened. Van Praagh was failing in his reading of a gentleman named Peter, who was poker-faced and obviously skeptical (without feedback Van Praagh's hit rate drops significantly). After dozens of misses Van Praagh queried: "Who is Charlie?" Peter sat there dumbfounded, unable to recall if he knew anyone of significance named Charlie, when suddenly the woman sitting in back of him—a complete stranger—blurted out "Charlie was our family dog." Van Praagh seized the moment and proclaimed that he could see Charlie and this woman's Dad taking walks in heaven together. Apparently Van Praagh's psychic abilities are not fine-tuned enough to tell the difference between a human and a dog.

The highlight of the *20/20* piece, however, was a case of hot reading. On a break, with a camera rolling, while relaxing and sipping a glass of water, Van Praagh suddenly called out to a young woman named Mary Jo: "Did your mother pass on?" Mary Jo shook her head negatively, and then volunteered: "Grandmother." Fifty-four minutes later Van Praagh turned to her and said: "I want to tell you, there is a lady sitting behind you. She feels like a grandmother to me." The next day, when I was shown this clip, one of the line producers said, "You know, I think he got that on the break. Too bad we don't have it on film." After checking they discovered they did, so Van Praagh was caught red-handed. When confronted by *20/20* correspondent Bill Ritter with the video clip, however, Van Praagh demurred: "I don't cheat. I don't have to prove . . . I don't cheat. I don't cheat. I mean, come on. . . ." Interesting. No one said anything about cheating. The gentleman doth protest too much.

As an example of the power of the Belief Engine, even after we caught Van Praagh cheating, Barbara Walters concluded in the wrap-up discussion: "I was skeptical. I still am. But I met James Van Praagh. He didn't expect to meet me. He knew that my father's name was Lew—Lewis he said—and he knew that my father had a glass eye. People don't know that." Ritter, doing his homework on this piece to the bitter end, explained: "You told me the story yesterday and I told you I would look and see what I could find out. Within a few minutes I found out that your father's name was Lew and that he was very well known in show business. And this morning I was looking in a book and found a passage that says he was blind in one eye—an accidental incident as a child—and he had a glass eye. If I found that out, then he could have." While Walters flustered in frustration, Hugh Downs declared without qualification: "I don't believe him."

Where have we heard all this before? A hundred years ago, when mediums, seances, and spiritualism were all the rage in England and America, Thomas Henry Huxley concluded, as only he could in his biting wit, that as nonsensical as it was, spiritual manifestations might at least reduce suicides: "Better live a crossing-sweeper than die and be made to talk twaddle by a 'medium' hired at a guinea a seance."

## The Tragedy of Death

The simplest explanation for how James Van Praagh can get away with such an outrageous claim on such questionable techniques is that he is dealing with a subject the likes of which it would be hard to top for tragedy and finality—death. Sooner or later we all will face this inevitability, starting, in the normal course of events, with the loss of our parents, then siblings and friends, and eventually ourselves. It is a grim outcome under the best of circumstances, made all the worse when death comes early or accidentally to those whose "time was not up." As those who traffic in the business of loss, death, and grief know all too well, we are often at our most vulnerable at such times. Giving deep thought to this reality can cause the most controlled and rational among us to succumb to our emotions.

I experienced the full force of this reality on April 2, 1998. The events of that day prompted me to consider what I would say to someone who is grieving. The ABC television program *20/20* came to my home and office, then followed me to Occidental College to shoot some background footage in my critical-thinking course. I thought I would ask the students to respond to a question I routinely receive from journalists: "What's the harm in what James Van Praagh does?" The students had plenty to say, but one woman named Melissa told a personal story about how her Dad had died when she was ten and that she had never really gotten over it. She was sad that her father never got to see her play volleyball or basketball, or to see her graduate from high school. Her opinion of James Van Praagh was less than charitable, to say the least. She could not imagine how such a performance could make someone feel better about death. In a maturity beyond her years, she expressed her opinion that one does not really get over such a loss; one just learns to live with it: "When my dad first died I just wanted to get on with my life and not let it bother me too much, now I'm just trying not to forget him. Next year when I turn twenty I will have lived ten years with my Dad and ten years without him . . . so I guess that is when my life will begin . . . like a new chapter or something." At this point she was fighting back her tears. It was a very touching moment.

When I returned home I was preparing to send Melissa an e-mail expressing how tragic it must have been to lose her Dad at such a young age, when I read this e-mail from my sister:

> *I was thinking of Dad today on this 12th anniversary and how proud he would have been of you and all you have accomplished with your life. For some reason, I have really been missing him lately, more than I have in a long time and it's still so hard to be without him. I really hope there is a heaven, even though I know otherwise, but the thought of never seeing him again, ever, is almost too hard to bear.*
>
> *Love you, Tina.*

Our father died twelve years ago that day, April 2, 1986, and it is probably a good thing I had not realized that in class as it would have been very difficult to remain composed.

This was such a peculiar conjuncture of events that it prompted me to give some thought about what I would say to someone experiencing grief. Having watched James Van Praagh now for more than five years, I would imagine he might say something to this effect:

*It's okay Melissa, your Dad is here now in the room with us. He's telling me he loves you. He says he watches over you. He loves watching you play basketball and volleyball. He saw you graduate. He is with you always. Don't be sad. Don't cry. You will get to see him again. Everything is fine.*

My response to Melissa, and to everyone who has ever received a "reading" from Van Praagh, is as follows:

*First of all, no one knows if any of this is true, but even if it is, why would your loved one talk with this guy you don't even know? Why would he choose to make his appearance in some television studio or at some hotel conference room with hundreds of other people around? Why doesn't he talk to you instead? You're the one he loves, not this guy getting $40 a seat in a hall with 400 people, or $200 a private reading, or two million dollars for a book filled with this sort of drivel. Why do you have to pay someone to talk to your loved one?*

In the *St. Louis Post Dispatch* (March 1, 1998) Van Praagh called me a "rat fink." I take this as a compliment because to "rat" on someone is to tell the truth about them. In Mafia circles it means a crime has been exposed. On the *20/20* show Van Praagh offered this view of the difference between my work and his: "He makes his life beating people down, putting people down. I make my life healing and bringing people up. I'm not a circus act. I'm not a side show. It's God's work." By now nearly everyone in America has heard what James Van Praagh says to aching hearts. Here is what I might say. It is not God's work, but you judge who is putting people down or bringing them up. To Melissa, to my sisters, Tina and Shawn, and to my own daughter, Devin, should I die before my time, I close with this statement:

*I am sorry this happened to you. It isn't fair. It isn't fair at all. If I were you I would feel cheated and hurt; I might even be angry that I didn't get more time with my Dad. You have every right to feel bad. If you want to cry, you should. It's okay. It's more than okay. It's human. Very human. All loving, caring people grieve when those they love are gone. And all of us, every last one of us, will experience this feeling at some point in our lives. Sometimes we grieve very deeply and for a very long time. Sometimes we get over it and sometimes we do not. Mostly we get on with our lives because there is nothing else we can do. But loving, caring people continue to think about their loved ones no matter how far they have gotten on with their lives, because our lost loved ones continue to live. No one knows if they really continue to live in some other place—I suspect not—but we do know for sure, with as much certainty as any scientific theory or philosophical argument can muster, that our loved ones continue to live in our memories and in our lives. It isn't wrong to feel sad. It is right. Self-evidently right. It means we love and can be loved. It means our loved ones continue to live because we continue to miss them. Tears of sadness are really tears of love. Why shouldn't you cry for your Dad? He's your Dad and you love him. Don't let anyone try to take that away from you. The freedom to grieve and love is one of the fundamentals of being human. To try to take that freedom away on a chimera of feigned hope and promises that cannot be filled is inhuman. Celebrate your love for your Dad in every way you can. That is your right, your freedom, your humanness.*

Chapter 4

# Why People Believe in God

## *An Empirical Study on a Deep Question*

*The belief in God has often been advanced as not only the greatest, but the most complete of all the distinctions between man and the lower animals. It is however impossible, as we have seen, to maintain that this belief is innate or instinctive in man. On the other hand a belief in all-pervading spiritual agencies seems to be universal; and apparently follows from a considerable advance in the reasoning powers of man, and from a still greater advance in his faculties of imagination, curiosity and wonder.*

—*Charles Darwin,* The Descent of Man, *Vol. II, 1871, p. 395*

Several years ago I attended a most unusual conference at the Santa Monica Miramar Sheraton Hotel in Southern California, sponsored by the Extropy Institute. Founded in 1988, the "Extropians" are dedicated to studying "transhumanism and futurist philosophy; life extension, immortalism, and cryonics; smart drugs and intelligence-increase technologies; machine intelligence, personality uploading, and artificial life; nanocomputers and nanotechnology; memetics (ideas as genes); effective thinking and information filtering; self-transformative psychology; rational market-based environmentalism; and probing the ultimate limits of physics." Limited in scope Extropians are not.

Led by Max More and Tom Morrow (not surprisingly, these are pseudonyms), the Extropians, one might reasonably assume, are a bunch of kooks on the lunatic fringe. They are not. The Extropians are on the cutting edge between science and science fiction, fact and fantasy. Conference speakers included Massachusetts Institute of Technology (MIT) artificial intelligence guru Marvin Minsky, University of California–Los Angeles (UCLA) pathologist and aging expert Roy Walford, and University of Southern California (USC) fuzzy logic founder Bart Kosko. They presented a mix of hard, scientific facts and soft, fanciful hopes about the future. Slides illustrating data were blended with inspirational orations. After Kosko gave a fact-filled summation of the science of fuzzy logic, a gentleman named FM-2030 (his legal name) delivered a sermon that would have been the envy of Billy Graham. By the year 2030, FM explained, nation-states will be obsolete, money will be purely electronic, computers will have near-human intelligence, and it will be possible to achieve considerable life extension in the range of hundreds or thousands of years, if not actual physical immortality. I spent several hours with FM and found him to be a most fascinating man, globally conscious (he has no permanent residence), open to all peoples and cultures (he refused to identify his race or accent, simply stating that he is human), interested in any science or technology that can be used to the betterment of humanity (he is eagerly awaiting his global cellular phone number), and ceaselessly optimistic about the future (he figures he will make it to 2030, and thus into centuries and millennia to come). If there was anyone for whom I would say that hope springs eternal it would be FM-2030, Max More, and this colorful band of Extropians.

What is perhaps most striking about this group, however, is the quasi-religious nature of their beliefs, including an almost faithlike devotion to science as a higher power. Scientism is their religion, technocracy their politics, progress their God. They hold an unmitigated confidence that because science has solved problems in the past, it will solve all problems in the future, including the biggest one of all—death. Why not follow the curve of scientific progress to its ultimate end, they argue? Medical science has cured many of the world's major diseases—why not eventually all of them, including aging? They point optimistically to "Moore's law" (in 1965 Gordon Moore, founder of Intel, accurately predicted that the density of transistors on integrated circuits would double every eighteen months) and speak fondly of nanotechnology, where some day computers will be the size of cells, capable of being injected into our bodies to repair organs, maintain tissues and systems, and eradicate cancers and other destructive agents.

Since death is something most of us would like to transcend, we must be particularly skeptical of claims that play on this deepest of all human desires, be it religiously or scientifically based. It is doubtful that the Extropians are right in their prediction that one day we will live into the thousands of years, if not achieve actual immortality. But I must admit it is fun to think about and occasionally, in quiet moments, I wonder . . . what if they are right?

In his 1996 book *Leaps of Faith,* psychologist Nicholas Humphrey speculates that true believers are in search of "supernatural consolation." There is what he calls a "paranormal fundamentalism" among the faithful who maintain "an unshakable conviction that no matter what the evidence, 'there must be something there.'" I would take this a step further and suggest that for *all of us* it is tempting to believe that "there must be something there." For secular religions like Marxism, Something There is the force of linear history inexorably marching through the stages of economic development toward communism. For capitalists, Something There is the invisible hand gently guiding markets to produce higher-quality products at lower prices. For Extropians, Something There is the vision of a paradisiacal future of longevity, intelligence, health, and wealth, delivered on the wings of scientific imagination. For some, science, or more precisely *scientism,* is a secular religion in the sense of generating loyal commitments (a type of faith) to a method, a body of knowledge, and a hope for a better tomorrow. Perhaps seeing Something There is partly hard-wired in us all.

## SEEING THE PATTERN OF GOD

As we have seen, humans are pattern-seeking animals. Our brains are hard-wired to seek and find patterns, whether the pattern is real or not. Psychologist Stuart Vyse demonstrated this in his research with his colleague Ruth Heltzer, in an experiment in which subjects participated in a video game, the goal of which was to navigate a path through a matrix grid using directional keys to move the cursor. One group of subjects were rewarded with points for successfully finding a way through the grid's lower right portion, while a second group of subjects were rewarded points randomly. Both groups were subsequently asked to describe how they thought the points were rewarded. Most of the subjects in the first group found the pattern of point scoring and accurately described it. Interestingly, most of the subjects in the second group also found "patterns" of point scoring, even though no pattern existed and the points were rewarded randomly. We seek and find patterns because we prefer to view the world as orderly instead of chaotic, and it is orderly often enough that this strategy works. In an ironic twist, it would appear that we were designed by nature to see in nature patterns of our design. Those patterns have to be given an identity, and for thousands of years many of those identities were called gods.

In his 1993 book, *Fuzzy Thinking,* Bart Kosko suggests that belief in God may be something similar to what we see when we look at the pattern in the Kanizsa-square illusion. The experience, Kosko suggests, is not unlike "our vague glimpses of God or His Shadow or His Handiwork . . . an illusion in the neural wiring of a creature recently and narrowly evolved on a fluke of a planet in a fluke of a galaxy in a fluke of a universe." The neural wiring in our brain creates "neural nets," or the sequence of neurons and the gaps between neurons called synapses, that together operate in the brain to store memory and pattern information. "These God glimpses or the feeling of God recognition," Kosko intimates, "may be just a 'filling in' or *déjà-vu* type anomaly of our neural nets." The Kanizsa square works to create the illusion of a square that is not really there. The four little

*The Kanizsa-square illusion works by fooling the mind into thinking there is a square. All that is seen are four figures turned at right angles to create four false boundaries and a bright interior. Perhaps God is an illusion of the mind, generated by the false boundaries and bright interiors of the universe.*

Pac-man figures are turned at right angles to one another to create four false boundaries and a bright interior. But there is no square in this figure. The square is in our mind. There appears to be Something There, when in actual fact there is nothing there. As pattern-seeking animals it is virtually impossible for us *not* to see the pattern. The same may be true for God. For most of us it is very difficult not to see a pattern of God when looking at the false boundaries and bright interiors of the universe.

Do people see the pattern of God in the world and in their lives, and therefore believe in God for perfectly rational reasons? And if they do, does that pattern represent Something There or nothing there? Or are there other reasons people believe, such as an emotional need, a fear of death, a hope for immortality, an explanation for evil and suffering, a foundation for morality, parental upbringing, cultural influence, historical momentum, and so on? To find out I decided to do what I always do when I want to know why people believe something—ask. I started off by asking skeptics—defined simply as readers of *Skeptic* magazine and members of the Skeptics Society—if they believe in God, why or why not, and why they think other people do. I then asked a random sample of the American population (defined by a professional polling agency, which provided the database) the same set of questions. The results were most enlightening. But first we must consider another issue: Is the propensity to believe in God hard-wired, either genetically or in the brain?

## IS BELIEF IN GOD GENETICALLY PROGRAMMED?

The renowned British psychologist Hans Eysenck, not noted for timidity in commenting on controversial issues, rang in on the God Question with this quip: "I think there's a gene for religiosity and I regret that I don't have it." Is there a gene for religiosity? No, any more than there is a gene for intelligence, aggression, or any other complex human expression. Such phenomena are the product of a complex interactive feedback loop between genes and environment, where *many* genes code for a *range of reactions* to environmental stimuli. The relative role of genes and environment would be impossible to tease apart were it not for the natural experiment of identical twins separated at birth and raised in relatively different environments. Intuitively it seems as if something as culturally variable as religion would be primarily, if not completely, the product of one's environment. Indeed, as late as 1989, Robert Plomin concluded that "religiosity and certain political beliefs . . . show no genetic influence." So pervasive is this presumption, in fact, that behavioral geneticists have used religiosity as a control variable in their studies of twins, while exploring other variables that could possibly be strongly influenced by genetics.

This assumption is beginning to change. Behavioral geneticist Thomas J. Bouchard Jr. directed the famous "Minnesota twins" study, one of the best known and most extensive studies to date. Bouchard and his colleagues have attempted to cleave the relative influence of nature and nurture on a number of variables long thought to be primarily under the control of the environment, including personality, political attitudes, and even religiosity. Studying fifty-three pairs of identical twins and thirty-one pairs of fraternal twins reared apart, looking at five different measures of religiosity, the researchers found that the correlations between identical twins were typically double those for fraternal twins, "suggesting that genetic factors play a significant role in the expression of this trait." How significant? While admitting that their findings "indicate that individual differences in religious attitudes, interests and values arise from both genetic and environmental influences . . . genetic factors account for approximately 50 percent of the observed variance on our measures." That is to say, about one-half of the differences among people in their religious attitudes, interests, and values is accounted for by their genes. After offering a proviso that much more research needs to be done in this area, and that this single study must be replicated, the twin-study experts concluded: "Social scientists will have to discard the a priori assumption that individual differences in religious and other social attitudes are solely influenced by environmental factors." Nancy Segal, in her 1999 book on twins, *Entwined Lives,* points out that genes, of course, do not determine whether one chooses Judaism or Catholicism, rather, "religious interest and commitment to certain practices, such as regular service attendance or singing in a choir, partly reflect genetically based personality traits such as traditionalism and conformance to authority." Clearly the fact that identical twins reared apart are more similar in their religious interests and commitments than fraternal twins reared together indicates that we cannot ignore heredity in our search to understand why people believe in God.

Taken at face value, a 50 percent heritability of religious tendencies may sound like a lot, but that still leaves the other half accounted for by the environment. Given the range of variables that individuals encounter in their religious experiences, there is much research still to be conducted. Virtually all studies implemented over the past century have found strong environmental factors in religiosity, including everything from family to class to culture. In other words, even with a genetic component to religiosity we still must examine other variables.

## IS THERE A GOD MODULE IN THE BRAIN?

During the month of October 1997 the media had a field day when the renowned University of California–San Diego neuroscientist, Dr. Vilayanur Ramachandran, delivered a paper at the annual meeting of the Society for Neuroscience, entitled "The Neural Basis of Religious Experience." One reporter stood outside Ramachandran's office and declared, "Inside this building scientists have discovered the God module." Robert Lee Hotz for the *Los Angeles Times* reported: "In what researchers called the first serious experiment aimed at the neural basis of religion, scientists at the UC San Diego brain and perception laboratory this week said they found evidence of neural circuits in the human brain that affect how strongly someone responds to a mystical experience. As evidence of how brain cells and synapses might process spiritual stirrings, the experiment suggests a physical basis for a religious state of mind." Hotz followed up six months later in the *Los Angeles Times* with a deeper analysis of "the biology of spirituality," in which he explored just how far science might go with this line of research. "The issues are huge," explained Robert John Russell, director of the Center for Theology and Natural Science in Berkeley. USC neuroscientist Michael Arbib agreed: "We cannot approach theology without some sense of the intricacy of the human brain. A lot of what people hold as articles of faith are eroded by neuroscience." And Nancey Murphy, from the Fuller Theological Seminary in Pasadena, rationalized the problem to Hotz this way: "If we recognize the brain does all the things that we [traditionally] attributed to the soul, then God must have some way of interacting with human brains."

Specifically, what Ramachandran said was that an individual's religiosity may depend on how enhanced a part of the brain's electrical circuitry becomes: "If these preliminary results hold up, they may indicate that the neural substrate for religion and belief in God may partially involve circuitry in the temporal lobes, which is enhanced in some patients." Using electrical monitors on subjects' skin (a skin conductance response commonly used to measure emotional arousal) Ramachandran and his colleagues tested three types of "emotional stimuli": religious, violent, and sexual, in three populations: (1) temporal lobe epilepsy (TLE) patients who had religious preoccupations, (2) normal "very religious" people, and (3) normal nonreligious people. In groups 2 and 3 Ramachandran found skin conductance response to be highest to sexual stimuli, whereas in the first group the response was strongest to religious words and icons, significantly above the religious control group.

Ramachandran considered three possible (but not mutually exclusive) hypotheses to explain his findings: (a) that the mystical reveries led the patient to religious beliefs; (b) that the facilitation of connections between emotion centers of the brain, like the amygdala, caused the patient to see deep cosmic significance in everything around him or her that is similar to religious experiences; (c) that there may be neural wiring in the temporal lobes focused on something akin to religion. Research other than Ramachandran's tends not to support the first hypothesis, which leaves b and c the likeliest explanations of the findings. Psychiatric and neurological patients experiencing hallucinations, for example, do not necessarily exhibit religious propensities, but TLE patients, when shown religious words, as well as words with sexual or violent connotations, showed much higher emotional response to the religious words. Cautious not to offend, Ramachandran concluded with this disclaimer: "Of course, far from invalidating religious experience this merely indicates what the underlying neural substrate might be."

Related to Ramachandran's research, with implications for both supernatural and paranormal beliefs, is the work of Michael Persinger at Laurentian University in Sudbury, Canada. Persinger places a motorcycle helmet specially modified with electromagnets on the subject's head, who lies in a comfortable recumbent position in a soundproof room with eyes covered. The electrical activity generated by the electromagnets produces a magnetic field pattern that stimulates "microseizures" in the temporal lobes of the brain which, in turn, produces a number of what can best be described as "spiritual" and "supernatural" experiences—the sense of a presence in the room, an out-of-body experience, bizarre distortion of body parts, and even religious feelings. Persinger calls these experiences "temporal lobe transients," or increases and instabilities in neuronal firing patterns in the temporal lobe. These "transients" are not unlike the seizures studied by Ramachandran. How do they produce religious states? Our "sense of self," says Persinger, is maintained by the left hemisphere temporal cortex. Under normal brain functioning this is matched by the corresponding systems in the right hemisphere temporal cortex. When these two systems become uncoordinated, such as during a seizure or a transient event, the left hemisphere interprets the uncoordinated activity as "another self," or a "sensed presence," thus accounting for subjects' experiences of a "presence" in the room (which might be interpreted as angels, demons, aliens, or ghosts), or leaving their bodies (as in a near-death experience), or even "God." When the amygdala is involved in the transient events, emotional factors significantly enhance the experience which, when connected to spiritual themes, can be a powerful force for intense religious feelings.

Persinger got his start in this field when he began to explore the possibility that electromagnetic disturbances in the earth's crust during earthquakes may cause such anomalies as ball lightning and other unusual atmospheric phenomena. From there he thought that perhaps earthquakes generate weak magnetic fields that could cause individuals to experience such paranormal phenomena as alien abductions and out-of-body experiences. Having now studied more than 600 subjects in the past decade, Persinger speculates that such transient events may account for psychological states routinely reported as happening outside the mind. These events, he suggests, may be triggered by the stress of a near-death experience (caused by an accident or traumatic surgery), high altitudes, fasting, a sudden decrease in oxygen, dramatic changes in blood sugar levels, and other stressful events.

In my 1997 book, *Why People Believe Weird Things,* I recount in detail my own alien abduction experience triggered by 83 hours of sleeplessness and riding a bicycle 1,259 miles without stopping (as part of the nonstop transcontinental bike race called Race Across America). I was, therefore, curious to experience Persinger's research firsthand, which a trip to his laboratory for a television program on the paranormal allowed me to do. The effects, Persinger explained, are subtle for most subjects, dramatic for a few. His lab assistants strapped me into the helmet, hooked up the EEG and EKG machines (to measure brain waves and heart rate), and sealed me in the soundproof room. I initially felt giddiness, as if the whole process were a silly exercise that I could easily control. Then I slumped into a state of melancholy. Minutes later, still believing the magnetic field patterns were ineffectual, I felt like part of me wanted to have an out-of-body experience, but my skeptical/rational mind kept pulling me back in. It was then I realized that it was the magnetic field patterns causing these experiences, but that I was fighting them. I concluded that the more fantasy prone the personality, the more emotional/spiritual would be the experience. Persinger confirmed my informal hypothesis in a post-experiment debriefing. In a large population there will be a wide range of mental experiences, with the more fantasy prone people interpreting these as being *outside* the mind (demons, spirits, angels, ghosts, aliens, God), and the more rationally prone people interpreting these as being *inside* the mind (lucid, dreams, hallucinations, fantasies).

There is, in fact, a long history of research into the possibility of mental states being equated with the presence of God and other supernatural beings, dating back to the mid-nineteenth century. The classic work in the field was Alexandre Brierre de Boismont's 1859 *On Hallucinations: A History and Explanation of Apparitions, Visions, Dreams, Ecstasy, Magnetism, and Somnambulism.* Brierre de Boismont, a French medical doctor, examined the relationship between hallucinations and a number of conditions, including "morality and religion." It is in the nature of man, he explained, to form a "mental representation of objects." We do this, for example, when we call up "the recollection of a friend, a landscape, or a statue" but these images, without practice, are "indistinct and obscure" and "still inferior to the original." With practice, however, the representation becomes much more realistic (as tested in a series of experiments with artists' models, who appeared for a fixed duration, followed by the artists' rendition from memory). Such mental representations can be produced in extended meditation, so a hallucination must be the product of something resembling the normal process of mental representation and not a state of disease—physiology, not pathology: "I believe I am justified in concluding that the phenomena—apparently so dissimilar—of sensorial *perception* or *sensation;* of voluntary and normal mental *representation (memory, imagination, conception),* and of involuntary and abnormal mental representation *(illusions, hallucinations)*—result from the operation of one and the same psycho-organic faculty, acting under different conditions, and with apparent degrees of intensity." If intense enough, the hallucination seems *"exterior* and at *some distance* from the *ego,"* and thus "the person *sees* and *believes.*" Especially intense hallucinations, as produced through reverie and meditation, and with religious overtones, are particularly effective, where "everything concurred to favour the production of hallucinations—religion, the love of the marvellous, ignorance, anarchy, and the still lingering fear that the end of the world was at hand." When Martin Luther wrote: "It happened on one occasion that I woke up suddenly, and Satan commenced disputing with me," this was no literary trope. He was hallucinating, says Brierre de Boismont. The "ideas of Luther, exalted by perpetual controversy, by the dangers of his situation, by the fulminations of the church, and by continually dwelling on religious subjects, would naturally fall under the influence of the demon, which he saw everywhere, and to whom he attributed all the obstacles he encountered, and whom—like his contemporaries—he conceived interfered in all the affairs of life." For Brierre de Boismont, Satan is a socially constructed hallucination, the product of a mind trapped in a demon-haunted world.

Brierre de Boismont's early theories, constructed long before even a crude understanding of brain physiology was realized, have held up remarkably well. And if Ramachandran's and Persinger's research is corroborated, we might inquire further about the origin of temporal lobe–stimulated religiosity. Persinger proffers an evolutionary explanation: "The God Experience has had survival value. It has allowed the human species to live through famine, pestilence, and untold horrors. When temporal lobe transients occurred, men and women who might have sunk into a schizophrenic stupor continued to build, plan, and hope." Maybe, but Ramachandran is more cautious: "Whether the findings imply the existence of a religion or a 'God module' in the temporal lobes remains to be seen."

In fact, according to neuroscientist David Noelle, "the hypothesis that the neural mechanisms underlying religion form a distinct brain module was not really tested by these experiments. Reports of evidence for a 'God module' in the brain are, at best, premature." When you consider the fact that most studies show that more than 90 percent of the population believes in God, it would take a big stretch of the temporal lobe imagination to suggest that billions of people of all faiths the world over have experienced or are experiencing temporal lobe seizures or transients. A more reasonable hypothesis is that the handful of fanatic religious leaders throughout history, who report hearing the voice and seeing the face of, and even communicating with God, the devil, angels, aliens, and other supernatural beings, can perhaps be accounted for by temporal lobe abnormalities and anomalies. Their followers need a different explanation.

## GOD AS MEME

In his 1976 book, *The Selfish Gene,* Richard Dawkins proposed a cultural replicator to explain the transmission of ideas through culture: "We need a name for the new replicator, a noun that conveys the idea of a unit of cultural transmission, or a unit of *imitation*. 'Mimeme' comes from a suitable Greek root, but I want a monosyllable that sounds a bit like 'gene'. I hope my classicist friends will forgive me if I abbreviate mimeme to *meme*."

Dawkins did not develop the concept much further and there it lay dormant until mathematician Richard Brodie pushed the meme as a "virus of the mind" in 1996, physicist Aaron Lynch took it in the direction of a "thought contagion" in 1996, and cognitive psychologist Susan Blackmore developed it into a "meme machine" in 1997 and 1999. In countless lectures for the past two decades since his creation of the concept, Dawkins has strongly suggested that God is a meme and religion is a virus, and all of these authors have followed his lead by devoting entire chapters to the subject. Lynch, for example, suggests that the commandment to "honor thy father and mother" is a meme for children to imitate their parents (including their religious beliefs), and that dietary laws and holy days are memes to encourage commitment to one's religion, to spread other memes within that particular faith, and to protect one faith's memes against another faith's memes: " 'I am the Lord thy God. Thou shalt have no false gods before me' supremely realizes this competition-supressing advantage. Thus arises the archetype of modern monotheism, right at the top of the Ten Commandments." Blackmore argues that religious memes are like computer viruses that contain a "copy me" program not unlike those irritating chain letters and computer virus "warnings" that command you to "copy and distribute" the document—if you do, happiness and success will be abundant; if you do not, misery and failure will be your fate: "From an early age children are brought up by their Catholic parents to believe that if they break certain rules they will burn in hell forever after death. The children cannot easily test this since neither hell nor God can be seen, although He can see everything they do. So they must simply live in life-long fear until death, when they will find out for sure, or not. The idea of hell is thus a self-perpetuating meme."

There may be something to this "God as meme" argument in the sense that all religions employ techniques to increase their membership, to compete against other religions, and to perpetuate themselves into future generations. Of course, all organizations do this—if not, they would quickly go the way of the Neanderthals and eight-track tapes. And it might even be possible to test meme theory through comparing and examining the exceptions. Judaism, for example, has a rather weak "copy me" program: Members are not encouraged to proselytize and recruit new members; converting to Judaism requires considerable time, energy, and commitment; interfaith marriages (where the non-Jewish spouse may or may not convert) are discouraged; and an aura of exclusivity (instead of the usual inclusivity found in most religions) surrounds the faith. As a consequence, the number of Jews worldwide for the past half century (after the Holocaust decimated their numbers) has hovered around thirteen million. By contrast, Catholicism, with one of the most effective "copy-me" memes ever created, boasts of a membership roll in excess of one billion souls. No corporate marketing and advertising program has even come close to the Catholic church's nearly two-millennium-long campaign of recruitment and conversion.

This meme's-eye view is intriguing, but there are a number of logical and scientific problems outlined by cognitive psychologist James Polichak, including not providing a clear operational definition of a meme, not presenting a testable model for how memes influence culture and why standard selection models are not adequate, ignoring the sophisticated social science models of information transfer already in place, and circularity in the explanatory power of memes. Blackmore has addressed these and other criticisms in her 1999 book, *The Meme Machine,* but what remains especially troubling is the pejorative and hostile spin put on religious memes by the memeticists—corporations employ memes, musicians and authors compose memes, science itself is a meme, but religion is a *virus,* a *disease,* a *scourge* on humanity, which, as with AIDS or some stealthy computer virus that threatens to erase the entire contents of civilization's hard drive, we must rid ourselves of before it does us in.

There is, unfortunately, much historical evidence to support this perspective. From the Crusades' numerous attempts to cleanse the Holy Land of infidels (anyone who was not a proper Christian), to the Inquisition's efforts to purge society of heretics (anyone who dissented from Christian dogma), to the Counter Reformation's push to extirpate reforming Protestants from Catholic lands, to the Holy Wars of the late twentieth century that continue to produce death rolls in the millions, all have been done in the name of God and the One True Religion. However, for every one of these grand tragedies there are ten thousand acts of personal kindness and social good that go largely unreported in the history books or on the evening news. Religion, like all social institutions of such historical depth and cultural impact, cannot be reduced to an unambiguous good or evil; shades of gray complexity abound in all such societal structures, and religion should not be treated any differently than, say, political organizations. One could easily build a case that state-sponsored terrorism, revolutions, and wars make even these horrific religion-sponsored catastrophies appear mild by comparison. If God is a meme, so is King and President; and if religion is a virus, politics is a full-blown epidemic replete with copy-me memes such as nationalism, jingoism, and outright racism. Yet no memeticist would propose that we do away with the state. Why? Because the state is a complex social entity with countless nuanced beneficent effects that go along with the pernicious.

Belief in God may partially be explained through the influence of techniques described by memeticists, but memes do not get to the core of what is going on inside the mind of the believer. To reach into that we must ask believers *why* they believe.

## SCIENTISTS' BELIEF IN GOD

For those atheists who believe that the secularization thesis is more prescriptive than descriptive (that is, even though secular institutions are not replacing religion, they should), there is the problem of explaining why so many scientists believe in God. In 1997, the British science journal *Nature* published the results of a random sampling of 1,000 scientists (from the latest edition of *American Men and Women of Science*), comparing these findings to a similar study from 1916 by the psychologist of religion, James Leuba. As earlier in the century, approximately 40 percent of scientists proclaimed a belief in a personal God. (Of the 60 percent who said they do not believe, 45 percent were strong in their convictions of "personal disbelief," whereas 15 percent consider themselves agnostics.) Edward Larson and Larry Witham, who conducted the 1997 study, concluded: "The stereotype of scientists is that they tend to reserve judgment about things they don't know about. It turns out not only in history but about the same in our time, that scientists seem to know what they believe—or don't believe. Either they're a theist or a nontheist. There was not that great sea of doubt I would have expected."

Belief in immortality was a different story. Here we see a shift downward in belief by more than 10 percent, as well as a change in belief across fields. Eighty years ago Leuba found that biologists showed the highest rate of disbelief—almost 70 percent—whereas today physicists and astronomers were the biggest skeptics at close to 80 percent. Of all the sciences, Larson and Witham found that mathematicians are the most likely to believe in God, coming in at 45 percent. (See the graph on page 275 of Appendix II, showing the breakdown of belief between 1916 and 1996.)

Larson and Witham concluded that "religious Americans will doubtless be pleased to know that as many as 40 percent of scientists agree with them about God and an afterlife." This study, however, stirred up a hornet's nest among many scientists, who felt that the 40 percent figure was too high. Gerald Bergman, for example, surveyed the literature on the religious beliefs of scientists and concluded: "The level of commitment and strength of belief is not always easy to determine. Many scientists attend church for the sake of their families, and many are simply following the tradition in which they were raised." Since scientists do not speak with one voice, in a follow-up study Larson and Witham controlled for "eminence," or what their predecessor James Leuba called the "greater" scientists—those who held "superior knowledge, understanding, and experience." Leuba discovered that disbelief in God rose from 60 percent among the general scientific population, to 67 percent and 85 percent in two different samples among these "greater" scientists (defined as members of the National Academy of Sciences, an extremely exclusive body whose members must be voted in based on a stellar body of original research). Eighty years later, Larson and Witham found, even more than Leuba, that when eminence is controlled for, disbelief in God rose to 69 percent among biologists, and 79 percent for physicists. When "doubt" or "agnosticism" is factored in, actual belief in God among eminent scientists (averaged over all fields) drops to a paltry 7 percent. Why? Larson and Witham attribute the difference with Leuba not to the intervening years, but to the fact that their sampling of "greater" scientists was from the National Academy of Sciences, a considerably more "elite" group than Leuba's, which was taken from the standard (and not so selective) reference work of the time, *American Men of Science.*

It should be reemphasized that these figures are for Americans. The United Kingdom, Europe, and other developed nations of the world show lower levels of belief for both the general population and among scientists, and creationism is almost nonexistent outside of the United States (with some isolated pockets, such as in Australia and New Zealand). The University of Cincinnati political scientist, George Bishop, for example, reported that while about 45 percent of Americans reject evolution and accept a strictly literal interpretation of the Bible creation story, only 7 percent do in Great Britain and even less in Germany, Norway, Russia, and the Netherlands. In the seventeen developed nations he studied, Bishop found that Americans were the most likely to accept the Bible as "the actual word of God . . . to be taken literally, word for word," and the least likely to read the Bible as "an ancient book of fables, legends, history and moral precepts recorded by man." In his survey published in *The Public Perspective,* the journal of the Roper Center, Bishop noted that the groups most likely to endorse biblical literalism and reject evolutionary theory were women, older Americans, the less well-educated, Southerners, African Americans, and fundamentalist Protestants.

## WHY PEOPLE BELIEVE IN GOD

For years after the founding of the Skeptics Society in 1992, we were accused by the media and public of being an organization of atheists. Curious to know the level of religious disbelief in the society, I conducted a survey of members in 1995. The society is a highly educated group, a fifth of whom have Ph.D.s and almost three-quarters of whom are college graduates. With most members working in the sciences and other professional careers, I expected the survey to show an extremely low level of belief in God. The results were surprising. While the vast majority of this group reported being skeptical about such things as the paranormal, reincarnation, near-death experiences, immortality, and Satan, over a third thought it "very likely" or "possible" that there is a God. At the other end of the spectrum, to the question *Do you think there is a God (a purposeful higher intelligence that created the universe)?,* 35 percent said, "Very Likely" or "Possibly," while 67 percent said, "Not Very Likely," "Very Unlikely," or "Definitely Not" (some answered more than one category). Nor were skeptics as oppugnant toward religion as expected. For example, 77 percent said they believe that religion is "always" or "sometimes" a force for morality and social stability.

In retrospect, those who might best be described as "militant atheists"—whose behavior often resembles in intensity that of the fanatical believers they despise—appear to be a vocal minority. The 35 percent of skeptics who believe that God's existence is either very likely or possible is not too far off Larson and Witham's 42 percent of general scientists who profess belief. It is also in the range of a 1969 Carnegie Commission study of 60,000 college professors that revealed that 34 percent of physical scientists considered themselves "religiously conservative," and 43 percent said they attended church two to three times a month, the latter figure being not so different from the general population.

So, while the *majority* of skeptics and scientists do not believe in God, a surprisingly *large minority* do. The question is, why? Why do scientists and skeptics believe in God? For that matter, why does anyone believe in God? As we have seen already, the question is partially answered by how our brains and genes are wired. But only partially. Although estimates of a 50 percent influence by genes on religiosity sounds like a lot, we must remember that genes do not *determine* behavior so much as code for a *range of reactions* to the environment in a complex and always interactive feedback loop between the two. Therefore the environment still plays an extremely powerful role in the expression of genetic traits. What is that role?

In 1998, MIT social scientist Frank Sulloway and I conducted an empirical study to answer this question, along with the more general one of why people believe in God. We began with a more sophisticated follow-up survey of members of the Skeptics Society, which had doubled in size since 1995. The survey was divided into four parts that included family background, religious beliefs, reasons for belief or disbelief, and an essay question asking why people believe (or disbelieve), and why they think other people believe. We followed up this survey with another that was mailed to a random sample of Americans.

### The Skeptics Survey

Of the approximately 1,700 respondents to the Skeptics Society survey, 78 percent were men, 22 percent were women, and the average age was 49. Surprisingly, although twice the size of the first study, this group was just as well educated as the 1995 group, with over a fifth holding Ph.D.s and over three-quarters college graduates. (As we shall see, education plays a crucial role in religiosity.) Since the wording of the questionnaire had changed, the answers to the question *Do you believe there is a God (a purposeful higher intelligence that created the universe)?* varied slightly from the first survey.

In a similar question, 14 percent called themselves theists and 23 percent agnostics. As I pointed out in Chapter 1, there is a significant difference between having no belief in a God and believing there is no God, and for this reason we asked specifically where nonbelievers fell on this issue. We found 22 percent nontheists (no belief in God), while 32 percent said they were atheists (there is no God). In all, 18 percent, or almost a fifth, said, "Definitely yes" or "Very likely yes" there is a God. However, since only 14 percent called themselves theists, clearly some who think of themselves as agnostics also have some belief in God. In fact, 2 percent who called themselves agnostics also answered, "Definitely yes" or "Very likely yes" to the God question. It would seem possible then to believe in God while simultaneously having some doubts about His existence. All of this shows just how personal and subjective religious beliefs can be.

Interestingly, although 67 percent of our respondents attended church at least once a week while growing up, a startling 94 percent said they "never" or "almost never" attend church now. How can so many people believe in God yet not attend church? One answer is that although 70 percent of skeptics reported having no religious affiliation at present, 30 percent do, with Jews, Catholics, and Unitarians accounting for 20 of the 30 percent. So, while skeptics as a group are not religious in any traditional sense, a significant minority belong to religious organizations and also have some belief in God.

Another way to examine this question is to compare skeptics to the general population on various measures of religious conviction. To do so, Sulloway and I computed a correlation between the questions *How strong are your religious convictions?* and *Do you believe there is a God?* (A correlation is a statistical measure of the relationship between two variables, for example, height and weight—see Appendix I for an explanation of the statistical nomenclature; Appendix II includes the various formal statistics linked to the text by page numbers.) In both the skeptics group and the general public the correlation between responses to these two questions is statistically significant, though less so among skeptics. What does this mean? For most people, the relationship between how religious you feel and a belief in God is very tightly linked—one defines the other. For skeptics, the relationship is much less determined, belief in God does not necessarily define their religiosity. This makes sense: If religiosity is a part of human nature (as I shall argue it is in Chapter 7, and as we saw supported by studies of twins), those who lose their faith in God's existence may not lose the feeling of religiosity. Such individuals may still report feeling religious and feel that they belong to a religious group, especially one like Judaism or Unitarianism, where belief in God is not a requirement for membership. In other words, skeptics may be skeptical of God but still consider themselves religious in some nontraditional sense, defining terms such as *God* and *religion* in different ways than other people do.

However the pie is sliced, the percentage of believers among skeptics, while significantly lower than in the general population, is surprisingly high. The question, of course, is why? We sought to get at an answer through two separate questions: *In your own words, why do you believe in God, or why don't you believe in God?* and *In your own words, why do you think most other people believe in God?* The diversity of answers we received was staggering. Two categories predominated, however: those who primarily believe in God because they "see" a pattern of God's presence in the world (that is, for intellectual or "empirical" reasons), and those who believe in God because such belief brings comfort (that is, for emotional reasons). What was most interesting about these two answers is that they neatly cleaved between why people believe in God *themselves* (for intellectual reasons) and why they think *other people* believe in God (for emotional reasons). Moreover, this was true for both skeptics and the general public, and, as we shall see, this response tells us something very revealing about the psychology of religion.

Carefully reading through the diverse array of answers people gave, it quickly became apparent that these responses could be grouped into roughly ten reasons. The box below presents the most frequent reasons skeptics say they believe in God, why they think other people believe in God, and why they do not believe in God. (The specifics of this distribution can be found in Appendix II.)

### Why Skeptics Believe in God

1. Arguments based on good design/natural beauty/perfection/complexity of the world or universe. (29.2%)
2. Belief in God is comforting, relieving, consoling, and gives meaning and purpose to life. (21.3%)
3. The experience of God in everyday life/a feeling that God is in us. (14.4%)
4. Just because/faith/or the need to believe in something. (11.4%)
5. Without God there would be no morality. (6.4%)

### Why Skeptics Think *Other People* Believe in God

1. Belief in God is comforting, relieving, consoling, and gives meaning and purpose to life. (21.5%)
2. The need to believe in an afterlife/the fear of death and the unknown. (17.8%)
3. Lack of exposure to science/lack of education/ignorance. (13.5%)
4. Raised to believe in God. (11.5%)
5. Arguments based on good design/natural beauty/perfection/complexity of the world or universe. (8.8%)

**WHY SKEPTICS DO NOT BELIEVE IN GOD**

1. There is no proof for God's existence. (37.9%)
2. There is no need to believe in God. (13.2%)
3. It is absurd to believe in God. (12.1%)
4. God is unknowable. (8.3%)
5. Science provides all the answers we need. (8.3%)

Compare the top answers given to the first two questions about personal belief and others' belief, and note the ranked difference between intellectual reasons versus emotional reasons. Those skeptics who believe in God do so primarily because of the good design of the world, whereas this reason drops to number five for why they think other people believe. Emotional need and comfort are instead the top two reasons skeptics think other people believe in God.

Note also the overwhelming reason skeptics do not believe in God—there is no evidence for His existence. This was corroborated by the answers given to the question *To what extent do you believe there is concrete evidence or proof of God?* On a scale of 1 to 9, from *Not at All* to *Completely,* 77 percent of skeptics checked the lowest category. What makes this so interesting is that the number-one reason people offer for their belief in God is evidence of good design of the world. How can one set of people find no evidence for God's existence while another set finds quite the opposite? Both are observing the same world. The answer, as we shall see, lies in the psychology of belief.

## The General Survey

In 1998 Frank Sulloway and I also undertook a survey of a random sample of Americans (from a list provided by the same organization used by the most notable political, social, and cultural surveys conducted by social scientists and the media) about their religious attitudes and belief in God and, more importantly, *why* they believe. As with the skeptics, we inquired about family background, religious beliefs, reasons for belief or disbelief, and an essay question asking why people believe and why they think other people believe. We also added a section on personality to see if there were any characteristics especially related to religiosity.

In this survey we received responses from almost 1,000 people. The average age was forty-two, and 63 percent were men and 37 percent were women. Although less well educated than the skeptics group, this was a fairly credentialed population by national standards: 12 percent were Ph.D.s and 62 percent college graduates. Not at all surprising was the dramatic increase in belief in God from 18 percent in the skeptic survey to 64 percent in the general survey, with disbelief dropping from 70 percent for skeptics to only 25 percent for the general public. (The graphs on the bottom of page 275 and the top of page 276 of Appendix II show the rates of belief and disbelief.)

Most surveys show that over 90 percent of Americans believe in God, so this 64 percent figure is remarkably low in comparison. The explanation is most likely to be found in education levels. As it turns out, the people who completed our survey were significantly more educated than the average American, and higher education is associated with lower religiosity. According to the U.S. Census Bureau for 1998, one-quarter of Americans over twenty-five years old have completed their bachelor's degree, whereas in our sample the corresponding rate was almost two-thirds. (It is hard to say why this was the case, but one possibility is that educated people are more likely to complete a moderately complicated survey.) This confirms what other social scientists have found: Of the numerous variables influencing religious attitudes, education is one of the most powerful. Precisely what is that influence and what are some of the other variables that lead people to believe (or not) in God?

To answer these questions, we examined the correlation between a number of variables on which we collected data with several measures of religiosity (see the graphs on pages 276 to 279 of Appendix II). In examining our findings, it is important to remember that the results represent *tendencies,* not absolutes. It turns out that the three strongest predictors of religiosity and belief in God are being raised religiously, gender (women are more religious than men), and parents' religiosity. The three strongest predictors of lower religiosity and disbelief in God are education, age, and parental conflict. In other words, being male, educated, and older tends to make people less religious, while being female and raised by religious parents generally makes you more religious. However, people do not live in a psychological laboratory where variables can be perfectly controlled. All of these variables interact, and the effect of these interactions complicates the picture. For example, being raised religiously makes people more religious unless they have conflict with their parents, in which case the rebellious thing to do is to become less religious. Likewise, a correlation between attending church when growing up and parental conflict showed that this combination led to a significant reduction in current church attendance. That is, if church attendance was high in youth but a person experienced conflict with parents, then lowered church attendance later was an apparent consequence of this conflict.

How religious attitudes *change* is important in understanding why people believe or do not believe in God. For example, interest in science corresponds to lower religious intensity. (It should be noted also that interest in science is itself highly predicted by education, gender, personality, and background—being educated, male, conscientious, and open to experience is associated with greater interest in science, while being raised religiously is associated with reduced interest in science.) But interest in science is only part of the larger and more powerful variable of education. Becoming more educated and getting older both cause religious attitudes to decrease. Why? As people get older they invariably encounter other belief systems that broaden their intellectual horizons, either through formal education or life experience, causing them to realize that religious attitudes and belief in God are perhaps not as certain as they seemed at a younger age. But age has other effects, and interacts with religious intensity. For example, we asked people, *Was there some age when you began to seriously doubt your religious faith?* Tellingly, the less the religiosity, the earlier was the age that serious doubt occurred. This makes sense, of course, since religiosity and belief in God peak in the late teens and then decline gradually until the eighties, at which point there is slight increase as people begin thinking about the end of their lives. This finding is confirmed by other studies such as a comprehensive one by Chris Brand, who discovered that the young and the elderly showed the highest levels of religious belief and involvement.

Although many of the findings were expected, there were also some surprises. For example, socioeconomic status had no direct influence on religious beliefs. However, political beliefs certainly did, with conservatives being more religious and liberals less so. Thus, while the majority of both conservatives and liberals believe in God, if you are a political liberal you are less likely to believe. Why? Probably because most religions represent the status quo, and what conservatives wish most to conserve is the status quo. (Despite the rhetoric of "change" professed by members from one end of the political spectrum to the other, when conservatives advocate change in the system, it is almost always change back to an older form of conservatism. And the most extreme examples of this type typically come from what is accurately called the religious right.) Liberals are more in favor of change away from traditional institutions, and among these are society's mainstream religions. (The exception is Judaism, which has traditionally supported liberal causes since Jews themselves have historically been a part of oppressed groups in virtually all cultures in which they have found themselves.) Thus, the liberal, radical thing to do is to change your religious attitudes, which usually means either becoming less religious, or adopting marginalized religious beliefs, as in the counterculture's embracing of fringe cults in the 1960s and 1970s, or the adoption of New Age spiritual movements in the 1980s and 1990s.

This connection between religion and politics is corroborated by other studies. For example, during the greatest religious revolution in history—the Protestant Reformation—Sulloway, for example, found that supporters "were more likely to be young, laterborn, lower class, and low in professional status," characteristics that today we would use to describe political liberals. And in a fascinating study on the religious attitudes and voting patterns of members of the 96th United States Congress, sociologists found that what they termed the most "legalistic" and "self-concerned" congressmen were the most conservative, whereas the "people-concerned" and "nontraditional" congressmen were the most liberal. For example, legislation favoring civil liberties received nearly three times the votes from the nontraditional/liberal congressmen as it did from the legalistic/conservative congressmen. David Wulff, summarizing a sizeable body of literature on the subject, showed that this tendency extends to the population as a whole. Measuring "piety" as a function of religious affiliation, church attendance, doctrinal orthodoxy, and self-rated importance of religion, "researchers have consistently found positive correlations with ethnocentrism, authoritarianism, dogmatism, social distance, rigidity, intolerance of ambiguity, and specific forms of prejudice, especially against Jews and blacks." That is to say, greater religiosity was associated with higher scores for these personality traits—traits that are the very antithesis of political liberalism.

Since personality plays an important role in many human beliefs, we examined a number of characteristics to see if there was any influence on religiosity. What is personality? Personality is the unique pattern of relatively permanent traits that shapes an individual's thoughts and actions. We might contrast personality *traits* with situational *states,* that is, merely temporary reactions to environmental circumstances. Personality is our core being—the stuff of which we are made. It may be flexible, where we react differently in different situations, but it is only flexible within certain parameters determined by an interactive combination of nature and nurture, genes and environment, biology and psychology. The most popular theory today is known as the Five Factor Model. The "Big Five" personality dimensions include: *Openness to experience* (imaginative, idealistic, adventurous), *Extraversion* (friendly, warm, sociable), *Agreeableness* (forgiving, tender-minded, sympathetic), *Conscientiousness* (efficient, organized, ambitious), and *Neuroticism* (anxious, moody, defensive). Sulloway and I measured these five dimensions using a scale of 1 to 9 on adjectives describing each dimension. For example, to measure your agreeableness you would rank yourself from tender-minded (1) to tough-minded (9); or for openness you would rank yourself from unadventurous (1) to adventurous (9). Each of the five dimensions had two questions and scales.

The most consistent finding related to religious intensity involved *openness.* A higher ranking on the openness dimension was associated with lower levels of religiosity and higher levels of doubt. Moreover, openness was significantly correlated with change in religiosity, with higher openness scores being associated with lowered piety, as well as lower rates of church attendance. There was a modest association between birth order and openness, with laterborns scoring higher than firstborns. Sulloway has pointed out that laterborns tend to be more open to experience than firstborns because they must generally be more exploratory in finding a valued family niche and to compete for limited parental attention and resources. Not surprisingly, we found a strong correlation between openness and political liberalism. But we also discovered a significant correlation on the agreeableness (tough-minded—tender-minded) scale: We found that religious people are more tender-minded. But it should be noted that laterborns, when controlled for sex, socioeconomic status, education, age, and sibship (brother and sister) size, are more liberal than firstborns. Related to this is the finding that laterborns are more tender-minded than firstborns. So, overall, belief in God was significantly related to being conservative and being tender-minded, but because laterborns are more liberal and also more tender-minded than their elder siblings, these two predisposing factors will tend to cancel themselves out in the expression of religiosity.

In sum, people who score high in openness are less religious, more likely to entertain religious doubts, more likely to change their beliefs, and less likely to attend church. Why? Additional adjectives that correlate highly with openness to experience on the Personality Inventory we used offer some insight. These include: inventive, versatile, curious, optimistic, original, insightful, and unconventional. Consider what it means to be less religious and skeptical of God in a country in which 90 to 95 percent of the population are believers. To even arrive at this position one would have to be inventive, curious, and insightful. And to maintain this skepticism in the face of the possibility of great scorn being heaped by zealous believers would mean one would need to be optimistic and original. More than anything else, one would need to be unconventional. Religion and belief in God is, if nothing else, conventional. In fact, I would argue that it is *the* convention in our culture. With the possible exception of politics (and even this is probably a distant second), you would be hard pressed to find another convention that generates so much zealousness on the part of followers. To be pious—an adjective almost exclusively used to describe compliance in the observance of religion—means compliance to convention.

In order to probe deeper into the question of why people believe, we asked next a series of questions with similar wording, for example, *To what extent does* emotional comfort *contribute to your religious beliefs?* (followed by a 1 to 9 scale, from *not at all* to *completely).* Additional reasons for belief included "faith," "apparently intelligent design of the world," "without God there is no basis for morality," and "a desire for meaning and purpose in life." We also included two questions involving the undermining of religious belief: *To what extent does the existence of evil, pain, and suffering undermine your religious beliefs?* and *To what extent have scientific explanations of the world undermined your religious beliefs?* The final question was concerned with belief and evidence: *To what extent do you believe there is concrete evidence or proof of God?*

In analyzing the data, we lumped these questions into two groupings: (1) rational influences on belief (the apparent intelligent design of the world; without God there is no basis for morality; the existence of evil, pain, and suffering; and scientific explanations of the world); and (2) emotional influences on belief (emotional comfort, faith, and desire for meaning and purpose in life). The single strongest correlation we found was for gender: Men tended to justify their belief with rational reasons, while women tended to justify their belief with emotional reasons. This finding dovetails well with the other significant relationships we found, such as a positive correlation between education and rational arguments for God's existence, and a negative correlation between education and emotional arguments for God's existence (as education decreased, preferences for emotional arguments increased). There was also a significant relationship between openness and a tendency to prefer rational reasons for belief over emotional reasons. This was confirmed in the finding of a significant negative correlation between openness and a preference for emotional reasons for belief—low openness is associated with a higher preference for emotional reasons.

In other words, educated, open people, and men feel the need to justify their faith with rational arguments, whereas less-educated people, especially women, are comfortable with their faith being based on emotional reasons. One explanation for this outcome is that, in general, education causes a decrease in faith, so for those who are educated and still believe, there is a need to justify belief with rational arguments. Since most people come to their faith by being raised religiously or through personal experiences, rational arguments are not typically a part of this process. We should not be surprised, then, that there were significant negative correlations between rational arguments and being raised religiously as well as parents' religiosity. That is, if your faith is a deep one, going back to childhood, there is less need to justify it with

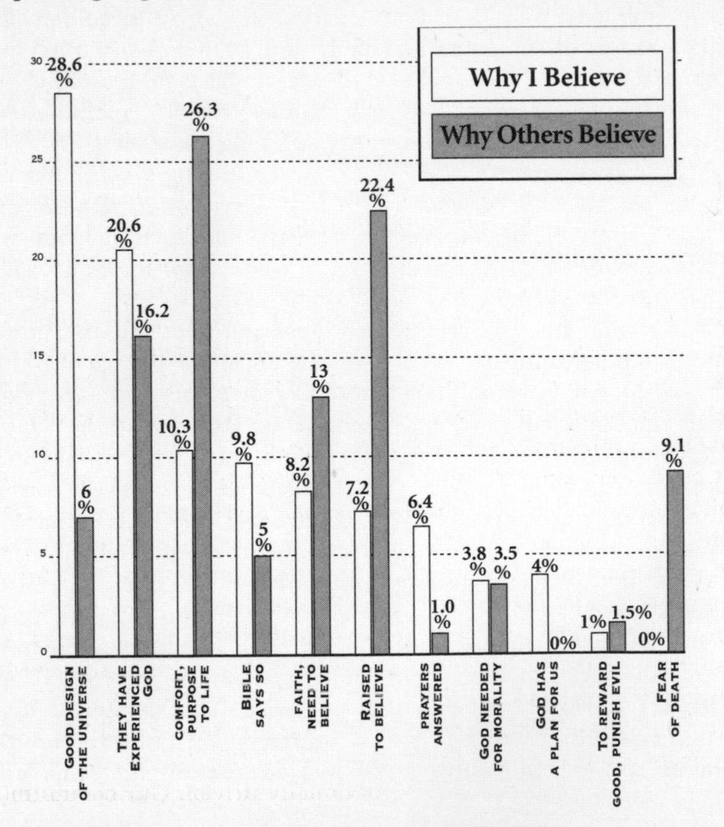

rational arguments. But these correlations, while significant, were weaker than for most we found in this study, indicating that education's even stronger role can override early-life experiences.

To give people an opportunity to say in their own words why they believe in God and why they think other people believe in God, we asked them exactly that. The graph on the facing page presents the most common reasons why people believe in God, and why they think other people believe in God.

## INTELLECTUAL AND EMOTIONAL REASONS TO BELIEVE

One of the most interesting results to come out of this study was that the intellectually based reasons for belief of "good design" and "experience of God," which were in first and second place in the first question of *Why do you believe in God?,* dropped to sixth and third place for the second question of *Why do you think other people believe in God?* Taking their place as the two most common reasons *other people* believe in God were the emotionally based categories of "comforting" and "raised to believe."

Why? One possible answer to this question is what psychologists call "biases in attributions." As pattern-seeking animals, we seek causes to which we can attribute our actions and the actions of others. According to attribution theory, we attribute the causes of our own and others' behaviors to either a situation or a disposition. When we make a *situational attribution,* we identify the cause in the environment ("my depression is caused by a death in the family"); when we make a *dispositional attribution,* we identify the cause in the person as an enduring trait ("her depression is caused by a melancholy personality"). Problems in attribution may arise in our haste to accept the first cause that comes to mind. But I suspect this is only part of the explanation. Social psychologists Carol Tavris and Carole Wade explain that there is, not surprisingly, a tendency for people "to take credit for their good actions (a dispositional attribution) and let the situation account for their bad ones." In dealing with others, for example, we might attribute our own good fortune to hard work and intelligence, whereas the other person's good fortune is attributed to luck and circumstance.

I would argue that there is an *intellectual attribution bias,* where we consider our own actions as being rationally motivated, whereas we see those of others as more emotionally driven. Our commitment to a belief is attributed to a rational decision and intellectual choice ("I'm against gun control because statistics show that crime decreases when gun ownership increases"); whereas the other person's is attributed to need and emotion ("he's for gun control because he's a bleeding-heart liberal who needs to identify with the victim"). This intellectual attribution bias applies to religion as a belief system and to God as the subject of belief. As pattern-seeking animals, the matter of the apparent good design of the universe, and the perceived action of a higher intelligence in the day-to-day contingencies of our lives, is a powerful one as an intellectual justification for belief. But we attribute other people's religious beliefs to their emotional needs. Here are just a few examples from the written portion of the surveys:

- A thirty-year-old male Jewish teacher with strong religious convictions (8 on a scale of 1 to 9), says he believes in God "because I believe in the Big Bang; and when you believe in the B.B., you have to ask yourself—'what came before that?' A creation implies a creator." (Aquinas's prime mover argument; see Chapter 5.) Yet, he goes on to explain: "I think that most people believe out of an emotional need, although there is a significant minority of rational (even skeptical!) believers such as myself."
- A fifty-one-year-old male with very strong religious convictions (9 on a scale of 1 to 9) but no formal religious membership writes that he believes in God based on his "personal experiences," but for others "belief in God provides emotional support and a belief structure that provides meaning, purpose, and rules of conduct for them. Many feel lost without believing something/someone more important than them runs their life rather than believing that they can and do create their reality and the universe."
- A sixty-five-year-old male Catholic with moderately strong religious convictions (7 on a scale of 1 to 9) gives the standard watchmaker argument: "To say that the universe was created by the Big Bang theory is to say that you can create *Webster's Dictionary* by throwing a bomb in a printing shop and the resulting explosion results in the dictionary." Nevertheless, other people believe in God because of a "sense of security" and "blind faith."
- A thirty-seven-year-old female Catholic with strong religious convictions (8 on a scale of 1 to 9) says she believes in God because "how else could you explain our origins? Only God could create a world and universe out of nothing. There are mir-

acles every day that science cannot explain." Others believe, she says, because it "gives hope."

- A forty-one-year-old male Baptist with very strong religious convictions (9 on a scale of 1 to 9) explains that he believes in God "due to the evidence of his magnificent creation and the extraordinary order of the universe," whereas other people believe because "without God there is no purpose for their lives or the universe."

There are many, many more examples. Morever, these data support Gallup polls taken in 1982 and 1991, where 46 percent of the public believe that "man was created pretty much in his current form at one time within the past 10,000 years," 40 percent believe that "man evolved over millions of years from less developed forms of life, but God guided the process, including the creation of man," but only 9 percent believe that "man evolved over millions of years from less developed forms of life. God had no part in the process." The Gallup polls did not ask *why,* but it seems obvious from our results that the answer is that people *see* God in the universe, in the world, and in their lives. Hardly anyone has heard of theologian William Paley and his eighteenth-century watchmaker argument for God, but they know this argument intuitively from their experiences. They also read about it from science popularizers like comet hunter David Levy, who told millions of readers of *Parade Magazine* that the "miracle of life" was due to the fact that the universe was "designed" for us, and that this is proved by such scientific facts as: (1) ice floats; (2) the night sky is dark; (3) protons and electrons have absolutely identical charges; (4) we have the right kind of Sun. There are perfectly rational, scientific explanations for these facts that have nothing whatsoever to do with life being "designed" or a "miracle" in any supernatural sense. But these counterarguments are also counterintuitive. The "feeling" one gets in studying the world and life is that it *seems* designed. And this is what people report about their perceptions and experiences.

Interestingly, the primary reasons people gave for *not* believing in God were also the intellectually based categories of "there is no proof for God's existence," followed by "God is a product of the mind and culture," "the problem of evil," and "science provides all the answers we need." For example, an eighteen-year-old Jewish male who considers himself an atheist, writes: "I don't believe in God because it is impossible for a being to be what God must be in order to be a god without being obvious and undeniable. In short, God is philosophically impossible and scientifically and cosmologically unnecessary." By contrast, and following the tendency to attribute to others emotional reasons for belief, he says other people believe in God because: "It's comforting. Additionally, some people find it easier to deal with problems if they believe it is 'God's will.' "

## ALL'S RIGHT WITH GOD IN HIS HEAVEN

In his 1781 classic work, *The Decline and Fall of the Roman Empire,* Edward Gibbon concluded his discussion of religion with this observation: "The various modes of worship which prevailed in the Roman world were all considered by the people as equally true; by the philosopher as equally false; and by the magistrate as equally useful." As we have seen, belief in God in the modern world is a function of a complex array of reasons that, while true for some people and false for others, certainly are equally useful. Consistently we find a fascinating distinction in belief attribution between why people think they believe in God and why they think *other* people believe in God.

This distinction was not lost on the psalmists of the Old Testament. To the choirmaster of Psalm 19:1, the author proclaims: "The heavens declare the glory of God; and the firmament showeth his handiwork." Yet in the psalm for the sons of Korah, Psalm 46:1–3, it is declared: "God is our refuge and strength, a very present help in trouble. Therefore will not we fear, though the earth be removed, and though the mountains be carried into the midst of the sea; Though the waters thereof roar and be troubled, though the mountains shake with the swelling thereof."

Are these not, in a way, two sides of the same coin? For believers, the heavens declare God's glory; for *other* believers He provides strength in their time of need. Or, as Robert Browning wrote in *Pippa Passes*: "God's in His Heaven—All's right with the world."

Chapter 5

# O Ye of Little Faith

## Proofs of God and What They Tell Us about Faith

*Faith has to do with things that are not seen, and hope with things that are not in hand.*
—Thomas Aquinas, *Summa Theologiae, LVII,* c. 1265

On Sunday, November 15, 1998, I debated God at the Church of the Rocky Peak, in Chatsworth, California, in the northwestern end of the San Fernando Valley. More precisely, I debated Dr. Doug Geivett, a professor of philosophy at the Talbot School of Theology and the author of such books as *Evil and the Evidence for God* and *In Defense of Miracles,* on the subject of "Does God Exist? Where Does the Evidence Point?" As a testimony to the interest in this subject, more than 1,500 seats were filled in this giant, modern church, with droves of students sitting on the floor in front of the dais, and standing room only at the back. The minister of the church, Dr. David Miller, was extremely accommodating to me, given that I was likely to be outnumbered in this venue. Was I ever. Miller asked for a show of hands of those who came specifically to support me. About fifty arms went up.

Dr. Geivett went first, presenting the standard arguments for God's existence, including: Big Bang cosmology is described by Genesis 1; the anthropic cosmological principle and the fine-tuned nature of the universe implies a creator; life has all the appearances of design; humans are moral and morality could only come from God, because if it did not, then no one would be moral; and the historical evidence supports the resurrection of Jesus. Geivett concluded his initial presentation by explaining that we are confronted here with an either-or choice: Either God exists or He does not; either the universe was created or it was not; either life was designed or it was not; either morality is natural or it is not; either Jesus was resurrected or he was not.

I opened my rebuttal by explaining that there are only two types of theories: those that divide the world into two type of theories, and those that do not. I explained that God's existence is an insoluble question, and then spent the majority of my time presenting evidence (as I do in Part II of this book) that belief in God and its expression through religion has all the earmarks of being a human creation and a social construction. In other words, I argued that humans made God, and not vice versa. In no way do I intend this belief to belittle religion or people's belief in any way. It is a testable hypothesis that I find reasonable and supported by the evidence from comparative mythology and world religions, evolutionary biology and psychology, and the anthropology, sociology, and psychology of religion. That man made God is every bit as fascinating as the reverse; and the evidence is even better.

I expected my debate opponent (and most of the audience), of course, to disagree, but I did not expect them to give me such a hard time about not addressing Dr. Geivett's "proofs" point by point. I touched on them briefly, but since I have always understood religious belief to be based on faith, the notion of "proving" one's faith seems oxymoronic. Nevertheless, in each of the three rebuttal segments, and in the question-and-answer period, my opponent reviewed the "proofs" over and over, demanding (along with the audience afterward) that I either refute them or accept God.

In Christian theology these arguments for God's existence are called apologetics, from the Latin *apologeticus*—"to speak in defense." I am quite familiar with them and began my study in the 1970s with the bestseller in the popular end of this genre, Josh McDowell's *Evidence That Demands a Verdict*, with its oft-quoted argument that Jesus was a liar, lunatic, or Lord. (Since Jesus could not be either of the first two, it is argued, he was, de facto, God incarnate.)

And for the past quarter century I have maintained an interest in apologetics because it is here where religion most closely intersects with science. Because we are rational, thinking beings, faith never seems to be enough for most of us. We want to *know* we are right, and in the Western world to know something is true is to prove it through reason or science, logic, and empiricism. Thus, arguments in favor of God's existence and the divine origin and authority of the Judaeo-Christian religion are couched in the language of science and reason, and there have been literally tens of thousands of books written along this vein.

But the question at hand is this: do any of these proofs actually *prove* God's existence? No. In fact, most of them are not so much proofs and arguments in favor of God's existence as they are "reasons to believe" (as one organization of modern Christian apologists is called) for those who already believe. The "God Question" remains as insoluble today as it ever was.

## PHILOSOPHICAL ARGUMENTS FOR GOD

As we saw in the previous chapter, the most common reason people give for believing in God is that there are arguments and evidence that lead them to that conclusion. Here are ten of the most commonly used philosophical arguments for God, and the problems with each. Since all of them are covered elsewhere in much greater depth (indeed, entire volumes are dedicated to each), I shall allocate more space to the scientific arguments that follow, which are, I think, more effective from the believer's perspective and thus require a more thoughtful response.

**1.**

*Prime Mover Argument.* This is the great Catholic theologian and philosopher St. Thomas Aquinas's first way to prove the existence of God as outlined in his great work, *Summa Theologica.* Everything in the universe is in motion. Nothing can be in motion unless it is moved by another. That something else must also be moved by yet another, and so on. But this cannot regress into infinity, "therefore it is necessary to arrive at a first mover, moved by no other; and this everyone understands to be God."

*Counterargument.* The universe is everything that is, ever was, or ever shall be. Thus, God must be within the universe or *is* the universe. In either case, God would himself need to be moved, and thus the regress to a prime mover just begs the question of what moved God. If God does not need to be moved, then clearly not everything in the universe needs to be moved. Maybe the initial creation of the universe was its own prime mover.

2. *First Cause Argument.* This is Aquinas's second way. All effects in the universe have causes. The universe itself must have a cause. But this cause-and-effect sequence cannot be regressed forever, so there had to be a first cause, a causal agent who needed no other cause. "Therefore it is necessary to admit a first efficient cause, to which everyone gives the name of God."

   *Counterargument.* This is essentially the prime mover argument rephrased. And, as in the problem with the prime mover argument, God must be within the universe or *is* the universe. In either case, God would himself need to be caused, and thus the regress to a first cause just begs the question of what caused God. If God does not need a cause, then clearly not everything in the universe needs a cause. Maybe the universe itself does not need a cause. Perhaps, as cosmologist Alan Guth suggests in his 1997 book *The Inflationary Universe*, it just sprang into existence out of a quantum vacuum, uncaused.

3. *Possibility and Necessity Argument.* Aquinas's third way argues that in nature it is possible for things to be or not to be. But not everything could be in the realm of the possible, for then there could be nothing. If there were at one time nothing, then the universe could not have come into existence. "Therefore we must admit the existence of some being having of itself its own necessity, and not receiving it from another, but rather causing in others their necessity. This all men speak of as God."

   *Counterargument.* Why is it not equally plausible that God, like the universe, is possible but not necessary? As Stephen Hawking likes to ask in his books and lectures: "Why does the universe bother to exist at all? Why should there be something rather than nothing?" No one knows. It is entirely possible that the universe, including God, did not need to come into existence. The problem here is that the human mind is incapable of conceiving of nothing (in the universal sense), and therefore this argument falls into what Martin Gardner calls a mysterian mystery—it is not just unknown, it is unknowable with the minds we possess. Evolution provided us with a big enough brain to ask such profound questions, but not big enough to answer them. This is an argument neither for nor against God.

4.
*The Perfection/Ontological Argument.* Aquinas argued in his fourth way that there are gradations from less to more good, true, and noble. "There is then, something which is truest, something best, something noblest. Therefore there must also be something which is to all beings the cause of their being, goodness, and every other perfection. And this we call God." This is known as the ontological argument and was first presented by St. Anselm, the eleventh-century archbishop of Canterbury, who in his *Proslogion* defined God as "something than which nothing greater can be conceived." Even a "fool" can understand this, says Anselm (referencing Psalm 14:1, "the fool hath said in his heart, There is no God"):

*For if it is actually in the understanding alone, it can be thought of as existing also in reality, and this is greater. Therefore, if that than which a greater cannot be thought is in the understanding alone, this same thing than which a greater cannot be thought is that than which a greater can be thought. But obviously this is impossible. Without doubt, therefore, there exists, both in the understanding and in reality, something than which a greater cannot be thought.*

Reversing the argument, Anselm says it is equally impossible to think of God as nonexistent:

*For something can be thought of as existing, which cannot be thought of as not existing, and this is greater than that which can be thought of as not existing. Thus, if that than which a greater cannot be thought can be thought of as not existing, this very thing than which a greater cannot be thought is not that than which a greater cannot be thought. But this is contradictory. So, then, there truly is a being than that which a greater cannot be thought—so truly that it cannot even be thought of as not existing.*

*Counterargument.* Anselm's somewhat confusing logical twists and turns are examples of the word play often found in these arguments. What does it mean to be perfect? Obviously no human can know, yet we are the creators of the concept itself. We can envision some maximal level of perfection and then argue that God must be above this, but what does that mean? No *human* can possibly know. And why couldn't an argument antithetical to Anselm's be made?: There is then, something which is falsest, something worst, something ignoblest. Therefore there must also be something that is to all beings the cause of falsity, badness, and ignobility. And this we call God.

Further, continuing the Flatland analogy from Chapter 1, why couldn't God, who is said to be omniscient, conceive of something even greater than perfection? The square thought the sphere the highest state of being, until he became a cube and realized there could be still higher states. The upper boundary of perfection is simply defined as such by our admittedly limited human mind. Does it not seem reasonable to argue that whatever state of "perfection" we might imagine, there could conceivably be a higher state? As with the concept of infinity, whatever number the mind can create as the seemingly largest, you can always add one to it. Why stop at God?

As for it being impossible to think of God as nonexistent, it seems equally impossible to think of nonexistence at all. Is it really possible to conceive of absolutely nothing—no galaxies, no stars, no planets, no life, no molecules, no atoms, no space, no time, no energy—no anything? The concept is epistemologically void. It neither proves nor disproves anything.

5. *The Design/Teleological Argument.* Aquinas's fifth way deals with "the governance of things." Since "natural bodies act for an end," and yet lack knowledge, they must have been designed. "Therefore some intelligent being exists by whom all natural things are ordered to their end; and this being we call God." Modern design arguments are more sophisticated and involve the intricacies of design in nature, such as symbiotic relationships between organisms like insects and flowers, or the apparent "anthropic" design of the cosmos—that is, it is precisely suited for the evolution of life.

*Counterargument.* Design arguments from nature are untenable by the simple fact that nature is not as beautifully designed nor as "perfect" as believers would have us think. The python's hind legs—unarticulated bones buried in flesh and totally useless—are indications of quirky and contingent evolution, not divine creation. Similarly, the whale's flipper—complete with useless humanlike upper arm, forearm, hand and finger bones—is obviously the evolutionary by-product of mammalian evolution, not the handiwork of a divine Geppetto. The anthropic cosmological principle will be dealt with below in the section on scientific arguments for God, but suffice it to say that any universe with the configuration that gives rise to pattern-seeking animals will appear designed, and those universes with laws that do not lead to life will not appear designed.

6.

*The Miracles Argument.* The miracles of the Bible, as well as those of modern times, cannot be accounted for by science or natural law, therefore they must have as their cause a higher power. This higher power is God. C. S. Lewis defined a miracle as "an interference with Nature by supernatural power." In fact, Lewis admits, "unless there exists, in addition to Nature, something else which we may call the supernatural, there can be no miracles."

*Counterargument.* A miracle—as so well displayed in Sidney Harris's cartoon in which a scientist inserts the phrase "and then a miracle occurs" in the middle of a long string of equations—is really just a name for something we cannot explain. This is the "God of the gaps" argument, but as soon as we are able to fill the gap with an explanation, it is no longer a miracle. If Jesus' walking on water is shown to be nothing more than a desert mirage, or the exaggerated tale of enthusiastic proselytizers, it is no longer a miracle. Additionally, how could you ever "prove" a miracle? It seems rather unlikely that one could prove that Jesus suspended the laws of nature that determine the surface tension of water (or of gravity, or whatever). The point of miracles is to inspire the faithful with religious reverence. One does not prove a miracle; one believes a miracle on faith, which is exactly how religion should be believed.

7.

*Pascal's Wager Argument.* At the age of thirty-one the French mathematician and philosopher Blaise Pascal had what he termed a "mystical experience" that changed his life. Not content to rest his belief entirely in the mystical (and knowing this would never convince fellow skeptics such as René Descartes), he formulated what has become known as Pascal's wager. If we wager that God does not exist and he does, then we have everything to lose and nothing to win. If we wager that God does exist and he does, then we have everything to gain and nothing to lose. Pascal was not naive enough to believe that people would then just place the bet, or that God would just accept the gamblers into His heavenly casino. He realized that belief comes through action so he argued that you also needed to go through the motions by attending Mass and taking the sacraments, and in time you would come to really believe. In the jargon of social psychology, you would shift from "conformity" to "internalization," incorporating God into your core of deeply held beliefs.

*Counterargument.* First, this is not actually a *proof* of God since Pascal himself admitted that one still needs faith. Second, believing in God and going through the motions of attending church, praying, taking the sacraments, and so forth, is not a case of "nothing to lose." There is plenty to lose, including the time and effort it takes to do all this when one could be doing something else. Finally, what if there were some other higher intelligence, even more powerful than God, and His sacraments included some of the more earthly pleasures? Not only would you be missing out on these, you might be eternally punished for placing the wrong wager or choosing the wrong God. This may sound unlikely, but from a purely objective point of view it is no more illogical than the existence of a Judaeo-Christian God.

**8.** *The Mystical Experience Argument.* This is the ultimate close encounter with God himself, directly and experientially: "I know God exists because I have experienced him." Bill Wilson, founder of Alcoholics Anonymous, reported just such an experience when he got on his knees and said, "If there is a God, let him show himself now." Wilson describes what happened next:

*Suddenly the room lit up, with a bright white light. I was caught up in an ecstasy for which there are no words to describe. It seemed to me in the mind's eye that I was on a mountain, that a wind, not of air, but of the spirit was blowing and then it burst upon me that I was in another world of consciousness. All about me and through me was a wonderful presence and I thought to myself, "so this is the God of the preachers."*

Such mystical experiences and conversions are not uncommon in history. Constantine's "vision" at the Milvian Bridge, preceding his victory over Maxentius in A.D. 312, cemented the Christian religion into his worldview and into our world. Augustine heard voices telling him, "Pick it up, read it; pick it up, read it!" upon which "I got to my feet . . . to open the Bible and read the first passage I should light upon." The passage told him to sell his belongings and give the money to the poor. This he did, and as he notes in his *Confessions,* "as the sentence ended, there was infused in my heart something like the light of full certainty and all the gloom of doubt vanished away." John Calvin reported in his *Commentary on the Psalms* that he had "a sudden conversion." Martin Luther was reportedly struck to the ground by a lightning bolt and cried in terror: "St. Anne, help me! I will become a monk."

*Counterargument.* As we saw in the previous chapter, these experiences are most probably the result of temporal lobe seizures or some other aberration in brain physiology. But it is the weakest of the so-called proofs of God, since not only is it not really a proof, it relies on personal experience, which by definition cannot be shared with others. I made the argument myself when I was a born-again Christian, and tried it out on my philosophy professor, Richard Hardison, who responded with a statement to our philosophy class that provides a potent refutation:

*The goals of a society that you have valued, and the achievements of the people that you have respected, have depended on objectivity. Even the occasional mystic who impressed you, stepped out of his mysticism when he made the analysis that you read. His very communication, by the nature of communication, was objective. Mystical "truths" by their very nature, must be solely personal. They can have no possible external validation. Nor can they produce any possible communication with those who do not share the particular mysticism. There is a fundamental flaw in all mysticisms: the mystic often seeks external support of his position and in the process, denies his mysticism.*

**9.**

*Fideism, or the* Credo Quia Consolans *Argument.* Of all the philosophical arguments for God, perhaps this stands up the best since it does not attempt to be a proof at all. Instead it is quite honest in its admission of the personal nature of belief. It says simply: "I believe because it is consoling." In his book, *The Whys of a Philosophical Scrivener*, Martin Gardner defines and defends fideism at length. It is a pragmatic argument, taken from the philosophers William James, Charles Peirce, and Miguel Unamuno. At its core it says that (1) in issues of extreme importance to human existence, (2) when the evidence is inconclusive one way or the other, and (3) you must make a choice, it is acceptable to take a leap of faith. Martin Gardner, the skeptic of all skeptics, is a fideist. He even admits that atheists have slightly better arguments than theists. But for personal, emotional reasons he was willing to make the leap.

*Counterargument.* One flaw in this argument is that it is based on the philosophy of pragmatism, which states that knowledge is valid if it "works" for you. But this does not necessarily apply to all ideas, including God. Some things we really can know, based on external validation. Another flaw is that fideism reduces belief to personality type. As recent research into personality development shows, one's acceptance or rejection of ideas is as much a function of one's family dynamics and personality characteristics as it is of empirical evidence. If beliefs are going to be based on emotion rather than argument or evidence, it would seem to eliminate the need for reason and science altogether. Why draw the line at some belief just because it feels good? Why not just say that God is an unknowable concept, an unsolvable mystery, and go about your life without the need for proofs?

**10.**

*The Moral Argument.* Humans are moral beings and animals are not. Where did we get this moral drive? Through the ultimate moral being—God. Without God, without the highest of higher moral authorities, anything goes and there would be no reason to be moral.

*Counterargument.* The argument that we cannot be good without God is easily refuted through a simple and straightforward question: *What would you do if there were no God?* The question can be followed by an additional question that draws the denouement: *Would you commit deception, robbery, rape, and murder, or would you continue being a good and moral person?* Either way the argument is over. If the answer is that people would quickly turn to deception, robbery, rape, or murder, then this is a moral indictment of their character, indicating they are not to be trusted because if, for any reason, they turn away from their belief in God (and most people do at some point in their lives), the plug is pulled on their constraints and their true immoral nature is revealed; we would be well advised to steer a wide course around them. If the answer is that people would continue being good and moral, then apparently you *can* be good without God.

## SCIENTIFIC ARGUMENTS FOR GOD

Scientifically based arguments that claim to prove the existence of God fall in the gray borderlands between science and philosophy, physics and metaphysics, and lie mostly in the realm of cosmology, in the study of the fundamental laws of nature, or in the complexities of the biological world. The first two fall into what might be called "The New Cosmology," and the last might be thought of as "The New Creationism."

## THE NEW COSMOLOGY

Most of the new cosmological arguments for God's existence are made by creationists such as Hugh Ross, whose series of books on *The Creator and the Cosmos, Creation and Time*, and *Beyond the Cosmos* argue "how the greatest scientific discoveries of the century reveal God." Ross's books are published by and for Christians, and are specifically written such that, as noted on the book jacket of the first installment, "whether you're looking for scientific support for your faith or new reasons to believe," these works "will enable you to see the Creator for yourself." Ross is, in fact, the president of Reasons to Believe, a nonprofit Christian corporation whose purpose, as gleaned from its name, is to provide believers with reasons to reinforce their faith. Among the strongest, he argues, are those from cosmology.

Many non-Christians also find cosmological arguments compelling. It may not be the God of Abraham in focus in the Hubble telescope, but behind the laws of nature, outside the large-scale structure of the universe, and inside the small-scale structure of the atom, lurks a higher intelligence, a spark of divinity. At the politically conservative American Enterprise Institute, for example, English literature scholar Patrick Glynn penned *God: The Evidence*, a more sophisticated presentation than Ross's but at the core presenting a similar set of arguments: The anthropic principle implies a creator, religious belief leads to greater physical and mental wellness, and near-death experiences prove there is an afterlife. Although Glynn is calling for "the reconciliation of faith and reason," he abandons the latter because "reason has proved an imperfect guide to the ultimate truths about the physical world, let alone the ultimate truths about the universe and human life." In the end, "reason rediscovers and reconstructs . . . what Spirit already knows." Of course, Glynn is using reason to bolster what *his* spirit already knows—that God exists. Rather than a reconciliation of faith and reason, it is faith in search of reasons to believe. In addition to this being a pointless exercise since faith cannot be proved, his reasons are not sound. The anthropic principle only implies that there is order in the universe (more on this later); religious belief may or may not lead to greater physical and mental wellness, but if it does, it is for perfectly understandable reasons, such as a social support system that encourages healthier living; and near-death experiences no more prove there is an afterlife than do hypnosis, hallucinations, or other altered states of consciousness.

The lengths some will go to in the endeavor to prove their faith strains credulity. Physicist Gerald Schroeder, in his 1997 book *The Science of God*, offers perhaps the most painfully contorted attempt to squeeze modern science into the Bible. According to Schroeder, modern scientists have discovered what ancient Jewish scholars always knew: Genesis describes the large-scale sequence of evolutionary change (sea creatures to land animals to mammals to man); the six days of creation perfectly match the description of the creation of a fifteen-billion-year-old universe (in relativistic time one day is equal to a couple of billion years); and medieval Kabalists like the Jewish scholar Nahmanides somehow got it all right. "With the insights of Albert Einstein," says Schroeder, "we have discovered in the six days of Genesis the billions of years during which the universe developed." How can a day be as a billion years? "The million-million-factor difference between our local perception of time and Genesis cosmic time is an average for the six days of creation. As discussed, it derives from the approximate million-millionfold stretching of light waves as the universe expanded." Faith and reason are reconciled, Schroeder concludes: "Genesis and science are *both* correct. When one asks if six days or fifteen billion years passed before the appearance of humankind, the correct answer is 'yes.'"

The fatal flaw in this argument is that the universe's age is only known within a factor of 2 (one often sees figure ranges reported such as ten to twenty billion years). This means that the days of Genesis, if defended scientifically, could have been anywhere from three to nine days. Since Schroeder argues that it must be six days (because, de facto, like everyone in this genre he begins with the assumption that the Bible must be true), the jig is up if the (still inconclusive) scientific evidence comes in at a figure at odds with Genesis.

A deeper and more troubling problem in this and other like-minded books is that Genesis is neither correct nor incorrect, because it is not a book of cosmology. Genesis is a *cosmogony*—a mythic tale of origins—and like all cosmogonies (for example, Egyptian, Hindu, Greek, Roman, Inuit, Polynesian, Mayan, Native American) it is neither true nor false because these evaluative terms are reserved for statements of fact, not myths and stories. Sure, if you stretch your imagination and play fast and loose with both the story and the science, you can find gross similarities between myth and nature. Comparing Genesis time to cosmic time is like comparing Taoism to quantum mechanics—the fact that they both speak of wholeness and integration means nothing more than that the author has found linguistic and conceptual similarities. But these comparisons do not prove anything, other than that the human mind is adept at finding and matching patterns.

Even those who do not consider themselves religious in any traditional way are attracted to some of these arguments for what they might imply about the possible existence of some sort of higher intelligence or human spirituality. In *Skeptics and True Believers,* physicist and astronomer Chet Raymo offers a very measured and reasonable discussion of the relationship between science and religion. Raymo considers himself "a thoroughgoing Skeptic who believes that words like *God, soul, sacred, spirituality, sacrament,* and *grace* can retain currency in an age of science, once we strip them of outworn overlays of anthropomorphic and animistic meaning. Like many others in today's society, I hunger for a faith that is open to the new cosmology—skeptical, empirical, ecumenical, and ecological—without sacrificing historical vernaculars of spirituality and liturgical expression." Along similar lines, Bruce Mazet, who has no belief whatsoever in the anthropomorphic Judaeo-Christian God, presented in the pages of *Skeptic,* "A Case for God." Mazet reviewed the fine-tuned universe argument in which the likelihood of the conditions for life to arise are astronomically small. He noted that there are counterarguments, such as that trillions of universes might have popped into and out of existence, one of which happened to have the right conditions for life (ours). The problem, Mazet notes, is that "there is no evidence whatsoever that this infinite number of hypothetical universes exist, and according to the cosmologists who postulate these hypothetical universes, there is no means by which to obtain any such evidence." Therefore, Mazet concludes, "I suggest that if it is acceptable to postulate the existence of hypothetical universes, then it is acceptable to postulate the existence of God."

That certainly sounds reasonable. After all, what is good for the cosmologist is good for the theologian. Let's examine what leading scientists are actually saying about God and cosmology, and consider how we might address these new cosmological arguments for God's existence.

**1.**

*Stephen Hawking's God.* When cosmologists deal with the beginning of the universe they are only a small step removed from Aquinas's prime mover and first cause arguments. After all, to ask such questions as: "What was there before the Big Bang?" or "Why should there be something rather than nothing?" is not so distant from "What was God doing before He created the universe?" or "What is God's purpose for the universe?" Stephen Hawking, in his quest to understand the origin and fate of the universe, admits his work often falls in that shadowland between science and religion, physics and metaphysics, as he told an ABC *20/20* reporter:

*It is difficult to discuss the beginning of the Universe without mentioning the concept of God. My work on the origin of the Universe is on the borderline between science and religion, but I try to stay on the scientific side of the border. It is quite possible that God acts in ways that cannot be described by scientific laws. But in that case one would just have to go by personal belief.*

In his book, *A Brief History of Time*, Hawking closes with this now oft-quoted line: "If we find the answer to that, it would be the ultimate triumph of human reason—for then we would know the mind of God." This was an unfortunate choice of words because in his position as one of the world's leading cosmologists Hawking is eminently quotable, and people have read this to mean the Judaeo-Christian God. According to his biographers Michael White and John Gribbin, although Hawking is not an atheist, he clearly does not believe in a personal God. Shortly after *A Brief History of Time* was released, in December 1988, the actress Shirley MacLaine asked Hawking at a luncheon if he believes that a God created the universe. In his characteristic economy of words, Hawking's machine voice answered "No." Similarly, in a BBC television production called *Master of the Universe*, Hawking waxed theological about his cosmology: "We are such insignificant creatures on a minor planet of a very average star in the outer suburbs of one of a hundred thousand million galaxies. So it is difficult to believe in a God that would care about us or even notice our existence." Indeed, in his chapter, "The Origin and Fate of the Universe," in *A Brief History of Time*, where he presents his no-boundary model of the cosmos, Hawking concluded that the universe may have no beginning or end, and thus no need for God:

*The idea that space and time may form a closed surface without boundary also has profound implications for the role of God in the affairs of the universe. With the success of scientific theories in describing events, most people have come to believe that God allows the universe to evolve according to a set of laws and does not intervene in the universe to break these laws. However, the laws do not tell us what the universe should have looked like when it started—it would still be up to God to wind up the clockwork and choose how to start it off. So long as the universe had a beginning, we could suppose it had a creator. But if the universe is really completely self-contained, having no boundary or edge, it would have neither beginning nor end: it would simply be. What place, then, for a creator?*

It is a difficult concept for the human mind to grasp, but Michael White and John Gribbin, in their biography of Hawking, make this analogy: Imagine walking all the way to the North Pole of the Earth. For the entire trip there you are heading north, but the moment you pass the pole you are now heading south. Similarly, imagine the universe as an expanding sphere beginning with the Big Bang, and that you are in a time machine traveling backward toward that initial point. For the entire trip you are heading back in time, but the moment you pass the starting point you are now heading forward in time. There is no beginning and no end—no boundaries. The universe always was, always is, and always shall be.

Whatever Hawking may mean when he speaks of God, he certainly does not mean the personal Judaeo-Christian God who created the universe and cares about us.

2. *Paul Davies's God.* Mathematical physicist Paul Davies is a believer in God and winner of the million-dollar Templeton Prize for "progress in religion." In his book, *The Mind of God*, Davies reviews all the philosophical and scientific arguments for God's existence, concluding that "belief in God is largely a matter of taste, to be judged by its explanatory value rather than logical compulsion. Personally I feel more comfortable with a deeper level of explanation than the laws of physics. Whether the use of the term 'God' for that deeper level is appropriate is, of course, a matter of debate." If one of the great believing scientists of our age says that God's existence cannot be proved, it would seem that some weight should be given to the position that belief in God is a matter of personality and emotional preference, also known as *faith*.

3. *Frank Tipler's God.* Cosmologist Frank Tipler's answer to the God Question, while a theistic one, begins with a premise unlike that of most theists. In his books *The Anthropic Cosmological Principle* and especially *The Physics of Immortality*, subtitled

*Modern Cosmology, God and the Resurrection of the Dead,* Tipler presents and defends his Omega Point Theory: The laws of nature and the configuration of the cosmos from atoms to galaxies is such that if you tweaked any of the parameters even slightly (and this often means a change many places after the decimal point in a number describing some aspect of nature), our universe, and we, could not exist in anything remotely similar to what we experience. Since "the Universe must have those properties which allow life to develop within it at some stage in its history," it does, and here we are. In other words, the universe had to be just so in order for us to be here, and the chances of it being just so are so small that it would have to have been made by some supreme being. More than this, says Tipler, "intelligent information-processing must come into existence in the Universe, and, once it comes into existence, it will never die out," so we must and we will take control of our universe and all other possible universes. In the process of doing this we "will have stored an infinite amount of information, including all bits of knowledge which it is logically possible to know." This, says Tipler, "is the end." It is the Omega Point—the all-knowing and all-powerful being (God as computer?)—that not only has the power but also the desire to resurrect everyone who ever lived or could have lived.

I have spoken to a number of cosmologists and physicists about Tipler's theory, and the conclusions are generally the same. Caltech theoretical physicist Kip Thorne, for example, found nothing wrong with Tipler's physics but concluded that his if-then leaps of logic between the steps of what must occur in order to reach the Omega Point were far too speculative to be meaningful; too much "hand-waving" between steps. John Casti, from the Santa Fe Institute, agreed with Tipler's speculations on how intelligent life could colonize the galaxy and, like Thorne, had no beef with Tipler's physics, but he concluded that each step in Tipler's chronology leading up to the universal resurrection could be broken down into further steps to the point where the probability of all these contingencies coming together was so unlikely that he does not know what value such a theory could have.

One of Tipler's most enthusiastic supporters, on the other hand, is the highly regarded German theologian Wolfhart Pannenberg, from the Institute for Fundamental Theology at the University of Munich. In a lecture given at the Innsbruck Conference in June 1997, Pannenberg concluded: "Tipler is justified in claiming that his statements on the properties of the Omega Point correspond to Biblical assertions on God. The God of the Bible is not only related to the future by his promises, but he is himself the saving future that constitutes the core of the promises: 'I shall be who I shall be.' " Yet even Pannenberg must go beyond Tipler's physics to admit that God is not *just* in the future: "In hidden ways he is already now the Lord of the universe which is his creation, but it is only in the future of the completion of this universe, in the arrival of his kingdom that he will be fully revealed in his kingship over the universe and thus in his divinity."

I even had the opportunity to ask Stephen Hawking's opinion of Tipler's theory during his 1998 visit to Caltech. Hawking's lecture dealt with something he calls the "pea instanton," a particle of space/time resembling a wrinkly pea, out of which the universe sprang into existence. As this universal "pea" expanded, the wrinkles were pushed out, leaving the relatively smooth universe we observe today. In Hawking's opinion, the question of the closed or open nature of the universe (Tipler's theory demands a closed universe) depends on the model applied to the question, which means that the universe can be both closed and open, not unlike how light can be both particle and wave. Without ever mentioning God, Hawking skirted that metaphysical line in discussing the Omega Point and the Anthropic Principle, so I inquired:

*You've been talking about the Omega Point and the Anthropic Principle. What is your opinion of your cosmologist colleague Frank Tipler's book,* The Physics of Immortality, *and his theory that the Omega Point will reach back from the far future of the universe into the past to reconstruct every human who ever lived or who ever could have lived in the ultimate Holodeck?*

Hawking composed his answer for about a minute, then his now-familiar computer voice responded: "My opinion would be libelous." Tipler responded to this charge as follows:

*All I do in my work is accept the logical consequences of the known laws of physics: quantum mechanics, relativity, and the Second Law of Thermodynamics. I'm not proposing any new laws of physics, just asking people to accept the logical consequences of the laws they claim to accept. Libeling the Omega Point Theory is equivalent to libeling the known laws of physics. Almost all contemporary theology* still *presupposes the truth of Aristotelean physics. This being the case, scientists naturally suppose theology is nonsense, or in a separate realm from science. With the almost unique exception of Pannenberg, theologians encourage them in this latter opinion. Only if theology is kept separate can it retain its Aristotelean physical basis.*

*The reality that the ancients were trying to capture in the word "soul" is expressed by defining the soul to be a computer program being run on the human brain. With this redefinition, we can keep the religious concept, and make it consistent with the facts. But most importantly, the redefinition makes the scientist realize that immortality is perfectly possible: there's no physical reason why a program cannot exist forever. Some of the programs now coded in our DNA have been around billions of years. Keeping the old definition makes Hawking want to libel a person whose book's central postulate is that the biosphere can go on forever. Is postulating the immortality of the biosphere an evil postulate? Shouldn't we at least try to make it so? Should a person who tries to figure out how to use the known physical laws to make the biosphere immortal be ostracized from scientific society?*

*Similarly for the word "God." If He is identified with the Omega Point, then the key religious meanings of "God" are retained, with science and religion integrated. As he wrote at length, the German theologian Wolfhart Pannenberg agrees that the Omega Point is in all essentials the God of the Bible. It's easier for a German theologian to come to this conclusion than an English speaker. God's Name, given in Exodus 3:14, was translated by Martin Luther as "ICH WERDE SEIN, DER ICH SEIN WERDE"—"I WILL BE WHAT I WILL BE." Failing to make this change of definition, which is to say, failing to give up Aristotelean physics, makes it difficult to accept the consequences of modern physics. These require the universe to terminate in its ultimate future in an Omega Point, a state of infinite knowledge, and infinite power.*

Certainly it is time to reject Aristotelian physics and with it the ancient and medieval concepts of God and soul. And with Tipler's narrow definition of life as information processing machines (with DNA coding for our anatomy and physiology and neurons coding for our thoughts and memories), it is conceivable the short-lived and fragile carbon-based, protein-chain life forms could be reconstituted into something more durable and long-lasting, such as silicon chips. A human life, by this analysis, is a "pattern" of information, and silicon can store the pattern much longer than protein, and there may be other future technologies we cannot yet conceive that could hold the integrity of the pattern still longer, perhaps approaching infinity, and thus immortality. As for God's future tense, *The Interpreter's Bible* notes that the common translation of Exodus 3:14 is "I AM WHO I AM," with a secondary alternative of "I AM WHAT I AM," and a tertiary translation of "I WILL BE WHAT I WILL BE." Richard Elliott Friedman, Professor of Hebrew and Comparative Literature at the University of California, San Diego, told me: "Tipler and Luther are simply wrong. God is not a future-tense verb in biblical Hebrew." Case closed. As for the Omega Point, Tipler says it is transcendent to time, but his God is the future c-boundary of the universe that acts back in time, not the personal anthropomorphic God who cares about us that most people think of when they think about God.

Why must the God conclusion be drawn from science? Why not speculate on the possibility of space travel, human occupation of the galaxy and eventually other galaxies, machine intelligence, and even the far future of the universe, without trying to tie it into some ancient mythic Hebrew doctrine created by and for people living on the margins of the Mediterranean nearly 4,000 years ago? What are the chances that this agrarian society, constructing myths and stories whole cloth out of traditions that preceded them sometimes by as much as a thousand years (and rewritten and reinterpreted to fit their social and cultural needs, as all myths are), just happened to anticipate one interpretation of late twentieth-century cosmology? Much more likely is that Tipler is pushing a particular rendition of modern cosmology and physics—one that is by no means shared by his colleagues—into and beyond the borderline between science and religion. It may be that someday science will reduce all religious and metaphysical questions to the equations of physics, but we are so far from that stage that wisdom would seem to dictate that we leave the God conclusion out of science altogether.

**4.**

*God and the Cosmologists.* After reading Tipler's book I thought perhaps I was missing something and that I better read what other cosmologists, astronomers, and physicists were thinking about the relationship of science and religion. According to David Deutsch, whom Tipler quotes in support, there is no God in his cosmos. Deutsch believes Tipler may be right about the Omega Point's future existence, and that it is conceivable we could all be resurrected in the far future of the universe, but, he concludes: "Unfortunately Tipler himself . . . makes exaggerated claims for his theory which have caused most scientists and philosophers to reject it out of hand." Deutsch points out that Tipler's Omega Point not only differs from everyone else's version of God, there are additional problems:

*For instance, the people near the omega point could not, even if they wanted to, speak to us or communicate their wishes to us, or work miracles (today). They did not create the universe, and they did not invent the laws of physics—nor could they violate those laws if they wanted to. They may listen to prayers from the present day (perhaps by detecting very faint signals), but they cannot answer them. They are (and this we can infer from Popperian epistemology) opposed to religious faith, and have no wish to be worshiped. And so on. But Tipler ploughs on, and argues that most of the core features of the God of the Judaeo-Christian religions are also properties of the omega point.*

Where Tipler and Davies see God in the cosmos, Deutsch and others do not. For example, in Alan Guth's well-received book, *The Inflationary Universe*, there is no mention of God or religion whatsoever. In his final chapter, "A Universe ex Nihilo," Guth concludes:

*While the attempts to describe the materialization of the universe from nothing remain highly speculative, they represent an exciting enlargement of the boundaries of science. If someday this program can be completed, it would mean that the existence and history of the universe could be explained by the underlying laws of nature. That is, the laws of physics would imply the existence of the universe. We would have accomplished the spectacular goal of understanding why there is something rather than nothing—because, if this approach is right, perpetual "nothing" is impossible.*

For Lee Smolin, in his 1997 *The Life of the Cosmos*, "the present crisis of modern cosmology is also an opportunity for science to finally transcend the religious and metaphysical faiths of its founders." Smolin's multiverse model includes an evolutionary mechanism where, like its biological counterpart, natural selection chooses from a variety of "species" of universes, each containing varying forms of laws of nature. Some of those universes with laws of nature like ours will be "selected" for intelligent life, which at some point in its evolution develops big enough brains to consider such questions of origins. Beyond that, Smolin admits, questions about ultimate existence and purpose "are in the class of really hard questions, such as the problem of consciousness or the problem of why there is in the world anything at all, rather than nothing. I do not see, really, how science, however much it progresses, could lead us to an understanding of these questions."

Maybe our universe simply popped into existence out of the quantum fluctuation of the vacuum of some larger multiverse. Maybe our universe is just one of those things that happened for no reason at all.

## THE NEW CREATIONISM

On the heels of the new cosmology is the new creationism, but with a far more activist agenda in working to see Genesis taught in public schools. In the twentieth century, creationists have employed three strategies to achieve this end: (1) banning the teaching of evolution, (2) demanding equal time for Genesis with Darwin, and (3) the demand of equal time for "creation-science" with "evolution-science." All three of these strategies were defeated in court cases, starting with the famed 1925 Scopes "Monkey Trial" and ending with the 1987 Louisiana trial, which went all the way to the United States Supreme Court where it was overturned by a vote of 7 to 2. This ended what I have called the "top-down" strategies of the creationists to legislate their beliefs into culture through public schools.

With these defeats they turned to "bottom-up" strategies of mass mailings to schools with creationist literature, debates at schools and colleges, and enlisting the aid of mainstream academics like University of California–Berkeley law professor Phillip Johnson and Lehigh University biochemist Michael Behe, and even roping in the conservative commentator William F. Buckley, whose PBS show, *Firing Line,* hosted a debate in December 1997, where it was resolved that "evolutionists should acknowledge creation." The debate was emblematic of a new creationism, employing new euphemisms such as "intelligent-design theory," "abrupt appearance theory," or "initial complexity theory," where it is argued that the "irreducible complexity" of life proves it was created by an intelligent designer, or God. In Behe's book, *Darwin's Black Box,* the biochemist, who has become something of a cult hero among creationists, explains this phrase: "By *irreducibly complex* I mean a single system composed of several well-matched, interacting parts that contribute to the basic function, wherein the removal of any one of the parts causes the system to effectively cease functioning."

Consider the creationists' favorite example of the human eye, a very complex organ that is, we are told, irreducibly complex—take out any one part and it will not work. How could natural selection have created the human eye when none of the individual parts themselves have any adaptive significance? There are four answers that refute this argument.

1. It is not true that the human eye is irreducibly complex, so that the removal of any part results in blindness. Any form of light detection is better than none—lots of people are visually impaired with any number of different diseases and injuries to the eyes, yet they are able to utilize their restricted visual capacity to some degree and would certainly prefer this to blindness. The creationists' "irreducible complexity" argument is an either-or fallacy. No one asks for partial vision, but if that is what you get, then like all life forms throughout natural history, you learn to cope in order to survive.

2. There is a deeper answer to the example of the evolution of the eye, and that is that natural selection did not create the human eye out of a warehouse of used parts lying around with nothing to do, any more than Boeing created the 747 without the ten million halting jerks and starts from the Wright Brothers to the present. Natural selection simply does not work that way. The human eye is the result of a long and complex pathway that goes back hundreds of millions of years to a *simple eyespot* where a handful of light-sensitive cells provides information to the organism about an important source of the light—the sun; to a *recessed eyespot* where a small surface indentation filled with light-sensitive cells provides additional data in the form of direction; to a *deep recession eyespot* where additional cells at greater depth provide more accurate information about the environment; to a *pinhole camera eye* that is actually able to focus an image on the back of a deeply recessed layer of light-sensitive cells; to a *pinhole lens eye* that is actually able to focus the image; to a *complex eye* found in modern mammals such as humans. And this is just part of the story—how many other stages of eye development were lost to the ravages of time because there was an organ that did not fossilize well?

We can also use the human eye as an example of bad design. The configuration of the retina is in three layers, with the light-sensitive rods and cones at the bottom, facing away from the light,

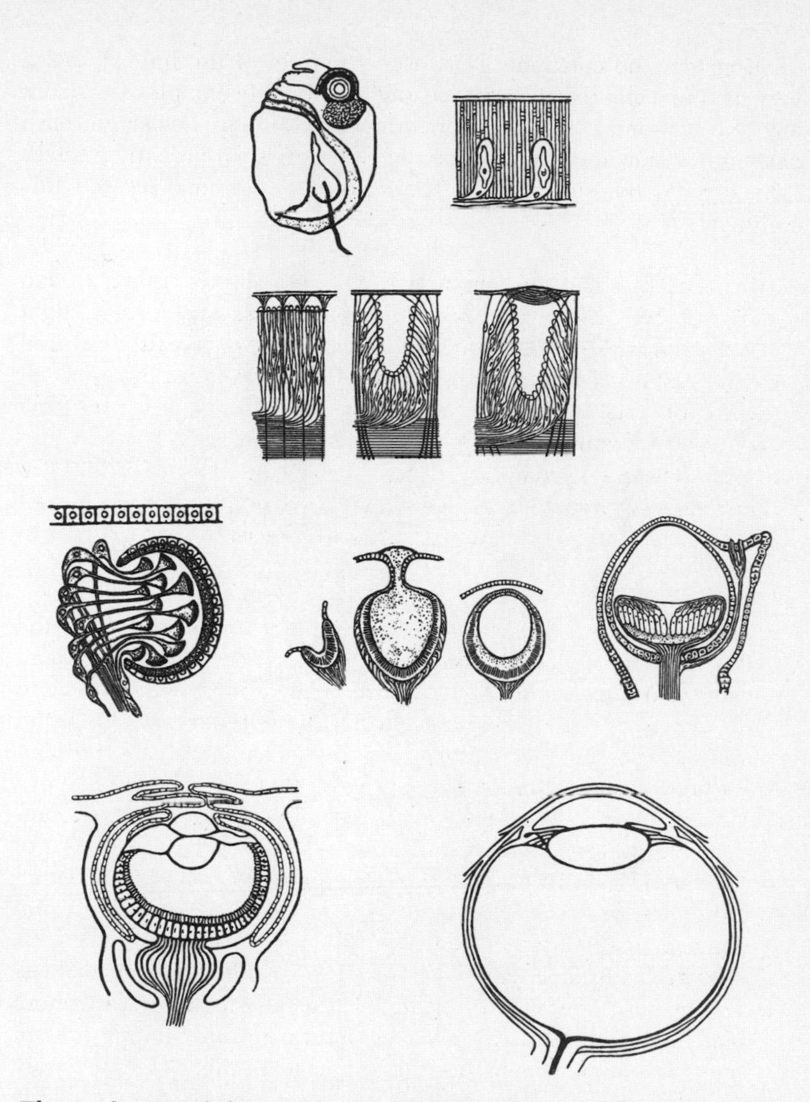

*The evolution of the eye from a simple eyespot to the complex eye, which has occurred independently at least a dozen times in natural history, shows that the eye is neither irreducibly complex nor intelligently designed. It was constructed by natural selection in fits and starts over hundreds of millions of years from available parts and systems already in use.*

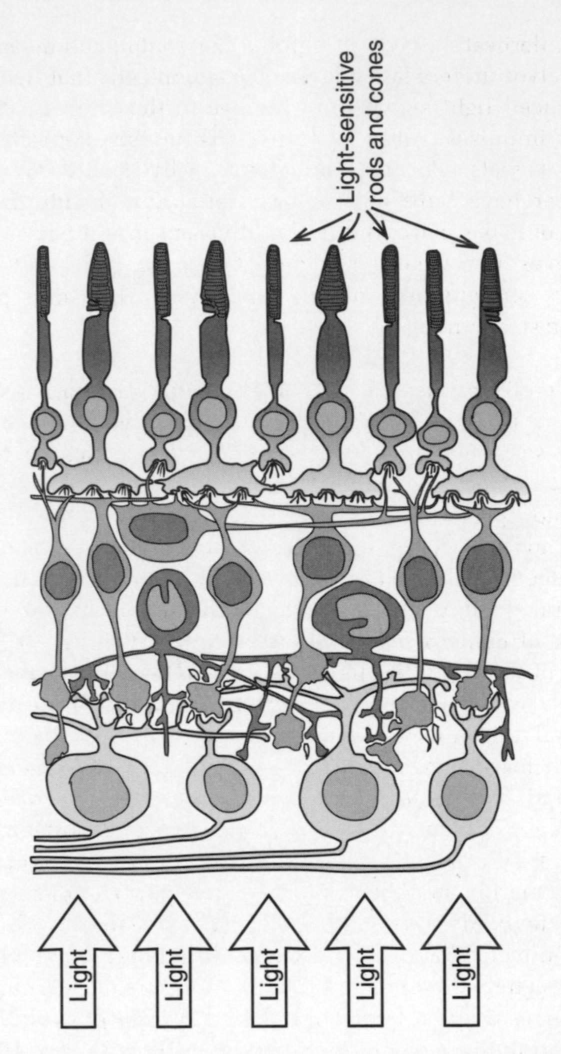


*The anatomy of the human eye shows that it is anything but "intelligently designed." It is built upside down and backwards, with photons of light having to travel through the cornea, lens, aqueous fluid, blood vessels, ganglion cells, amacrine cells, horizonal cells, and bipolar cells, before reaching the light-sensitive rods and cones that will transform the signal into neural impulses. From the rods and cones, the impulses are then sent to the visual cortex at the back of the brain for processing into meaningful images.*

and underneath a layer of bipolar, horizontal, and amacrine cells, themselves underneath a layer of ganglion cells that help carry the transduced light signal from the eye to the brain in the form of neural impulses. And this entire structure sits beneath a layer of blood vessels. For optimal vision, why would an intelligent designer have built an eye backwards and upside down? This does not make sense. But it would make sense if natural selection built eyes from whatever materials were available, and in the particular configuration of the ancestral organism's preexisting organic structures.

**3.**

The "intelligent design" argument, similar to Aquinas's fifth way to prove God, also suffers from the fact that the world is not always so intelligently designed! Look at the animal on the following page. It is *Ambulocetus natans*, a transitional fossil between the quadrupedal land mammal *Mesonychids* and the direct ancestor of modern whales, the *Archaeocetes. Ambulocetus natans,* say the paleontologists who discovered it, swam "by undulating the vertebral column and paddling with the hindlimbs, combining aspects of modern seals and otters, rather than by vertical movements of the tail fluke, as is the case in modern whales." First of all, why would God, in His infinite wisdom and power, create a mammal that appears midway between a land mammal and a modern marine mammal, that combines the movements of both land and marine mammals, and, most uniquely, paddles with hind limbs obviously well designed for land locomotion? For that matter, why would He create air-breathing, warm-blooded, breast-feeding marine mammals only moderately well "designed" for living in the oceans, when he could have just stuck with the much more efficient fish design? Finally, on a larger scale, why would God design the fossil record to look like descent with modification was the result of hundreds of millions of years of evolution, rather than sprinkling geological strata willy-nilly with, say, trilobites in Cretaceous strata, and a T-Rex or two alongside some Neanderthal fossils? The fossil record screams out evolution, not creation.

**4.**

When Michael Behe defines irreducible complexity, he concludes: "An irreducibly complex system cannot be produced directly (that is, by continuously improving the initial function, which continues to work by the same mechanism) by slight, successive modifications of a precursor system, because any precursor

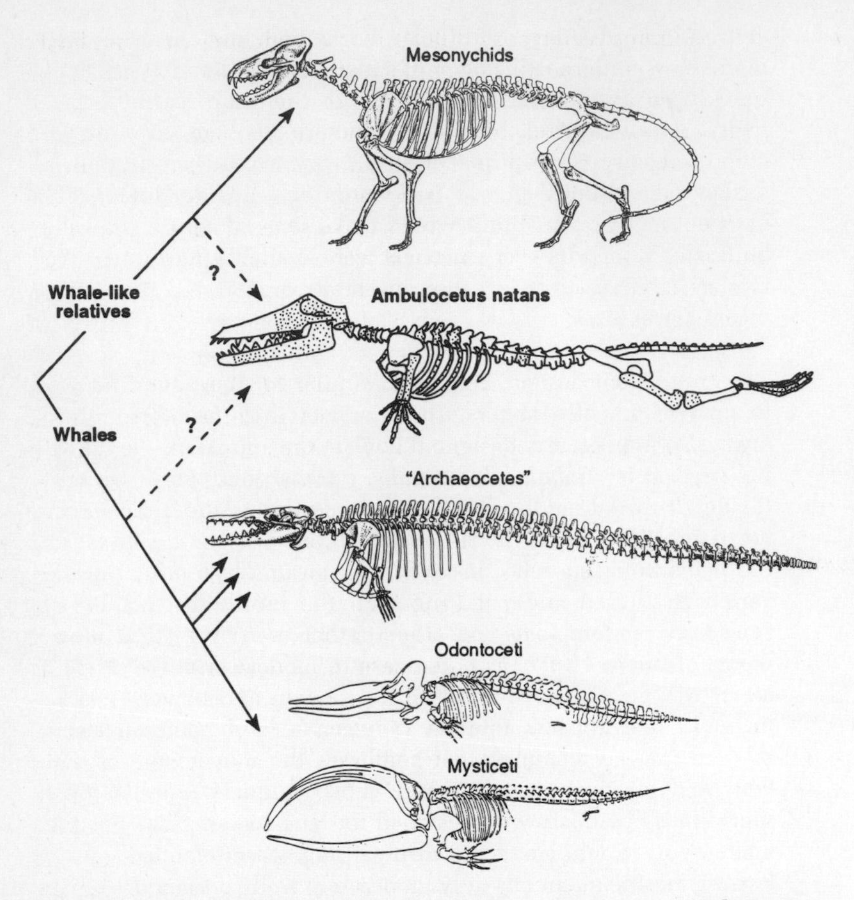


*The frequent rallying cry of creationists and other antievolutionists demands proof of the existence of "just one transitional fossil." The discovery of* Ambulocetus natans, *a transitional fossil between the land-based* Mesoynchids *and the marine mammal* Archaeocetes, *the direct ancestor of modern whales, answers that demand. This fossil record has all the earmarks of an evolutionary process of Darwinian "descent with modification" rather than a creationist "abrupt appearance."*

to an irreducibly complex system that is missing a part is by definition nonfunctional." Philosopher Robert Pennock has pointed out that this last phrase employs a classic fallacy of bait-and-switch logic—reasoning from something that is true "by definition" to something that is proved through empirical evidence. Creationists counter the above arguments about the eye by redefining what constitutes an eye, reducing its complexity until they get one that does not work. This is not allowed in the rules of right reasoning.

The new creationists have also mounted an attack on the very foundations of science—its philosophical naturalism (sometimes called methodological naturalism, materialism, or scientism). This is the belief that life is the result of a natural and purposeless process in a system of material causes and effects that does not allow, or need, the introduction of supernatural forces. The argument against naturalism is trumpeted by University of California–Berkeley law professor Phillip Johnson, a self-proclaimed "philosophical theist and a Christian" who believes in "a Creator who plays an active role in worldly affairs." In his book *Darwin on Trial*, Johnson claims that scientists unfairly define God out of the picture by saying, essentially, "we are only going to examine natural causes and shall ignore any supernatural ones."

This is a fallacy of fuzzy definitions. What does Johnson mean by *super*natural? Cosmologists who find God in the anthropic principle are both theists and naturalists. Supernatural simply means a lack of knowledge about the natural. We might as well call it *ignatural*. To medieval Europeans the weather was caused by supernatural forces; they abandoned that belief when natural forces were understood. This is, once again, the "God of the Gaps" argument, which is what philosophers call "arguments from ignorance." The rules of logical reasoning do not allow the following: "You cannot explain *X*, therefore *Y* must be the cause," or, to cut to the chase, "Science cannot explain all life, therefore God must be the cause." Of course, just as naturalism allows us to tell creationists that they cannot "prove" God through science, we cannot "disprove" God through science. After all, as the anthropologist Eugenie Scott cleverly notes, an "omnipotent God by definition can do anything it wants, including interfering in the universe to make it look exactly like there is no interference!"

Even if we did allow creationists to make the gaps argument, it is easily countered. Although in their public debates and published works creationists replace "God" with such obfuscating phrases as "abrupt appearance" and "intelligent design," their true colors fly when you attend their church services and monitor their Internet chat rooms. There is no question in anyone's mind that when creationists argue for an intelligent designer they mean God, and it is almost always the Judaeo-Christian God and all that goes with it. But why must an intelligent designer be God? Since creationists like William Dembski argue that what they are doing is no different from what the astronomers do who look for intelligent design in the background noise of the cosmos in their search for extraterrestrial intelligent radio signals, then why not postulate that the design in irreducibly complex structures such as DNA is the result of an extraterrestrial experiment? Such theories have been proffered, in fact, by some daring astronomers and science fiction authors who speculated (wrongly, it appears) that the Earth was seeded with amino acids, protein chains, or microbes billions of years ago, possibly even by an extraterrestrial intelligence. Suffice it to say that no creationist worth his sacred salt is going to break bread or sip wine in the name of some experimental exobiologist from Vega. And that is the point. What we are really talking about here is not a scientific problem in the study of the origins of life, it is a religious problem in dealing with the findings of science.

Finally, at the core of the new creationists' argument is the arrogant and indolent belief that if *they* cannot think of how nature could have created something through evolution, it must mean that scientists will not be able to do so either. (This argument is not unlike those who, because *they* cannot think of how the ancient Egyptians built the pyramids, assume these structures must have been built by Atlantians or aliens.) It is a remarkable confession of their own inabilities and lack of creativity. Who knows what breakthrough scientific discoveries await us next month or next year? The reason, in fact, that Behe has had to focus on the microscopic world's gaps is that the macroscopic gaps have mostly been filled. They are chasing science, not leading it. Also, sometimes we must simply live with uncertainties. A scientific theory need not account for *every* anomaly in order to be viable (this is called the *residue problem*—we will always have a "residue" of anomalies). It is certainly acceptable to challenge existing theories, and call for an explanation of those anomalies. Indeed, this is routinely done in science. (The "gaps" that creationists focus on have all been identified by scientists first.) But it is not acceptable in science to offer as an alternative a nontestable, mystical, supernatural force to account for those anomalies.

## THE BIBLE CODE

A classic example of the misapplication of science in the service of God and religion can be found in Michael Drosnin's 1997 book, *The Bible Code*, that skyrocketed up the *New York Times* bestseller list, received full-page reviews in both *Time* and *Newsweek*, was sold to Warner Brothers for a possible television movie, was the subject of an entire episode of *Oprah*, and is being utilized by the Aish HaTorah's Discovery Seminars as proof to doubting Jews that God exists and that the Bible tells the absolute truth. Because of its cultural impact and importance, and for how similar its approach is to God and the Bible, it is worth examining its claims more closely to reveal the deeper flaw in all such arguments—the negation of faith.

It turns out God is not a mathematician, physicist, or cosmologist; God is a cryptanalyst and computer programmer. According to Drosnin, a former journalist for the *Wall Street Journal*, the Bible is actually an encrypted code book filled with meaningful portents of newsworthy events: Yitzhak Rabin's assassination, and John and Robert Kennedy's too; Netanyahu's election; comet Shoemaker-Levy's collision with Jupiter; Watergate; the Oklahoma City bombing and Timothy McVeigh; the asteroid that killed the dinosaurs; an earthquake in California; and, of course, just in time for the soon-to-come millennium madness, the end of the world in the year 2000.

Do not bother dusting off your old King James Bible. You will not find any of these revelations there. You need a Hebrew Bible, specifically, the Torah—Genesis, Exodus, Leviticus, Numbers, and Deuteronomy. *The Bible Code* is based on the work of Eliyahu Rips, an Israeli mathematician and computer expert who, along with two other authors (Doron Witztum and Yoav Rosenberg), published an article in 1994 in the prestigious academic journal *Statistical Science*. It is a peer-reviewed journal, but the editors made it clear they were publishing it because it was an interesting statistical phenomenon and "a challenging puzzle," not because they endorsed it.

Rips eliminated the spaces between all the words in the entire Torah, converting it into one continuous strand of 304,805 letters (which is how the Torah was allegedly dictated to Moses by God). With this strand Rips utilized an equidistant letter sequencing (ELS) computer program: Start with the first letter of Genesis and then enter a "skip-code" program by taking every *n*th letter, where *n* equals whatever number you wish—every 7th letter, 19th letter, 3,023th letter, or whatever it takes to find meaningful patterns. If

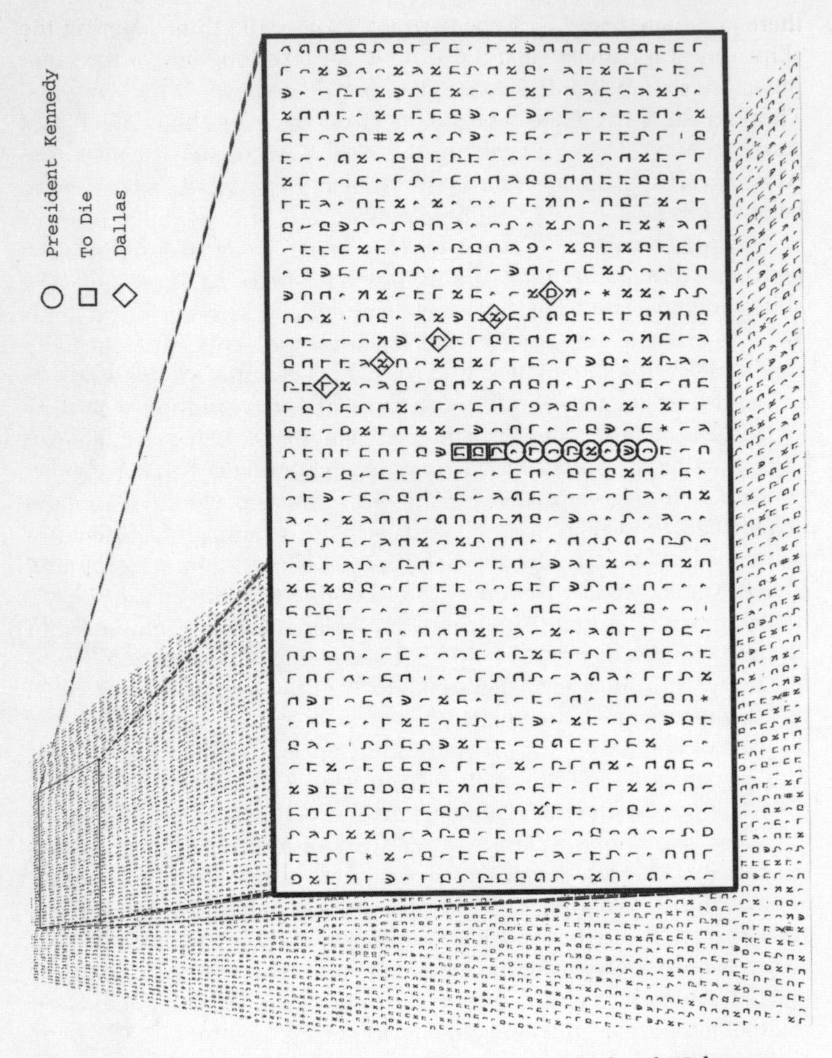


*Page 108 from* The Bible Code *presents a block of Hebrew type allegedly predicting the assassination of President Kennedy in Dallas ("Kennedy" is in circles, "Dallas" in diamonds, "to die" in squares). Since these sequences depend on how the letters align themselves vertically, and this in turn depends on the margin width, which has been arbitrarily set by humans, the "divine" nature of the code quickly disappears. This "margins problem" is one of many in* The Bible Code.

there are none, begin with the second letter, or the third, altering the skip *n* until a pattern emerges. It does not take long before the computer finds it: "Hitler," "Nazi," "Kennedy," "Dallas," "Pearl Harbor." They are all there. How can this be? The only way this ancient text could "know" these future events is if it were the work of the Almighty Himself, thus the code becomes a form of evidence for believers. Is it?

There are numerous flaws in *The Bible Code* that reinforce the point that humans are pattern-seeking animals who have a remarkable ability to find patterns even when none exist.

1. *The Margins Problem.* Look closely at the block of Hebrew type on the facing page that Drosnin claims has special significance—a field of Hebrew letters purporting to show the name *Kennedy* (in the circles), positioned near the word *Dallas* (in diamonds), adjacent to *to die* (in squares). The obvious problem is that the margin widths determine the type flow. Reduce or expand the margins and those alignments would disappear. The row widths, Drosnin explains, were determined by the skip-code *n*. An *n* of 10 means each row would be 10 letters long. An *n* of 4,772 would be 4,772 letters long. But why should this be? What is so special about a margin–skip-code correlation? Nothing. Since it is humans doing the margin and skip-code selecting, not God, this reveals the source of the pattern.

2. *The Vowels Problem.* Since ancient Hebrew is written without vowels, they are added *after* the skip-code program is run. If it were English, for example, *RBN* could be Rabin, or Ruben, or Rubin, or Robin. Bible scholar Ronald Hendel, for example, explained: "The same word may be spelled with a vowel letter in one sentence and without that vowel letter in the next sentence. As a result of these differences, every known ancient Hebrew manuscript of the Bible—including every ancient manuscript of the traditional Masoretic text—has a different number of letters." This is fatal for a skip-code computer program. Additionally, even though Hebrew is read from right to left, the Bible decoders also look left to right, up to down, down to up, and diagonally in any direction. If you have a name or word in mind ahead of time, just search to find it. Or you can look at the letter sequences and then find a meaningful name or word. Seek and ye shall find.

3.

*The Falsifiability Problem.* One of the tenets of science is falsifiability. In order to determine if something is true or not, there must be a way to test it, or falsify it. Drosnin provided one such test when he told *Newsweek* (June 9, 1997): "When my critics find a message about the assassination of a prime minister encrypted in *Moby Dick* I will believe them." Australian math professor Brendan McKay did just that, finding in *Moby Dick* no less than thirteen assassinations of public figures, several of them leaders of countries and even prime ministers. The results of the experiment that falsifies *The Bible Code* are revealed in the *Moby Dick* code:

O R WI T H A WH **I** T E P

N A H A B Y O U N **G** M A N

K L E S HI S G R **A** N D D

D S Y E T I N G E **N** E R A

**T H E B L O O D Y D E E D**

E R M WH A L E S **H** E A D

T T O I MP O S S **I** B L E

Indian Prime Minister Indira Gandhi was assassinated on October 31, 1984, a bloody deed to be sure.

Brendan McKay did not stop with *Moby Dick*. He also found "Hear the law of the sea" in the United Nations' *Convention on the Law of the Sea*, and fifty-nine words related to Hanukkah in the Hebrew translation of *War and Peace*, including "miracle of lights" and "Maccabees." The odds against all fifty-nine, he calculated, are more than a quadrillion to 1. Are we to believe that Tolstoy's hand was directed by God?

Similarly, in their book *The Signature of God*, which predates *The Bible Code* by two years, authors Grant Jeffrey and Yacov Rambsel report that they found the phrase "Yeshua is my Name" ("Jesus is my Name") with an ELS $n = 20$ in Isaiah 53, which some interpret as the prophecy of Jesus' coming. But others found that the phrase "Muhammad is my name" occurs twenty-one times, and "Koresh is my name" appears no less than forty-three times! Should we have listened to David Koresh's ramblings more closely?

4. *The Biblical Origins Problem.* Drosnin claims that "all Bibles in the original Hebrew language that now exist are the same letter for letter." All serious scholars of the Bible know this is utter nonsense. Richard Elliott Friedman, in his classic work, *Who Wrote The Bible?*, traces the multiple sources and authors that went into the construction of the Torah. In his latest research, carefully documented in *The Hidden Book in the Bible*, Friedman examines the oldest Hebrew documents to reveal that within the cacophony of biblical voices lies a single prose masterpiece that, in the editing process, had been fractured into what we know as the Old Testament. Our Bible is anything but a letter-by-letter transcription from ancient Hebrew. Ronald Hendel adds: "We do not have the original Hebrew version of the Old Testament, and all ancient manuscripts of the Hebrew Bible that we do have differ in the number of letters." Most biblical scholars now believe that the Torah was written by more than one individual, thus accounting for the different styles, the two different creation stories in Genesis, and other inconsistencies, and that there was a "redactor," or editor, who coalesced the multiple writings into one tome. Biblical archaeologist Gerald Larue also notes that even allegedly original biblical documents are anything but—quotes may be from memory or a compilation of several sources, errors are faithfully reproduced from one manuscript to another, and Hebrew letters look enough alike that names and words can be easily confused with others that are similar. Concern for accuracy and preserving the original text of the Bible came nearly 1,500 years after the originals were dictated (itself an oral tradition known for generating inaccuracies). All of these problems undermine the belief that the Torah was written by Moses, as inspired by God. Without this foundation, the Bible as an encrypted code of prophecies falls apart, and with it the claim that it provides evidentiary proof of God's existence.

5. *The Translation Problem.* In reading Drosnin's book in English, it is reasonable to wonder what we are losing in the translation from Hebrew. Ronald Hendel points out that the phrase "assassin that will assassinate" near Rabin's name is more properly translated as "murderer who murders inadvertently." Can you have an inadvertent assassination? Hendel identifies other translation howlers by Drosnin: "After the death of Abraham" (Genesis 25:11) is rendered as "after the death (of) Prime Minister"; "[Jacob] set it

up as a standing stone" (Genesis 31:45) is rendered as "shooting from the military post"; and "Which she [Rebekah] has made" (Genesis 27:17) is rendered as "fire, earthquake."

6. *The Prediction–Free Will Problem.* In *The Bible Code* Drosnin tells the dramatic story that he tried to warn Rabin a year before his assassination. In his claim that the Bible Code predicts such future events, Drosnin has unknowingly wedged himself into an insoluble paradox. Consider the implications: Say Rabin took the warning seriously and changed his schedule and was not assassinated. Would this mean that humans are more powerful than God, or that some statistician can rerun the universe to produce a different outcome? Does this mean that biblical prophecies are self-fulfilling prophecies, or that they are not prophecies at all, but warnings? Drosnin tries to solve this problem through an awkward blend of pop-science, pseudoscience, and hand-waving that is typical of most of the modern arguments for God. In his last chapter—"The Final Days"—Drosnin says the Bible Code predicts that the end of the world will occur in 2000, or 2006, or it will be delayed until a later date, or it might not happen at all. Some prediction! He gets around this problem by applying chaos theory, Heisenberg's uncertainty principle, and Feynman's quantum physics: "There isn't just one real future, there are many possible futures." In fact, he concludes, "the Bible Code revealed each of them." None of this works. Remarkably, after 178 pages of breathtaking revelations about biblical prophecies, Drosnin confesses that the Bible does not actually predict anything: "It is not a promise of divine salvation. It is not a threat of inevitable doom. It is just information." Even Rips has cut the tether in a public statement: "I do not support Mr. Drosnin's work on the codes, or the conclusions he derives. I did witness in 1994 Mr. Drosnin 'predict' the assassination of Prime Minister Rabin. For me, it was a catalyst to ask whether we can, from a scientific point of view, attempt to use the codes to predict future events. After much thought, my categorical answer is no." Does the Bible code prove God's existence? The categorical answer is no.

## THE REAL MEANING OF ARGUMENTS FOR GOD

All of this emphasis on proving God's existence is as if to say: "See, modern science supports what we have been saying all along—there really is something unique and special about the Bible." Is there? There is. The Bible is one of the greatest works of literature in the history of Western thought. It is a book of myth and meaning, moral homilies and ethical dilemmas, poetry and prose. Few works have been so influential to so many people over so many millennia. In an epilogue Drosnin admits: "I'm not religious. I don't even believe in God." It shows. Drosnin, like the creationists, has taken a beautiful book of literature and ruined it by trying to turn it into a book of science.

Science and religion are, at present, largely separate spheres of knowledge divided by, more than anything else, a difference in methodologies. Science is a process of inquiry aimed at building a testable body of knowledge constantly open to rejection or confirmation; its "truths" are provisional, fluid, and changing. Religion is the affirmation of a set of beliefs aimed at providing morals and meaning; its truths are final, confirmed by faith. Because we live in the Age of Science and no longer the Age of Faith, temptations abound to use science to bolster faith. Such attempts at reconciling science and religion always fail for the fundamental reason that religion ultimately depends on faith. The whole point of faith, in fact, is to believe regardless of the evidence, which is the very antithesis of science. "Now faith is the substance of things hoped for, the evidence of things not seen" (Hebrews 11:1). "For we walk by faith, not by sight" (II Corinthians 5:7). "Blessed are they that have not seen, and yet have believed" (John 20:29).

William Jennings Bryan ended his famous "Address to the Jury in the Scopes' Case" (published posthumously as it was never delivered during the trial and he died two days later), after pages of text marshalling the evidence for God and against evolution, with a plea to "sing that old song of triumph," faith:

> Faith of our fathers! living still
> In spite of dungeon, fire, and sword!
> O, how our hearts beat high with joy,
> Whene'er we hear that glorious word:
> Faith of our fathers, holy faith!
> We will be true to thee till death.

O, ye of little faith. Why do you need science to prove God? You do not. These scientific proofs of God are not only an insult to science; to those who are deeply religious they are an insult to God.

# Part II

# Religion and Science

**Thomas Paine,**
***The Age of Reason, III,* 1794**

*The world is my country, all mankind are my brethren, and to do good is my religion.*

Chapter 6

# In a Mirror Dimly, Then Face to Face

## *Faith, Reason, and the Relationship of Religion and Science*

*For now we see in a mirror dimly, but then face to face. Now I know in part; then I shall understand fully.*
—I Corinthians 13:12

There is a certain predictable, expected pleasure in making a discovery that comes at the end of a long and ordered journey, especially when that discovery is the goal of the trek itself. Discoveries made by accident, with no jaunt planned or purpose in mind, also generate their own unique pleasures, reserved for those rare occasions when contingent sequences include us in their wanderings. As a minimalist example of the latter, I once encountered the following coincidence at the home of a friend. During a quiet moment alone I grabbed for the nearest piece of reading material and happened to pull down a 1954 edition of *The Story of the Starry Universe,* part of the Popular Science Library's series of illustrated books of science for the general reader. Flipping to the final page to see what prognostications were being made for the future, I read that V-2 rockets were being hurled into space with scientific instruments (instead of the warheads of a decade prior), so that the stars might be studied from above the ultraviolet filter of the ozone layer. The research was so new it was not even published yet, but the authors boldly speculated:

> *Scientists are even talking about the possibility of sending rockets completely outside of the earth's atmosphere and causing them to move in an approximately circular orbit, permanent satellites of the earth for special laboratory studies. It has been estimated that perhaps ten years or so will elapse before such a ladder to the skies will have been perfected.*

My contingent gem came later that day in the same room, when I opened the paper to view the magnificent newly released color photographs from the Hubble Space Telescope—our "ladder to the skies."

It may have taken four decades instead of the estimated one, but the prize was well worth the wait. In science, as in most cultural productions, time frames rarely match expectations. But there is no disputing the fact that science changes faster than religion. Compare this 30-year discrepancy to the 360-year abyss between Galileo's 1633 indictment for the heretical support of Copernicus's heliocentrism and Pope John Paul II's acquittal of him in his April 1993 address to the Pontifical Biblical Commission; or the 137-year gap between Darwin's 1859 *Origin of Species* and Pope John Paul II's acceptance of evolution as a viable theory in his October 1996 address to the Pontifical Academy of Sciences.

As Popes go, John Paul II is relatively progressive in embracing science and its underpinnings of logic and empiricism. He is both broadly and deeply read, and sensitive to the relationship between faith and reason, religion and science. In his 1993 address he explained that "it is necessary to determine the proper sense of Scripture, while avoiding any unwarranted interpretations that make it say what it does not intend to say," and in order to do so "the theologian must keep informed about the results achieved by the natural sciences." His 1996 address, entitled *Truth Cannot Contradict Truth,* was written to update and revise Pope Pius XII's 1950 Encyclical *Humani Generis,* in which Catholics were told that there is no conflict between reason and faith when dealing with the theory of evolution: "The Teaching Authority of the Church does not forbid that, in conformity with the present state of human sciences and sacred theology, research and discussions, on the part of men experienced in both fields, take place with regard to the doctrine of evolution, in as far as it inquires into the origin of the human body as coming from pre-existent and living matter—for the Catholic faith obliges us to hold that souls are immediately created by God." For Pius XII, however, evolution as a theory was still up in the air and could one day be proved false. Thus, while there had been no opposition to provisionally accepting evolution (of the body only), if it turned out wrong, Catholics have lost nothing:

> *However, this must be done in such a way that the reasons for both opinions, that is, those favorable and those unfavorable to evolution, be weighted and judged with the necessary seriousness, moderation and measure. . . . Some however, rashly transgress this liberty of discussion, when they act as if the origin of the human body from pre-existing and living matter were already completely certain and proved by the facts which have been discovered up to now and by reasoning on those facts, and as if there were nothing in the sources of divine revelation which demands the greatest moderation and caution in this question.*

With this level of equivocation on the part of his predecessor, John Paul II felt it necessary to bring his over one billion followers up to date on the outcome of half a century of scientific research. The verdict is now in, the Pope explained in 1996, evolution happened:

> *Today, almost half a century after the publication of the Encyclical, new knowledge has led to the recognition of more than one hypothesis in the theory of evolution. It is indeed remarkable that this theory has been progressively accepted by researchers, following a series of discoveries in various fields of knowledge. The convergence, neither sought nor fabricated, of the results of work that was conducted independently is in itself a significant argument in favor of the theory.* [Note: John Paul II's address was published on October 22, 1996, as a translation into English from French. In the November 19, 1996, edition of *L'Osservatore Romano,* the editor, Father Robert Dempsey, explained that the paper's original translation was overly literal and that instead of "more than one hypothesis," John Paul II's intent was to say that the theory of evolution is "more than a hypothesis" (*plus qu'une hypothèse,* where the indefinite article *une* should be read as *"a"* not *"one"*).]

John Paul II showed the depth of his reading in the evolutionary sciences by his awareness of the plurality of levels of evolutionary analysis: "And, to tell the truth, rather than *the* theory of evolution, we should speak of *several* theories of evolution. On the one hand, this plurality has to do with the different explanations advanced for the mechanism of evolution, and on the other, with the various philosophies on which it is based." It is in these philosophies where the "Church's Magisterium is directly concerned with the question of evolution, for it involves the conception of man: Revelation teaches us that he was created in the image and likeness of God." Since "truth cannot contradict truth," and since both the Bible and the theory of evolution are true, how does John Paul II reconcile the existence of body and soul? He finds a solution in Aristotle and Aquinas, in their belief that the body and soul are ontologically separate. Evolution created the body, God created the soul:

> *With man, then, we find ourselves in the presence of an ontological difference, an ontological leap, one could say. However, does not the posing of such ontological discontinuity run counter to that physical continuity which seems to be the main thread of research into evolution in the field of physics and chemistry? Consideration of the method used in the various branches of knowledge makes it possible to reconcile two points of view which would seem irreconcilable. The sciences of observation describe and measure the multiple manifestations of life with increasing precision and correlate them with the time line. The moment of transition to the spiritual cannot be the object of this kind of observation, which nevertheless can discover at the experimental level a series of very valuable signs indicating what is specific to the human being. But the experience of metaphysical knowledge, of self-awareness and self-reflection, of moral conscience, freedom, or again, of aesthetic and religious experience, falls within the competence of philosophical analysis and reflection, while theology brings out its ultimate meaning according to the Creator's plans.*

Catholics, says the Pope, can have faith and reason, religion and science.

## A THREE-TIERED MODEL OF RELIGION AND SCIENCE

Implied (but not directly stated) in John Paul II's address is his division of knowledge into types: empirical (science), reason (philosophy), and faith (religion). The Pope's blending of these epistemologies places him squarely in the second tier of what is here proposed as a three-tiered model of the relationship between science and religion.

1. *Conflicting-Worlds Model.* This "warfare" model of science and religion, in its modern incarnation dates back to the 1874 publication of John William Draper's *History of the Conflict between Religion and Science,* and the 1896 publication of Andrew Dickson White's *A History of the Warfare of Science with Theology in Christendom,* for three-quarters of a century considered the definitive histories of the relationship. In his preface Draper explained the difference between two ways of knowing: "Faith is in its nature unchangeable, stationary; Science is in its nature progressive; and eventually a divergence between them, impossible to conceal, must take place." In his introduction, White explained that his book grew out of a lecture entitled "The Battlefields of Science," that carried this unqualified thesis: "In all modern history, interference with science in the supposed interest of religion, no matter how conscientious such interference may have been, has resulted in the direst evils both to religion and to science, and . . . all untrammelled scientific investigation, no matter how dangerous to religion some of its stages may have seemed for the time to be, has invariably resulted in the highest good both of religion and of science."

   Both Draper and White presented simplified histories of the alleged war through such prominent events as the discovery of the earth's sphericity, Galileo's heresy trial, and the 1860 Huxley–Wilberforce debate over evolution. In our own century the most famous case study in the *conflicting-worlds* model is the 1925 Scopes trial, where the relationship was forced into a courtroom out of which a winner and loser emerged. The monument in front of the Rhea County Courthouse where the trial was held in Dayton, Tennessee, presents the case as a conflict, but gets the outcome wrong—Scopes was found guilty and fined $100 by the judge, allowing the Tennessee Supreme Court to overturn the conviction on the grounds that the jury, not the judge, should have imposed the fine. With that, there was no conviction to appeal, the case was over, and the anti-evolution law remained on the books until 1967. Never was the conflict model so evident in practice, and clearly distorting what really happened.

   Among the holders of the conflict model today are fundamentalist Christians and many creationists who reject, bend, shape, or distort science until it fits their theology. Mathematician and philosopher William Dembski, for example, is a fellow of the Discovery Institute's Center for the Renewal of Science and Culture in

*A roadside sign commemorating the Scopes trial. The 1925 trial is practically a monument to the conflicting-worlds model of religion and science.*

Irving, Texas, where he argues that what believers need to do is "rather than look for common ground on which all Christians can agree, propose a theory of creation that puts Christians in the strongest possible position to defeat the common enemy of creation, to wit, naturalism." Since science is based on the philosophy of naturalism, it is the "common enemy" to be defeated; stronger fight'n words were never spoken.

**2.**

*Same-Worlds Model.* In the last couple of decades this position has become popular among mainstream theologians, religious leaders, and believing scientists, who have moved beyond the pugnacious conflicting-worlds model, and hope for an integrative conciliation. Religion and science, faith and reason, they argue, are two ways of examining the same reality. As modern science progresses to a greater understanding of the natural world, we are discovering that the wisdom of the ancients neatly matches the findings of modern scientists. Sometimes figuratively (as in day-age models where a biblical day represents a geological epoch), sometimes literally (where scientific findings are interpreted as supporting, point by point, biblical passages read nonmetaphorically), most residing on this tier are believers who work mightily to read into these ancient writings the findings of modern science, or to read into scientific theories biblical stories. The German theologian Wolfhart Pannenberg, and his scientific counterpart, the mathematical physicist Frank Tipler, meet at this level, arguing that theology and cosmology are rapidly converging into one sphere of knowledge. Two signs from the wall of the museum at the Institute for Creation Research in Santee, California, demonstrate the confusion implicit at this tier. In the first sign, creation-

**Science And Religion**

Religion and science are not separate spheres of study, as some say. Both involve the real world of human life and observation. If both are true, they must agree.

In fact, true science supports the Biblical worldview. There are many facts of science revealed in the Bible and no proven scientific errors.

However, science does not support false religions (e.g., atheism, evolutionism, pantheism, humanism, etc.).

**Does God Exist?**

The Bible (e.g., Genesis 1:1) simply assumes the existence of God as self-evident. Science, as such, cannot prove God exists, but neither can it prove there is no God.

The atheistic scientist Isaac Asimov (author of more books on science than any other scientist) does not believe in God, but he has admitted: "I don't have the evidence to prove that God doesn't exist" (*Free Inquiry*, 1982).

*The Institute for Creation Research: A monument to confusion on the relationship between religion and science.*

ists claim that "religion and science are not separate spheres of study," and that "if both are true, they must agree." The implication is that if they do not agree, one must be right and the other wrong—as explained at the end of the text where the science of evolution is rejected as another false religion along with atheism, pantheism, and humanism. In the second sign (next to the first on the wall), God's existence is claimed to be "self-evident" and science can neither prove nor disprove God's existence, the implication being that science and religion *are* separate spheres. Well, which is it? You cannot have it both ways.

John Paul II's 1996 *Truth Cannot Contradict Truth* falls squarely in this tier as he argues here, and in his 1998 *Fides et Ratio,* that faith and reason can work together toward the same goal of understanding the universe and our place in it. At first blush that sounds reasonable, but as we shall see, trying to mesh these two radically divergent methods of understanding the world into one worldview does not work.

**3.**

*Separate-Worlds Model.* Residents on this tier are still in the minority in their belief that science and religion are neither in conflict nor in agreement, but, in Stephen Jay Gould's apt phrase (adopted from the Pope's 1996 address), are "nonoverlapping magisteria." Science philosopher Michael Ruse agrees and notes: "If you want evolution plus souls, that is your option, and if you want evolution less souls, that is also your option. Either way, evolution is untouched. . . . More than this, together with the Pope, I believe that his tradition is right in feeling that evolution—even evolution through selection—is no barrier to faith. Were I a Catholic, I would positively welcome Darwin as an ally." Anthropologist Eugenie Scott also believes the worlds of science and religion should be kept separate, especially in the classroom, for three very practical reasons: "Using the classroom to indoctrinate students to any belief or nonbelief is, first of all, a violation of the First Amendment of the Constitution's establishment clause; second, it will be misleading to students who will have difficulty separating science as a way of knowing from personal philosophy; and third, it is bad strategy for anyone concerned about the public understanding of evolution."

To clarify further the similarities and differences among these three tiers, it might be useful to make a distinction between the two primary purposes of religion and belief in God: (1) an explanation for the natural world in the form of cosmogony myths, and

(2) a guide to human life and an institution for social cohesiveness in the form of morality myths. Clearly, modern cosmology has displaced ancient cosmogonies in the minds of all but a tiny handful of young-earth creationists. Most believers have now abandoned the six-thousand-year-old young-earth model in favor of one that more closely parallels the tenets of deep geological time. This process of displacement has been under way for the past four centuries and continues to this day, with a few holdouts from the *same-worlds* tier struggling to squeeze the square peg of science into the round hole of religion. Evolutionary biology and the study of the chemical origins of life have also paved new roads into the ancient question of life's origination, to the point where these types of religious myths are now obsolete. And, most dramatically, modern cosmology has presented us with theories so unlike anything described in any ancient myths (black holes, wormholes, quantum foam, inflationary cosmology), that a *separate-worlds* model really is the only viable alternative.

The distinction may not be so clear as we move into the human realm, but it is there nonetheless (at least for now). Although some progress has been made since the Enlightenment to ground moral values in nonreligious, metaphysical concepts such as "rights," and to construct a secular system by which one can live a meaningful and moral life without any belief in God, we are a long way from finding agreement among scientists and philosophers about whether, say, abortion is moral or immoral; whether lying is permissible in certain circumstances; whether we have free will or are determined; how to operationally define good and evil, especially about such subjective matters as meaning and purpose of human existence. Scientists have opinions on these questions, of course, but there is no consensus (and considerable disagreement) among them, to such an extent that these matters are rarely even dealt with in the scientific literature, let alone agreed upon. But the best reason to keep science and religion separate is because they employ radically different methods. Science is not a "thing," but a "process"—more than a body of knowledge, science is a method for obtaining answers to questions about the natural world. Religion, in its second mode, deals with matters about which science has little to say.

To that end, the separate-worlds model is better for science because religion, by definition, deals with subjects beyond our scope and practice. But the separate-worlds model is also better for religion because science is constantly changing and thus it is dangerous to attach religious doctrines to scientific theories, which may go out of date in a matter of years. If Stephen Hawking's no-boundary universe is true, for example, then there is no beginning, no end, and no need for a creator. Catholic cleric and professional astronomer Guy Consolmagno, a scientist at the Vatican Advanced Technology Telescope in the high desert of southeast Arizona, summarized this position well when he explained why he believes in God: "It's not because of the beauty or symmetry or design of the universe that I see in my science, even though all of those things can lead me to appreciate the God I already believe in. It's not because some particular scientific theory is true or false, but because truth and falseness themselves are important. And because, the last time I asked God if he existed, His reply was, 'Last time I looked, I did.'" Faith, not reason, religion, not science, is the proper domain of God's existence.

## *FIDES ET RATIO*

Two years after his 1996 address on evolution, John Paul II released his thirteenth Encyclical Letter—*Fides et Ratio of the Supreme Pontiff to the Bishops of the Catholic Church on the Relationship between Faith and Reason.* Coming in at no less than 35,000 words, divided into 7 chapters and 108 numbered subchapters, and featuring a weighty 132 scholarly endnotes, it was a significant expansion of his commentary on evolution. By any standards *Fides et Ratio* is an impressive work of scholarship. It begins poetically: "Faith and reason are like two wings on which the human spirit rises to the contemplation of truth." Faith and reason, the Pope points out, both must be employed in addressing the most fundamental questions about human existence: "Who am I? Where have I come from and where am I going? What is there after this life?"

To answer these questions, to become "ever more human," we begin with philosophy, "one of noblest of human tasks." Philosophers employ logic and reason to yield "genuine systems of thought" as well as "a body of knowledge which may be judged a kind of spiritual heritage of humanity." That heritage can be seen in "certain fundamental moral norms which are shared by all," and thus "the Church cannot but set great value upon reason's drive to attain goals which render people's lives ever more worthy." However, "the positive results achieved must not obscure the fact that reason, in its one-sided concern to investigate human subjectivity, seems to have forgotten that men and women are always called to direct their steps towards a truth which transcends them," giving rise "to different forms of agnosticism and relativism which have led philosophical research to lose its way in the shifting sands of widespread scepticism." John Paul II then launches an attack on "undifferentiated pluralism" where "all positions are equally valid, which is one of today's most widespread symptoms of the lack of confidence in truth."

Most scientists would join the Pope in voicing their concerns for the decay of knowledge and truth standards, and the attacks from postmodern deconstructionists who claim that science is nothing more than a socially constructed myth. But these same scientists would soon part company with John Paul II when he turns, not to more rigorous philosophical standards for reason or empirical guidelines for science, but to faith: "Philosophy and the sciences function within the order of natural reason; while faith, enlightened and guided by the Spirit, recognizes in the message of salvation the 'fullness of grace and truth' (cf. John 1:14) which God has willed to reveal in history and definitively through his Son, Jesus Christ." John Paul II, of course, is not proffering a radical new epistemology. According to the First Vatican Council, which he cites: "There exists a twofold order of knowledge, distinct not only as regards their source, but also as regard their object. With regard to the source, because we know in one by natural reason, in the other by divine faith. With regard to the object, because besides those things which natural reason can attain, there are proposed for our belief mysteries hidden in God which, unless they are divinely revealed, cannot be known." Thus, John Paul II concludes: "The truth attained by philosophy and the truth of Revelation are neither identical nor mutually exclusive."

But what does it mean to attain truth through revelation? Here the argument becomes circular. Once you have decided that there is a God, it follows that "by the authority of his absolute transcendence, God who makes himself known is also the source of the credibility of what he reveals. By faith, men and women give their assent to this divine testimony. This means that they acknowledge fully and integrally the truth of what is revealed because it is God himself who is the guarantor of that truth." In other words, God's revelations are true because they come from God. What we are to do, then, is apply reason as far as it will go, then take the leap of faith. Why? Because that is the only way to truly understand these divine truths: "Thus the world and the events of history cannot be understood in depth without professing faith in the God who is at work in them." The self-evident truth of

God's existence leads to the inevitable conclusion that His revelations are true by definition. Reason cannot reveal the nature of these truths or of God, therefore we must have faith. Yet it is with faith that we come to believe in God in the first place, thus closing the circle of this circular argument. If there is no God, of course, or if God is not the omniscient, omnipotent, or omnibenevolent god of Abraham, then faith goes out the window as a viable epistemological system.

In reading *Fides et Ratio,* one fluctuates between awesome respect for the deep learning and wisdom of John Paul II, and befuddlement as to how so great a mind can so contradict himself in one document. On the one hand, he reflects modernity and liberalism when he writes that "inseparable as they are from people and their history, cultures share the dynamics which the human experience of life reveals," that "cultural context permeates the living of Christian faith," and when he warns missionaries that the cultures of other peoples should be respected and preserved because "no one culture can ever become the criterion of judgment, much less the ultimate criterion of truth with regard to God's Revelation." On the other hand, in arguing for "the Magisterium's interventions in philosophical matters," he continues to get himself entangled in logical knots, such as when he says "how inseparable and at the same time how distinct were faith and reason." Either faith and reason are inseparable or they are distinct. They cannot be both. Yet that is precisely what John Paul II wants: "Therefore, reason and faith cannot be separated without diminishing the capacity of men and women to know themselves, the world and God in an appropriate way. There is thus no reason for competition of any kind between reason and faith: each contains the other, and each has its own scope for action."

One problem of trying to have both faith and reason within the same sphere is in using the language of reason to describe the process of faith. When John Paul II wants to have it both ways, his language is not only circular but fuzzy as well: "Faith sharpens the inner eye, opening the mind to discover in the flux of events the workings of Providence." If we are going to mix reason and faith, then it is reasonable to ask what it can possibly mean to have an "inner eye" that opens the mind. St. Augustine, to whom John Paul II turns for clarification, is no help in his equally tautological and woolly reasoning: "To believe is nothing other than to think with assent. . . . Believers are also thinkers: in believing, they think and in thinking, they believe. . . . If faith does not think, it is nothing. If there is no assent, there is no faith, for without assent one does not really believe."

Another problem in wedding religion and science is in dealing with subjects appropriate in one sphere but not in the other. This forces one into the uncomfortable position of simultaneously embracing and rejecting science. For example, John Paul II eloquently expresses "my admiration and in offering encouragement to these brave pioneers of scientific research, to whom humanity owes so much of its current development, I would urge them to continue their efforts without ever abandoning the sapiential horizon within which scientific and technological achievements are wedded to the philosophical and ethical values which are the distinctive and indelible mark of the human person." Yet shortly before this praise, he noted with distress that "Scientism is the philosophical notion which refuses to admit the validity of forms of knowledge other than those of the positive sciences; and it relegates religious, theological, ethical and aesthetic knowledge to the realm of mere fantasy." Only those in the conflicting-worlds or same-worlds tiers would so categorize these forms of knowledge. As we saw, science cannot solve such problems, so in holding the same-worlds model the Pope is forced to lay siege to science because it "consigns all that has to do with the question of the meaning of life to the realm of the irrational or imaginary." Such questions can only be "irrational" when inappropriately treated as subjects of rational analysis. When kept in their appropriately separate worlds such questions cannot produce conflict or paradox.

At the beginning of *Fides et Ratio,* John Paul II references I Corinthians 13:12, to make the point that reason without faith leaves one's perception and comprehension faint and fragmentary: "For now we see in a mirror dimly, but then face to face. Now I know in part; then I shall understand fully." That vision and understanding, as we have seen, can only be achieved when these two *different* methods are employed in these two *different* worlds.

## MORAL COURAGE AND NOBILITY OF SPIRIT

I witnessed a poignant example of the power of religion in the second mode (a moral guide to human life and an institution for social bonding) while on a trip to the South in late 1998 to visit a close friend on the eve of his campaign for election to the United States Senate. Michael Coles was running on the Democratic ticket, and on the Sunday morning before the election I joined him and the other Democratic candidates as they visited six different black Baptist churches in and around the Atlanta area. (Since 85 percent of the black vote goes to the Democratic party, these visits were to answer the question asked by a brochure being distributed at one of the churches, entitled "The Black Church Vote: Will God hold us accountable for who governs?") Among the churches we attended were the Shiloh Missionary Baptist Church, the Ray of Hope Christian Church, and, most movingly, the late Martin Luther King, Jr.'s Ebenezer Baptist Church. The statue of a father holding his newborn child to the sky, adjacent to the church and the Martin Luther King, Jr. Center for Nonviolent Social Change, is emblematic of this second mode of religion captured in the epigram beneath the statue: "Dedicated to the memory of Dr. Martin Luther King, Jr. for his moral courage and nobility of spirit."

I had always heard (but never witnessed first-hand) that the black religious experience is qualitatively different from that of the white. I cannot speak for all white churches, of course, but having attended a wide variety of services throughout my life, I have seen nothing quite like the élan expressed in these houses of worship. These services go far beyond mere Hosannas and Amens, with the fellowship's vocal repartee to the songs, hymns, prayers, and sermons every bit as much

*Martin Luther King, Jr.'s Ebenezer Baptist Church and the monument to his moral courage and nobility of spirit.*

a part of the service as anything planned for the day. "Say it preacher" . . . "Oh yeah brother" . . . "That's right sister" . . . and hundreds of other rejoinders came bursting forth from around a room filled with energy and anima. You would have to be made of wood not to feel a spiritual presence there; thus only the tiniest amount of faith, and only a modicum of the willing suspension of disbelief is necessary to "get into the spirit" of the experience. The specific content of the services was not necessary to understand the power of religion on this most foundational level—the human experience of moral courage and nobility of spirit.

At one of the churches that morning, during a rare quiet moment between modes of spiritual expression, Michael leaned over and whispered to me: "You know, the church is the only social institution in the last four centuries that has not let these people down." Although religion has played its own ugly role in the ghastly history of slavery, and anything can and has been justified in the name of God and religion, including murder, war, rape, and slavery, Coles is right. Through centuries of slavery, decades of corrupt reconstruction, years of Jim Crow American apartheid, and overt and covert racism at all levels of our society, religion has remained steadfast by the side of African Americans, providing a safe haven where they might enjoy (however fleeting) a sense of freedom from the physical or psychological chains that bound (and, in many ways, still bind) them.

The anthropologist Anthony Wallace estimates that over the course of the past 10,000 years humans have constructed no less than 100,000 religions. God is alive and well, not only in the past, but in the present. Most people believe in a god of some kind, and if the historians, anthropologists, and archaeologists are right, almost everyone who ever lived believed in one god or another. Religion evolved for the two modes of myth making and social bonding. To those scientists, skeptics, and humanists who believe that science and humanism will one day replace these two functions of religion, E. O. Wilson wrote the following reality check in his Pulitzer prize–winning book, *On Human Nature*:

> *Skeptics continue to nourish the belief that science and learning will banish religion, which they consider to be no more than a tissue of illusions. . . . Today, scientists and other scholars, organized into learned groups such as the American Humanist Society and Institute on Religion in an Age of Science, support little magazines distributed by subscription and organize campaigns to discredit Christian fundamentalism, astrology, and*

> *Immanuel Velikovsky. Their crisply logical salvos, endorsed by whole arrogances of Nobel Laureates, pass like steel-jacketed bullets through fog.*

*Skeptic* magazine, with its circulation of 40,000, would probably fit into this category when compared to various religious publications whose circulation numbers run well into the hundreds of thousands or even millions. We might answer Wilson by explaining that we only take on Christian fundamentalists and their related brethren when they cross over into our turf by trying to use science to prove articles of faith which, by definition, cannot be proved. Indeed, some of this book is aimed at just this sort of invasion. But sometimes we go beyond what our science can really say about some of the great and enduring questions traditionally addressed by religion, particularly the big three addressed by the Pope: "Who am I? Where have I come from and where am I going? What is there after this life?"

Cosmology and evolutionary biology provide a good answer to the first question and a partial answer to the second—we are star stuff and biomass, we evolved by descent with modification from our ancestors, but no one knows where we are going. About the third question, science knows a lot about the process of dying, has adequate explanations for phenomena such as near-death experiences, struggles when it comes to defining death, but can say nothing about what happens *after* we die, other than "No one knows."

As for the second mode of religion's purpose in the moral and social realm, humanists have worked to present viable secular alternatives. But as Wilson observed, we are small in number and, compared to religion, largely impotent as a social force. Studies show, for example, that following the 1992 Los Angeles riots it was religion that helped rebuild the looted and torched neighborhoods, not business, not government, and certainly not the humanists. Perhaps it is because religion has a 10,000-year head start on these other social institutions, or perhaps it is because that is what religion does best. Only time will tell. But the notion that religion will soon fall into disuse would seem to be belied by the data of both science and anecdotal observation. In this sense, at least for now, the separate-worlds model emerges as the only possible description of the relationship of religion and science. While scientists may manifest commendable moral traits, or act with admirable social consciousness, they do so as an expression of their humanity, not their science. Science has never trafficked, and likely never will, in the business of moral courage and nobility of spirit.

# The Storytelling Animal

## *Myth, Morality, and the Evolution of Religion*

*I have given the evidence to the best of my ability; and we must acknowledge, as it seems to me, that man with all his noble qualities, with sympathy which feels for the most debased, with benevolence which extends not only to other men but to the humblest living creature, with his god-like intellect which has penetrated into the movements and constitution of the solar system—with all these exalted powers—Man still bears in his bodily frame the indelible stamp of his lowly origin.*

—Charles Darwin, The final paragraph of *The Descent of Man,* Vol. II, p. 405, 1871

During the broadcast of the 1997 Cable ACE awards on VH-1, the rock star Madonna was called upon to present an award. After slinking her way down a flight of stairs in a tightly wrapped full-length skirt and stiletto heels, she approached the podium and announced that she did *not* wish to talk about Princess Di, or the paparazzi who hounded her, or the tabloids that exploited her, or the public who worshiped her to death. After adroitly making her point, while simultaneously denying she wanted to make it, Madonna instead suggested that we look for the deeper cause of Princess Di's tragic death—our fascination with gossip and other people's personal lives, especially when it is none of our business. Ignoring the simple and obvious fact that both Diana and Madonna, like most celebrities, depend and thrive upon the very obsession they pretend to hate, Madonna was asking that humans, who are by nature storytelling animals, quit telling stories about their favorite subject—other humans. Why does no one ever discuss the *true* cause of Diana's death—just one more case of speeding and drunk driving? Because that would make her just another boring statistic—over 25,000 people are killed every year in America alone due to drunk driving. Nothing interesting about that fact—no juicy gossip, no web of sex and deceit, no assassination cabals, and, most of all, no evil villains. Every story needs a hero and a villain. Princess Diana died an ignoble death, and that does not make for a very interesting story.

What does make for an interesting story? Why do we tell stories and so enjoy hearing them? Cognitive neuroscientist Michael Gazzaniga, in his book *The Mind's Past,* argues that we are all storytellers, in the sense that we take the facts of our everyday experience and weave them into a narrative, from which we spin doctor our self-image. "The spin doctoring that goes on keeps us believing we are good people, that we are in control and mean to do good." Gazzaniga calls the brain mechanism that carries out this task the *interpreter,* "probably the most amazing mechanism the human being possesses." Dreams serve as another example, since at least some appear to be random firings of neural impulses that are hung together by our internal storyteller—recall dreams you have had that include disparate elements and people never found together in reality, yet make perfect sense in a dream narrative. This is the power of the pattern-seeking, storytelling animal.

What has this to do with religion and the belief in God? We have recognized two primary purposes of religion: (1) The creation of stories and myths that address the deepest questions we can ask ourselves: Where did we come from? Why are we here? What does our ultimate future hold?; (2) The production of moral systems to provide social cohesion for the most social of all the social primates. God(s) figures prominently in both these modes as the ultimate subject of mythmaking and the final arbiter of moral dilemmas and enforcer of ethical precepts. Why did this capacity to tell stories, create myths, construct morality, develop religion, and believe in God evolve?

## THE HOW AND THE WHY: IN SEARCH OF DEEPER ANSWERS

In the seventh century B.C.E. the Greek philosopher Archilochus penned one of the pithiest yet most thoughtful epigrams when he observed: "The fox knows many things, but the hedgehog knows one great thing." Twenty-two centuries later the nineteenth-century British philosopher William Whewell described science as employing a fox-like method to arrive at a hedgehoglike conclusion, that he called a *consilience of inductions*, or what might also be called a *convergence of evidence*. Bringing into focus numerous theories, models, and data from disparate and unconnected fields, each one of which converges to a similar conclusion, together allows us to increase the confidence in our theory. We know, for example, that evolution happened not by any one fossil or organism, but by tens of thousands of bits of data from unrelated fields, all of which converge to a single conclusion; paleontology, geology, comparative anatomy, comparative physiology, molecular genetics, population genetics, zoology, botany, biochemistry all independently point to an evolutionary history of life on Earth. Together they converge to an inescapable focal point of scientific truth.

We can engage this convergence method to understand how and why religion and belief in God evolved in human societies. (This book, in fact, has been doing just that, using the convergence and comparative methods from the various behavioral and social sciences such as neurophysiology, behavior genetics, cognitive and social psychology, anthropology, sociology, history, and archaeology.) A survey of the extensive body of literature on religion shows that this is one of the most complex of all human social phenomena, not explicable in terms of an overriding hedgehog theory. We need to take a foxlike approach, yet look for a consilience of evidence to see what these disparate fields of thought might reveal. There is no question that different cultures express religious behavior in many different and unique ways. But is there something *underneath* these diverse expressions?

To get at the deeper question of the purpose of religion, I begin with evolutionary theory and the distinction biologists make between *how* questions and *why* questions. *How* questions are concerned with *proximate* causes—the immediate or nearest cause or purpose of a structure or function—the "how does it work?" type of question. "How is it that fruit taste good?" A physiologist might answer "because it stimulates the sweet receptors on the tongue." This is a proximate answer. Evolutionary biologists are also concerned with *ultimate* causes—the final cause or end purpose of a structure or function—the "why does it exist?" type of question. "Why does fruit taste good?" An evolutionary biologist might answer "because there was a natural selection for taste bud receptors and brain modules to produce a pleasurable sensation with certain food substances that are both scarce and healthy." We can go even deeper and ask what drives the selection process, and postulate that those organisms for whom healthy foods tasted good ate more of them, had less disease, lived longer, and thus left behind more offspring. Since differential reproductive success is the ultimate result of natural selection, and natural selection is the primary driving force behind evolution, we have reached the deepest level of causality in answering our question.

If questions about anatomy and physiology can be answered at the deeper evolutionary level, what about behavior? In the late 1970s the field of ethology—the study of animal behavior from an evolutionary perspective—came of age. John Alcock's *Animal Behavior: An Evolutionary Approach* and Irenäus Eibl-Eibesfeldt's classic text, *Ethology: The Biology of Behavior,* demonstrated that behaviors are not just the result of reinforced learning in response to changing environments but also the product of millions of years of evolution. A herring gull chick, for example, pecks at a red dot on its mother's beak. Its mother then regurgitates her food for the chick to eat. The chick did not "learn" this behavior by trial and error in its own lifetime but inherited it from the evolutionary history of the species. The chick was born "knowing" that when it sees a red dot it should peck at it. The mother, in turn, was born "knowing" that when a chick pecks at its beak, it should regurgitate its food.

Of course, compared to herring gulls and other simple organisms, human behavior is vastly more complex and influenced by learning and the environment. Nevertheless we are animals, and no less than any other organism on earth we are the product of evolution. In order to fully understand human behavior we must also address our own ultimate "why" questions from an evolutionary perspective. What humans have done in the past 13,000-year history of civilization is nothing short of miraculous, but we must not discount the orders of magnitude of deeper time that preceded the age of civilization, when the human animal was shaped over the course of hundreds of thousands of years as Pleistocene hunter-gatherers, and over the course of millions of years as primates, and tens of millions of years as mammals.

In the past decade the field of ethology has been joined by the emerging discipline of evolutionary psychology. Where evolutionary biologists focus on the effects of the physical environment, evolutionary psychologists concentrate on the influence of the *social* environment. Primates are extremely social mammals, and the human primate is, arguably, the most social of all. In their introduction to the field's most influential text, anthropologists Jerome Barkow and John Tooby, and psychologist Leda Cosmides explain: "The central premise of *The Adapted Mind* is that there is a universal human nature, but that this universality exists primarily at the level of evolved psychological mechanisms, not of expressed cultural behaviors. In this view, cultural variability is not a challenge to claims of universality, but rather data that can give one insight into the structure of the psychological mechanisms that helped generate it."

The Harvard evolutionary biologist Edward O. Wilson has done just this in such works as *Sociobiology, On Human Nature,* and most recently in *Consilience: The Unity of Knowledge.* Wilson argues "that the etiology of culture wends its way tortuously from the genes through the brain and senses to learning and social behavior. What we inherit are neurobiological traits that cause us to see the world in a particular way and to learn certain behaviors in preference to other behaviors." Wilson says these complex interactions between genes and learning are guided by *epigenetic rules,* which "comprise the full range of inherited regularities of development in anatomy, physiology, cognition, and behavior. They are the algorithms of growth and differentiation that create a fully functioning organism."

A simple example of an epigenetic rule is one-trial learning—taste aversion being the most obvious example. Pairing a food or drink substance with violent nausea, for example, will produce an aversion to that substance for some time to come (red wine once did me in and for over a decade I could not drink even a small amount of it). This is an evolved mechanism for avoiding toxic foods—the learning needs to take place immediately in one trial—there is no margin for error, no time for a gradual learning sequence. Language is a much more complex epigenetic rule in which we all learn the language to which we are exposed as infants, but the basic rules of language were learned over the past hundred thousand years by our ancestors. Moving beyond basic ethological concepts of innate mechanisms, Wilson shows how genes and culture interacted in our evolutionary history in complex ways he calls *gene-culture coevolution.* Forget the nature–nurture debate with its artificially imposed percentages assigned to each component (for example, 40 percent genes, 60 percent environment). We are well past such facile delineations (a process that itself may be a product of an epigenetic rule that directs us to cleave a continuous nature into bivariate categories in order to simplify our complex world). Humans, says Wilson, are products of both biological and cultural evolution so inextricably interwoven that the two cannot be separated:

> *Culture is created by the communal mind, and each mind in turn is the product of the genetically structured human brain. Genes and culture are therefore inseverably linked. But the linkage is flexible, to a degree still mostly unmeasured. The linkage is also tortuous: Genes prescribe epigenetic rules, which are the neural pathways and regularities in cognitive development by which the individual mind assembles itself. The mind grows from birth to death by absorbing parts of the existing culture available to it, with selections guided through epigenetic rules inherited by the individual brain.*

The fear of and fascination with snakes, so common in peoples around the world, is produced by an epigenetic rule with obvious survival significance. But how that fear of and fascination with snakes is uniquely expressed depends on the culture in which the individual was raised. Snake stories, myths, and narratives all differ, depending on the culture, but the focus on snakes themselves is hard-wired. Wilson shows how:

> *Some individuals inherit epigenetic rules enabling them to survive and reproduce better in the surrounding environment and culture than individuals who lack those rules, or at least possess them in weaker valence. By this means, over many generations, the more successful epigenetic rules have spread through the population along with the genes that prescribe the rules. As a consequence the human species has evolved genetically by natural selection in behavior, just as it has in the anatomy and physiology of the brain.*

This line of reasoning leaves behind the pejorative accusations and false dichotomies of biological and environmental determinism, with extremists on the political left and right accusing each other of employing Darwinian models to justify certain social or political agendas. The epigenetic rules that guide gene-culture coevolution are so complex and interactive that such name-calling tells us as much about the name-caller's agenda as that of the accused. Culture, says Wilson, evolves "in a track parallel to and usually much faster than genetic evolution." But, he notes, "The quicker the pace of cultural evolution, the looser the connection between genes and culture, although the connection is never completely broken."

## FROM PATTERN-SEEKING TO STORYTELLING

Humans are pattern-seeking animals who seek and find causal relationships in our physical and social environments. The process is called learning. As we have seen, sometimes we get it right (*Type 1 and 2 Hits*—not believing a falsehood and believing a truth) and sometimes we get it wrong (*Type 1 and 2 Errors*—believing a falsehood and rejecting a truth). But we do much more than this. We do not just process environmental data like a computer, spewing out cold, hard facts. We tell stories about it. Humans are storytelling animals.

In his book *How to Argue and Win Every Time,* the mediagenic attorney Gerry Spence explains that one of the reasons he is so successful is that he does not speak to the jury like a lawyer, with all the legalese and law-school language spouted by most Ivy League–trained attorneys. Spence talks to them conversationally. He tells them stories. In one case, he began his closing statement with a story about a cocky young man who wanted to show up his wiser elder. His plan was to capture a small bird in his hand, approach the old man and ask him if the bird was alive or dead. If the old man said "dead," he would let the bird go. If the old man said "alive," he would crush the life out of the bird. Either way he would show up the old man. So the young man captured a small bird, approached the old man, and asked him if it was alive or dead. "The bird's life," replied the old man, "is in your hands." Spence says he won that case because the jury understood that the story was a metaphor for the life of his client, which they held in their hands. His point was not that telling good stories wins court cases. It was that humans can relate to stories better than they can to pure logic or objective facts. It is simply easier to keep track of a complex argument if it includes people, places, and events rather than propositions, syllogisms, and symbolic logic.

Psychologists, in search of ultimate *why* answers to human behavioral questions, have discovered this fact about storytelling as well. Through a series of clever experiments Peter Wason discovered that when students are presented with traditional problems in logic, which they normally have a difficult time in solving, they improve significantly if these same problems are presented in the form of a story, especially a story involving people and relationships in which the students are to detect cheating and rule breaking in social contracts. Cosmides and Tooby review subsequent experiments that corroborate Wason's findings, demonstrating that "human reasoning is well designed for detecting violations of conditional rules when these can be interpreted as cheating on a social contract." They conclude that this is the result of an evolved mechanism because "social exchange behavior is both universal and highly elaborated across all human cultures—including hunter-gatherer cultures—as would be expected if it were an ancient and central part of human social life."

Anthropologist Misia Landau, in a fascinating study of the evolution of storytelling, believes that "the central claim of narratology is simply that human beings love to tell stories." But she goes further, arguing that stories are not just *about* our reality, they help *create* our realities:

> *Narrative, then, is . . . a defining characteristic of human intelligence and of the human species. Related to this assumption . . . is the idea that we have certain basic stories, or deep structures, for organizing our experiences. Each deep structure comes in many versions and in several different modes. For example, the Cinderella story is embedded not just in fairy tales but in novels, films, operas, ballets, and television shows. Some narratologists, stressing the central role of narrative in human experience, would further argue that we have not only different versions of stories but different versions of reality which are shaped by these basic stories.*

Origin myths among indigenous peoples, of course, neatly fit this description. Landau, however, goes on to show how *scientific* theories of human origins are no less susceptible to narrative bias. Was it bipedalism that gave rise to tool use, which generated big brains? Or was it tool use that led to bipedalism and then big brains? Were early hominids primarily hunters—man the killer ape, warlike in nature? Or were they primarily gatherers—man the vegetarian, pacifist in nature? More importantly, does the narrative change in response to empirical evidence, or does the interpretation of the evidence change as a result of the currently popular narrative? This is a serious problem in the philosophy of science: To what extent are observations in science driven by theory? Quite a bit, as it turns out, especially in history and the social sciences. And this fact supports the thesis that humans are primarily storytelling animals. The scientific method of purposefully searching for evidence to falsify our most deeply held beliefs does not come naturally. Telling stories in the service of a scientific theory does.

## FROM STORYTELLING TO MYTHMAKING

One night in the early 1970s a young couple was parked in a vacant lot high in the Hollywood hills overlooking the lights of Los Angeles. The young man told his date about a recently escaped one-armed convict who was known to be roaming those very foothills, killing parked young couples by slashing them with his arm hook. The girl got scared and insisted her date take her home at once. He did, and when he went to open her car door he discovered a hook dangling from the handle.

For decades now high school kids have been telling this story, along with another favorite, the vanishing hitchhiker: Driving along a country road you pick up a hitchhiker who gets in the backseat of your car and instructs you where to drop her (sometimes it is a he) off. When you arrive at the house (sometimes it is a graveyard) you discover that the girl has disappeared from your car. You then discover that the girl was killed (sometimes "disappeared") while hitchhiking on that very stretch of highway that same day the year before.

As Jan Harold Brunvand noted in his 1981 book about such urban legends, *The Vanishing Hitchhiker,* such stories are appealing because they contain three mythic elements: (1) a strong story; (2) a foundation in actual belief; (3) a meaningful message. (He presents no less than fifteen versions of the vanishing hitchhiker story!) Myths contain a staggering diversity of themes, including stories about life and death, birth and rebirth, adolescence and coming of age, love and marriage, the origin and end of the universe, moral dilemmas, the meaning of life, and all manner of human triumphs and traumas. A myth should not be thought of in terms of its veracity or lack thereof, as when we say that an urban legend is a myth, meaning it is not true. Urban legends, in fact, are a subspecies of myths; they are stories about our fears and anxieties, as in the hook-man and vanishing hitchhiker, or others like alligators living in New York City sewers. All cultures throughout the world, and all peoples throughout history have had myths. Long before there was the written word there was the spoken word, and with language humans told stories—stories about ourselves and our relationships, stories about our origin and our end, and stories about our world and our environment. These stories became myths.

What are myths, what do they mean, and what methods should we employ to understand them? The *Oxford English Dictionary*'s history of the word's usage is enlightening in this regard. A myth is "a purely fictitious narrative usually involving supernatural persons, actions, or events, and embodying some popular idea concerning natural or historical phenomena." In fact, the original Greek meaning of *mythos* was "word," in the sense of a final pronouncement, to be contrasted with *logos,* also "word," but one whose veracity may be disputed. The point of a myth is not whether it is true or false, but what it represents. The ancient world was rich in myths, but so too is the modern world. Science fiction, for example, is a genre of modern myth—*Star Trek* is filled with supernatural persons, actions, and events; it remains one of our most popular myths even in this, the Age of Science. Myths, science-fiction author Thomas Disch might say, are the dreams our stuff is made of. In his 1949 classic statement on the subject, *The Hero with a Thousand Faces,* Joseph Campbell describes the diversity of thought on these questions:

> *Mythology has been interpreted by the modern intellect as a primitive, fumbling effort to explain the world of nature (Frazer); as a production of poetical fantasy from prehistoric times, misunderstood by succeeding ages (Müller); as a repository of allegorical instruction, to shape the individual to his group (Durkheim); as a group dream, symptomatic of archetypal urges within the depths of the human psyche (Jung); as the traditional vehicle of man's profoundest metaphysical insights (Coomaraswamy); and as God's revelation to his children (The Church). Mythology is all of these. The various judgments are determined by the viewpoints of the judges. For when scrutinized in terms not of what it is but how it functions, of how it has served mankind in the past, or how it may serve today, mythology shows itself to be as amenable as life itself to the obsessions and requirements of the individual, the race, the age.*

Campbell's theory, as described in his 1972 *Myths to Live By,* is that myths serve four functions: (1) *mystical,* which "serves to awaken and maintain the individual sense of awe and gratitude in relation to the mystery dimension of the universe, not so that one lives in fear of it, but so that he recognizes that he participates in it"; (2) *explanatory,* or "an image of the universe which will be in accord with the knowledge of the time, the sciences and the fields of action of the folk to whom the mythology is addressed"; (3) *normative,* or to "validate, support, and imprint the norms of a given, specific moral order, that, namely of the society in which the individual is to live"; and (4) *guidance,* or "to guide him [the individual], stage by stage, in health, strength and harmony of spirit, through the whole foreseeable course of a useful life." This is a useful outline to help us get our minds around the varied culture of myths, but it is only answering *how* questions about myths at a *proximate* level. To know *why* humans need to experience the mystical, explain the world, create norms, or seek guidance, we need to consider myths from an evolutionary perspective.

A myth is a form of symbolic communication that invests stories not only with ordinary people and events but also with gods, supernatural beings, and extraordinary happenings, often unfolding in a place or time different from that of ordinary human experience. There are many themes and subjects embodied in myths: *origins* (cosmogony and creation), *eschatology* (end times and destruction), *heroes* (humans with special powers and experiences), *time and eternity* (ages of man, periods of history), *providence and destiny* (destiny, mastery over fate), *memory and forgetting* (prenatal existence, previous lives, collective unconscious), *higher beings* (celestial gods), *founders of religions, nations, and peoples* (Abraham, Moses, Buddah, Romulus and Remus, Siegfried), *kings and ascetics* (Arthur and Merlin), *transformation* (coming of age), *rebirth and renewal* (seasons and ages), and *messianic and millenarian* (second comings and new world orders).

If myths are to be explained on a deeper evolutionary level, then they must be universal for all peoples, including ourselves. Myths are not just someone else's story, or stories that come from far-off times or places. We have plenty of myths of our own. Marxism was a political myth, as was pure laissez-faire capitalism, both providing explanatory, descriptive, and, most importantly, normative mythic functions. Freudian psychoanalysis was a psychosocial myth, as was Skinnerian behaviorism, both serving to justify and control human behavior. Science fiction provides descriptive myths, often of dystopian or paradisiacal future states of the world. This evolutionary explanation of myths, in fact, is itself explanatory mythmaking. The fact that scientific reasoning and empirical data are employed to support the argument makes it no less a myth—a story for us, of our time, that provides meaning and purpose.

In this regard, science is a type of myth, in both function and typology. To some degree, cosmologists give us origin and eschatology myths, from the Big Bang to the Big Crunch. Historians provide hero myths, from Martin Luther to Martin Luther King, Jr. Archaeologists present time and eternity myths, from the Paleolithic Age to the Neolithic Age. Economists proffer providence and destiny myths, from total free market anarchism to pure communism. Psychologists furnish memory and forgetting myths, from recovered memories to repression. Astronomers and computer scientists supply higher beings myths, from extraterrestrial intelligences to artificial intelligences. Biblical archaeologists contribute founders of religions myths, from King David to Moses. Anthropologists produce transformation myths, from Samoan teenagers to Yanomamö warriors. Sociologists confer rebirth and renewal myths, from childhood to adulthood. And political scientists proffer messianic and millenarian myths, from the new president to the new world order.

Why do we continue telling stories and constructing myths today? Because the epigenetic rules for mythmaking still reside within us. Consider monsters and beasts as myths. From the earliest cave paintings to the present, monsters and beasts lurk at the interstices of the natural world, appear on the margins of our perception, dwell in the dangerous lands remote from human habitation, come out at night, or in our nightmares. In the light of E. O. Wilson's epigenetic rule for fear of snakes that generates narratives in the form of snake myths in cultures worldwide, we should not be surprised that the modern sciences of zoology and especially cryptozoology (the "science of hidden animals") are ripe with mythic tales. For millions of years hominids evolved alongside other primates and mammals in a rich and varied zoological world. The identification of other animals, and the anticipation of possibly dangerous cryptids would have produced not only Type 1 and 2 Hits but also numerous Type 1 and 2 Errors in our thinking. Thus, in our own time we have correctly identified such genuine cryptid surprises as the giant panda, the pygmy hippopotamus, the Komodo dragon, and the long-thought extinct coelacanth. The now-famous mountain gorilla was only discovered in 1903. The pygmy chimp (making us the "third chimpanzee" in Jared Diamond's apt phrase) was only recently identified as being a separate species. But the field of cryptozoology is ripe with pseudoscientific hoaxes, exaggerated descriptions, and ridiculous claims. These are Type 1 Errors in thinking. But we must be cautious not to commit a Type 2 Error in rejecting a new discovery. There may very well be new and possibly dangerous animals lying in wait. They may not be dangerous to those of us living in suburban America, but they certainly could have been to our paleolithic ancestors, and this is where an epigenetic rule that would generate mythic monsters could have evolved.

Or consider how an epigenetic rule might apply to the myths of dragons and werewolves. Dragons are the most common of all mythological creatures, usually portrayed as the hybrid of a serpent or crocodile, and constructed of any number of disparate mix-and-match parts such as the scales of a fish, the wings and occasionally the head of a bird, the forelimbs and sometimes the head of a lion, the ears of an ox, the feet of a tiger, the claws of an eagle, the horns of a deer, and the eyes of a demon. In the ancient world the dragon was a winged lizard or serpent, regarded as the enemy of mankind, and its overthrow is made to figure among the greatest exploits of the gods and heroes of mythology. The dragon is found in the myths of most peoples, where it has been worshiped as a god, endowed with both beneficent and malevolent attributes, combatted as a monster, or attributed supernatural power. It is mentioned thirty-one times in Judaeo-Christian scriptures, starting with the serpent in the Garden of Eden. Dragons are often associated with water and sometimes live in caves under lakes or in the ocean bottom. In medieval tales the dragon dried up rivers and caused drought, forcing inhabitants to pay an annual tribute of gold or fair maidens. Many heroes of mythology are dragon slayers: Marduk, Hercules, Apollo, St. Michael, St. George, Beowulf, King Arthur. Some mythologists conjecture that the male dragon slayer is a symbol that represents the shift from egalitarian societies to patriarchal societies. The real source of the dragon myth may be frilled lizards, or reptiles that spit a toxic venom. But more likely it comes from a prebiblical Babylonian myth of the prime female deity who was a dragon named Tiamat. She was associated with the flooding of the Tigris-Euphrates river system and the beginning of the growing season, and her ritual killing by Marduk is possibly the source of many of the dragon and dragon-slayer stories in the world.

Similarly, werewolves figure prominently in mythology. The peak of prosecutions for lycanthropy—the "condition" of a human taking on wolflike characteristics—was in the sixteenth and seventeenth centuries in France. The most famous case was that of Jean Grenier who, in 1603, boasted to three girls that he was a werewolf, telling them that a man "gave me a wolfskin cape; he wraps it around me, and every Monday, Friday and Sunday, and for about an hour at dusk every other day, I am a wolf, a werewolf. I have killed dogs and drunk their blood; but little girls taste better, and their flesh is tender and sweet, their blood rich and warm." Looking at this tale with the distance of almost 400 years, it is likely nothing more than male boasting and posturing; but since several children had been murdered at the time Grenier was fingered and convicted. Why a wolf? The earliest myths are associated with a ceremony of a man putting on a wolf's skin for protection from the cold, or to act as concealment when hunting for food. This mutated into the theme that the wearing of the skin passed on to the man great magical powers of strength, speed, and stealth, not just for hunting, but for exacting vengeance or gaining power over others. From here it was but a small step to changing the man into a wolf through the common mythic motif of *shapeshifting,* where creatures or objects can change into other creatures or objects, either at will or under special conditions. An evolutionary argument could also be made that, as pack hunters, wolves were a principal competitor to early humans in northern latitudes. Dogs, as loyal friends and noncompetitors to humans, do not generate such myths as wolves. (It also should be noted that werewolves did not have a monopoly on the genre. There were werebears, weretigers, werehyenas, werecrocodiles, and werejackals. Vampires were a type of werebat. Shapeshifting is found in countless myths, including the Burma-Assam tiger men who can share a tiger's body, or the leopard men of certain regions of Africa.)

Telling stories and constructing myths about animals have obvious survival significance to humans living in a paleolithic environment. Most simply and directly, it is a form of pedagogy and a medium of knowledge transfer of important information about the flora and fauna of the local ecology. A simple story can relay to a child that a particular food is poisonous or a certain animal is dangerous. A myth codifies this knowledge into the permanent record of a people's store of wisdom. Anthropologist Melvin Konner, for example, in his study of the !Kung San people of Africa, observed that their knowledge of the local ecology was "detailed and thorough enough to astonish and inform professional botanists and zoologists." And this knowledge, he noted, was often exchanged around the campfire in the form of storytelling and mythmaking:

> *[Their knowledge covered] everything from the location of food sources to the behavior of predators to the movements of migratory game. Not only stories, but great stores of knowledge are exchanged around the fire among the !Kung and the dramatizations—perhaps best of all—bear knowledge critical to survival. A way of life that is difficult enough would, without such knowledge, become simply impossible.*

As Wilson observed: "Storytelling may be central in language because, in simulating real experience, they bring into play all of the cognitive and emotional circuitry evolved to deal with real experience. In other words, narrative is the best mnemonic procedure; it maximizes rate of learning and understanding." It seems reasonable, therefore, to offer the following evolutionary explanation for myths: *Some individuals inherited an epigenetic rule for mythmaking, in this case myths related to animals, that enabled them to survive and reproduce better in the surrounding environment and culture than individuals who lacked these rules, thus spreading the rules. As part of gene-culture coevolution, myth culture was reconstructed by each generation collectively in the minds of individuals. When oral myths were supplemented by written myths, the culture of myth grew indefinitely large, but the fundamental influence of the epigenetic rules for myths remained constant. Since some myths survived and reproduced better than competing myths, this caused mythic culture to evolve in a track parallel to, and faster than, genetic evolution. This quicker pace of mythic cultural evolution loosened the connection between genes and culture, although the connection was never completely broken. Thus we witness the plethora of modern myths, and our fascination with them.*

## FROM MYTHMAKING TO MORALITY

One of the classic myths of medieval Europe is the story of Beowulf and the monster Grendel. The myth comes to us from a single manuscript, dated circa A.D. 1000, but probably derives from an oral tradition of the eighth century. In its nascent form it was without title. It was later named for the Scandinavian hero Beowulf, although there is no historical evidence that such a person ever lived. The myth has two parts. In the first part the evil monster, Grendel, devours Danish King Hrothgar's warriors and ravages his kingdom. Young Beowulf, a prince of the Geats of southern Sweden, hears about the monster through his noble uncle, Hygelac (who may have been a historical figure), and makes the king an offer to rid him of the Grendel monster. Meanwhile, the monster strikes at night, while everyone sleeps, stealing away numerous thanes (feudal lords) and devouring them in his keep. Beowulf sets a trap whereby Grendel grabs him one evening, but Beowulf, a mighty warrior, tears off Grendel's arm. The beast flees, and the next day the people follow the trail of blood to discover the deceased monster. But the next night Grendel's mother avenges her son, killing one of Hrothgar's earls. The next day the people once again trek to the keep of the Grendel monster and there discover many monsters and dragons of the sea. Beowulf arrives to wreak vengeance for the latest killing, and finds and slays Grendel's mother. In the second part of the story Beowulf assumes the throne when King Hrothgar dies, only to have a fire-breathing dragon ravage his land. Beowulf, now an old man, fights the dragon but is no match for the beast. With the aid of a young warrior named Wiglaf the dragon is defeated, but in the process Beowulf dies. His last words are uttered in desperation: "Dear Wiglaf, quickly now help me to see this old treasure of gold, the gladness of its bright jewels, curiously set, that I may yield my life the more easily and the lordship I have held so long."

What many scholars see in this myth, Campbell notes, is "the old Germanic virtues . . . of loyalty and courage, pride in the performances of duty, and, for a king, selfless, fatherly care for his people's good." But, we might ask, *why* does this myth contain ethical values of the Germanic code of loyalty to chief and tribe and vengeance to enemies? Might there be a deeper reason, one rooted in epigenetic rules pertaining to the human condition in a paleolithic community? How do we get from pattern-seeking, storytelling, and mythmaking, to religion?

As pattern-seeking animals, humans evolved speech as one of the earliest symbolic patterns, with sounds and words representing objects and events in the physical and social environment. But no one knows exactly *when* language evolved. The scientific evidence is sketchy at best. Cranial endocasts of hominids as old as those of *Homo habilis* and even *Australopithecus africanus* (dating several million years old) reveal the nooks and crannies of the exterior surface of the brain. Some of these may correspond to the distinctive language centers of the modern human brain, but whether they drove language in these ancient hominids is impossible to prove. In the Kebara 2 burial site in Israel there is evidence of language in Neanderthals in the form of a nearly complete hyoid bone—a free-floating bone attached to soft tissue in the larynx that anchors throat muscles involved in speech—found next to a Neanderthal mandible.

Equally important, however, is the question, why language? Language may have evolved for strictly adaptive purposes, giving our hominid ancestors a selective advantage in dealing with the physical and especially the social environment. On the other hand, all other modern primate species are social and hierarchical, yet not one of them has developed as complex a language system as ours. Perhaps language is, in part, a spandrel—a contingent by-product of an enlarged brain evolved for dealing with symbols and different components of language developed for other reasons and later employed in language and speech. Donald Johanson summarizes this as-yet-unsolved mystery:

> *Language evolution is probably intimately linked to brain evolution, and since our brain has been growing and reorganizing over the past 2 million years, it seems unlikely that language suddenly arose from some radical new mutation. Human brains could have been language-competent long before spoken languages appeared. The enlarging brain of early* Homo *no doubt was capable of complicated cognitive coordination and calculation and as such relied on and used skills important to language. Perhaps language evolved in tandem with our enlarging brain or was a cause, rather than a consequence, of brain enlargement during the Pleistocene.*

Whenever and however language evolved, from pattern-seeking to speech-making to storytelling to mythmaking, humans solved problems through language. Anthropologist Terrence Deacon goes so far as to invent a new species designation for us, *Homo symbolicus,* the hominid symbol user. Anthropologists studying modern hunter-gatherer societies, for example, have found that problems are often couched in the language of stories, myths, and other symbolic narratives, such as songs and poems. In his description of the Copper Eskimo, for example, anthropologist David Damas notes that "every man or woman in that group was said to have had his own compositions. Some of the subjects of the songs were man's impotence in the universe, hunger, songs of the hunt, songs of lust, the fear of loneliness, and death." The description is as applicable for suburban commuters in New York as it is for hunter-gatherers in Alaska.

Paleoanthropologists believe that we evolved in small hunter-gatherer (and scavenging) communities operating out of a home base and utilizing considerable cooperation and communication. The late archaeologist Glynn Isaac proffered the "home base hypothesis" from which hunting and gathering would have been conducted, with food substances brought back to a specific place where it was shared. Archaeologist Lewis Binford pushes for a "scavenging" model, where ancient hominids more likely would have taken what they could find from the remains of already hunted animals, rather than hunting themselves. Either way, anthropologist Robert Bettinger demonstrates how, compared with individuals, "groups may often be more efficient" not only "in finding and taking prey, particularly large prey" but also in coordinating the activities of individuals, who might otherwise unduly interfere with one another. Finally, as in the case of resource storage, foraging groups that pool and share resources have the effect of 'smoothing' the variation in daily capture rates between individuals." That is, as the group grows larger, "lucky" individuals share their take with "unlucky" individuals, and everyone benefits. Cooperation would have been as powerful a drive in human evolution as competition, if not more so. And communication is an essential tool of cooperation, so it makes sense that Paleolithic hunter-gatherers, as well as their modern counterparts, would have employed language to tell stories and solve problems.

How large were these communities? Most modern hunter-gatherer groups range in size from 50 to 400 residents, with a medium range of 100 to 200 people. Anthropologist Napoleon Chagnon, in his extensive studies of the Yanomamö people in the Amazon, found the typical group to be roughly 100 people in size, with 40 to 80 living together in the rugged mountain regions, and 300 to 400 members living together in the largest lowland villages. He has also noted that when groups get excessively large for the carrying capacity of their local environment (given their level of technology), they fission into smaller groups. Such bifurcations may also be a product of exceeding the carrying capacity of the *social* environment. Psychologist Robin Dunbar, in his book, *Grooming, Gossip and the Evolution of Language,* argues that the figure of 150 people in a typical group has a deeper evolutionary basis. It turns out that 150 is roughly the number of living descendants (wives, husbands, and children) a Paleolithic couple would produce in four generations at the birthrate of hunter-gatherer peoples—this is how many people they knew in their immediate and extended family. Archaeologists believe that early agricultural communities in the Near East 7,000 years ago typically numbered about 150 people. Even modern farming communities, like the Hutterites in Europe (and now Dakota and Canada), average about 150 people.

When groups get large they split into smaller groups. Why? According to the Hutterites, it is because shunning does not work as well in large groups, and shunning is a primary means of social control. Sociologists know that once groups exceed 200 people a hierarchical structure is needed to enforce the rules of cooperation and to deal with offenders, who in the smaller group could be dealt with through informal personal contracts and social pressure. Still larger groups need chiefs and a police force, and rule enforcement involves more violence or the threat of violence. Even in the modern world with a population of six billion people crowded into dense cities, people find themselves divided into small groups. In the Second World War, for example, the average size company in the British army was 130 men, in the United States army it was 223 men. The 150 average also fits for the size of small businesses, of departments in large corporations, and of efficiently run factories. A Church of England study, conducted in an attempt to balance the financial support provided by a large group and the intimacy of a small group, concluded that the ideal size for congregations was 200 or less. The average number of people in any given person's address book also turns out to be about 150 people.

It would appear that 150 is the number of people each of us knows fairly well. Dunbar claims that this figure fits a ratio of primate group size to their neocortex ratio; that is, the volume of the neocortex—evolutionarily the most recent regions of the cerebral cortex—to the rest of the brain. Extremely social primates need big brains to handle living in big groups, because there is a minimum amount of brain power needed to keep track of the complex relationships, in order to live in relative peaceful cooperation. Dunbar concludes that these groupings "are a consequence of the fact that the human brain cannot sustain more than a certain number of relationships of a given strength at any one time. The figure of 150 seems to represent the maximum number of individuals with whom we can have a genuine social relationship, the kind of relationship that goes with knowing who they are and how they relate to us. Putting it another way, it's the number of people you would not feel embarrassed about joining uninvited for a drink if you happened to bump into them in a bar."

Morality most likely evolved in these tiny bands of 100 to 200 people as a form of *reciprocal altruism,* or I'll scratch your back if you'll scratch mine. But as Lincoln noted, men are not angels. There are cheaters. Individuals defect from social contracts. Reciprocal altruism, in the long run, only works when you know who will cooperate and who will defect. In these small groups, cooperation is regulated through a complex feedback loop of communication among members of the community. (This also helps to explain why people in big cities can get away with being rude, inconsiderate, and uncooperative—they are anonymous and thus not subject to the normal checks and balances that come with seeing the same people every day.) In order to play the game of reciprocation you need to know whose back needs scratching and who you will trust to scratch yours. This information is gathered through telling stories about other people, better known as gossip. From an anthropologist's perspective, gossip is a tool of social control through communicating cultural norms, as Jerome Barkow observed: "Reputation is determined by gossip, and the casual conversations of others affect one's relative standing and one's acceptability as a mate or as a partner in social exchange. In Euro-American society, gossiping may at times be publicly disvalued and disowned, but it remains a favorite pastime, as it no doubt is in all human societies."

The etymology of the word *gossip,* in fact, is enlightening. The root stem is *godsib,* or *god* and *sib,* and meant "akin or related." Its early use, as traced through the *Oxford English Dictionary,* included "one who has contracted spiritual affinity with another," "a godfather or godmother," "a sponsor," and "applied to a woman's female friends invited to be present at a birth" (where they would gossip). (In one of its earliest uses in 1386, for example, Chaucer wrote: "A womman may in no lasse synne assemblen with hire godsib, than with hire owene flesshly brother.") The word then mutated into talk surrounding those who are akin or related to us, and eventually to "one who delights in idle talk," as we employ it today. Not surprisingly, we are especially interested in gossiping about the activities of others that most affect our *inclusive fitness,* that is, our reproductive success, the reproductive success of our relatives, and the reciprocation of those around us. Normal gossip is about relatives, close friends, and those in our immediate sphere of influence in the community, plus members of the community or society who are high ranking or have high social status. It is here where we find our favorite subjects of gossip—sex, generosity, cheating, aggression, violence, social status and standings, births and deaths, political and religious commitments, physical and psychological health, and the various nuances of human relations, particularly friendships and alliances. Gossip is the stuff of which not only soap operas but also grand operas are made. But why, in our culture, do we gossip about total strangers, namely celebrities? The probable reason is that the mass media make these figures so familiar to us that they seem *like* relatives, friends, and members of our community. Why would anyone care with whom Princess Diana slept or what her status was in the royal family? Because our Pleistocene brains are being tricked into thinking that Princess Diana is someone we personally know and care about.

## FROM MORALITY TO RELIGION

What has all this to do with religion? *Religion is a social institution that evolved as an integral mechanism of human culture to create and promote myths, to encourage altruism and reciprocal altruism, and to reveal the level of commitment to cooperate and reciprocate among members of the community.* That is to say, religion evolved as the social structure that enforced the rules of human interactions before there were such institutions as the state or such concepts as laws and rights. We would do well to remember that the history of the modern nation-state with constitutional rights and protection of basic human freedoms can be measured in mere centuries, whereas humans evolved as social primates over the course of millions of years, and human culture itself dates back at least 35,000 years, if not more. The principal social institution available to facilitate cooperation and goodwill was probably religion. An organized establishment with rules and morals, with a hierarchical structure so necessary for social primates, and with a higher power to enforce the rules and punish their transgressors, religion evolved as the penultimate effort of these pattern-seeking, storytelling, mythmaking animals. How and why did it evolve?

At the most fundamental level, blood is thicker than water, and Richard Dawkins' famous selfish-gene model accounts for altruism and cooperation among families and extended families. That is, the percentage of genes shared among various degrees of kinship will predict the amount of benefits we receive from a given relative (on average—families will vary, of course). Thus, we do not need religion and gods to enforce the rules in the immediate family where the ties are close. Most parents do just fine. But when we move out from the circle of extended families and into the community and society, we need other mechanisms to ensure that people are kind to one another.

Morality evolved over eons in the paleolithic environment where individuals cooperated and competed with one another to meet their needs. Individuals belonged to families, families to extended families, extended families to communities, and, in the last couple of centuries, communities to societies. This natural progression, which is now in its latest evolutionary stages of perceiving societies as part of the species, and the species as part of the biosphere, is illustrated in the Bio-Cultural Pyramid on the facing page.

The lower strata of the Bio-Cultural Pyramid depict the 1.5 million years over which our moral behavior evolved under primarily biogenetic control, and the middle layer the transition about 35,000

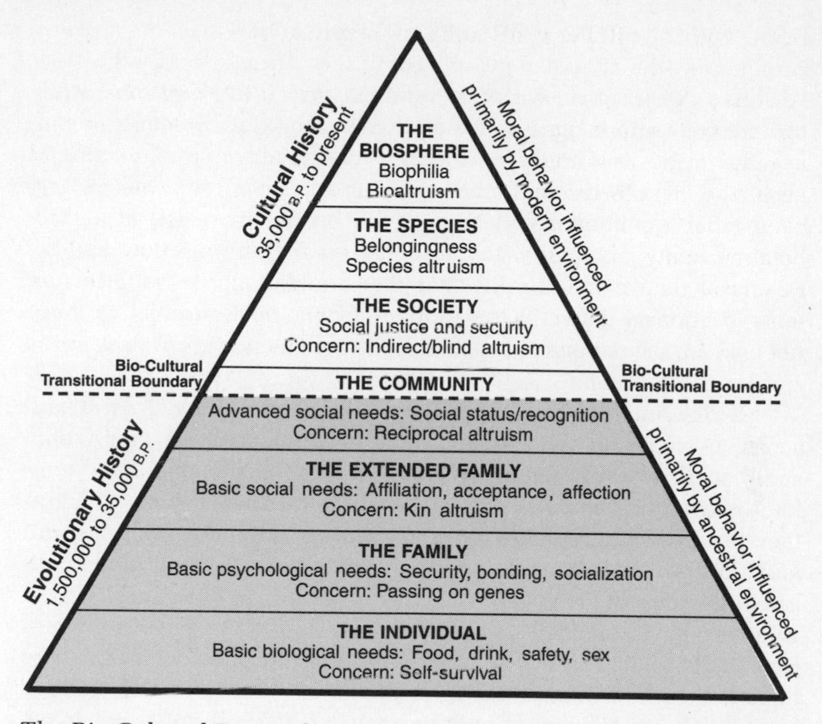


*The Bio-Cultural Pyramid: A model of the origin and development of ethical behavior.*

years ago when sociocultural factors increasingly assumed control in shaping our ethical precepts. Obviously this was a continuous process. There was no point at which an Upper Paleolithic Moses descended from a glacier-covered mountain to present The Law to his fellow Cro-Magnons.

Nevertheless, the semipermeable *bio-cultural transitional boundary* divides time and dominant source of influence, where the individual, family, extended family, and paleolithic communities were primarily molded by natural selection; whereas neolithic communities and modern societies were and are primarily shaped by cultural selection. Starting at the bottom of the Bio-Cultural Pyramid, the individual's need for survival and genetic propagation (through food, drink, safety, and sex) is met by way of the family, extended family, and the community. The nuclear family, however, is the foundation. Despite assaults on it in the second half of the twentieth century, the family remains the most common social unit around the world.

Even within extremes of cultural deprivation—slavery, prisons, communes—the two-parents-with-children structure emerges: (1) African slave families broken up retained their attachment and structure for generations through the oral tradition; (2) in women's prisons pseudofamilies self-organize, with a sexually active couple acting as "husband" and "wife" and others playing "brothers" and "sisters"; (3) even when communal collective parenting is the norm (e.g., Kibbutzim), many mothers switch to the two-parent arrangement and the raising of their own offspring. For this foundational social structure our evolutionary history is too strong to overcome. Conservatives need not bemoan the decline of families. They will be around as long as the species continues.

Moving up the Bio-Cultural Pyramid, basic psychological and social needs such as security, bonding, socialization, affiliation, acceptance, and affection evolved as mental programs to aid and reinforce cooperation and altruism, all of which facilitate genetic propagation through children. *Kin altruism* works indirectly—siblings and half-siblings, grand- and great-grandchildren, cousins and half-cousins, nieces and nephews, all carry portions of our genes. This is what is known as *inclusive fitness,* and applies to anyone who is genetically related to us. In larger communities and societies, where there is no genetic relationship, *reciprocal altruism* (if you scratch my back I will scratch yours) and *indirect altruism* (if you scratch my back now I will scratch yours later) supplements kin altruism. Inclusive fitness gives way to what we might call *exclusive fitness.* The natural progression of exclusive fitness may be the adoption of *species altruism* and *bioaltruism* (we will prevent extinction and destruction now for a long-term payoff), which Wilson argues in *Biophilia* may even have a genetic basis. But, Wilson confesses, this should probably still be grounded in self-interest arguments—*my* children and grandchildren will be better off in a future with abundant biodiversity and a healthy biosphere—since inclusive fitness is more powerful than exclusive fitness.

The width of the Bio-Cultural Pyramid at any point indicates the strength of ethical sentiment, and the degree to which it is under evolutionary control. The height of the pyramid at any point indicates the degree to which that ethical sentiment extends beyond our own genomes (ourselves). But the pyramid also shows that these two sets of sentiments are inversely related. The further a sentiment reaches beyond ourselves, the further it goes in the direction of helping someone genetically less related, and the less support it receives from underlying evolutionary mechanisms.

New research by philosopher Elliott Sober and biologist David Sloane Wilson, presented in their 1998 book *Unto Others: The Evolution and Psychology of Unselfish Behavior,* indicates that there may have been an additional selection component in human evolution that gave rise to cooperation and altruism, and that is a modified version of *group selection.* This is a volatile subject among evolutionary theorists because for the past thirty years, group selection has been next to creationism as the doctrine strict Darwinians most love to hate. From George Williams's 1966 book *Adaptation and Natural Selection* to Richard Alexander's 1987 book *The Biology of Moral Systems* to Richard Dawkins's several books throughout the 1990s, group selection was vilified as the pap of bleeding-heart liberals who couldn't deal with the reality of "nature red in tooth and claw." Michael Ghiselin's 1974 description summed up the Darwinian literalists perspective, especially the last line:

> *The economy of nature is competitive from beginning to end. . . . The impulses that lead one animal to sacrifice himself for another turn out to have their ultimate rationale in gaining advantage over a third. . . . Where it is in his own interest, every organism may reasonably be expected to aid his fellows. . . . Yet given a full chance to act in his own interest, nothing but expediency will restrain him from brutalizing, from maiming, from murdering—his brother, his mate, his parent, or his child. Scratch an "altruist," and watch a "hypocrite" bleed.*

Sober and Wilson, through a sophisticated mathematical model and series of logical arguments, and defining a group as "a set of individuals that influence each other's fitness with respect to a certain trait but not the fitness of those outside the group," demonstrate that "natural selection can operate at more than one level of the biological hierarchy." They show how "individual selection favors traits that maximize relative fitness within single groups," and that "group selection favors traits that maximize the relative fitness of groups." Of course, "altruism is maladaptive with respect to individual selection but adaptive with respect to group selection." Therefore, they conclude, "altruism can evolve if the process of group selection is sufficiently strong." For example, they cite William Hamilton's analysis of how consciousness might have provided a group selective advantage for certain human populations with regard to the ethical enforcement of rules: "Consider also the selective value of having a conscience. The more consciences are lacking in a group as a whole, the more energy the group will need to divert to enforcing otherwise tacit rules or else face dissolution. Thus considering one step (individual vs. group) in a hierarchical population structure, having a conscience is an 'altruistic' character."

Part of the problem in this debate is in how certain terms are defined, such as *altruism* and *cooperation,* and the tendency to force these categories into either-or choices for human actions. Humans are *either* altruistic *or* selfish. Humans are *either* cooperative *or* competitive. But *altruistic* and *cooperative* are not reified *things,* they are behaviors. And like all behaviors, there is a broad range of expression, from a little to a lot. Applying fuzzy logic can help clarify this complex human phenomenon, where we might assign fuzzy numbers to altruism or cooperation. Depending on the circumstances, someone might be, say, .2 altruistic and .8 nonaltruistic (or selfish), or .6 cooperative and .4 noncooperative (or competitive). Humans can be both altruistic and nonaltruistic, cooperative and noncooperative.

One problem with reciprocal altruism is this: How do I know that if I scratch your back you will scratch mine? I am more than willing to cooperate with unrelated members of my community, but only if I am reasonably certain that they are going to reciprocate. How can I find out who are the cooperators and who are the defectors? Gossip is one way. Past experience with my fellow community members is another. Combined, these give me enough information to make a decision (even if it is on an unconscious level) about whom I can trust.

In a way, daily life can be modeled by a game theory technique called the *Prisoner's Dilemma.* Two individuals who cooperated in committing a crime are caught, arrested, and offered the chance of a reduced sentence if one will rat out the other. The district attorney can convict both of them of a minor offense, but if one of them confesses, he can go free while the other rots in jail with a long sentence. What will they do? It depends on their respective reputations for being trustworthy. Let us simplify the game where each player gets one point if both cooperate, either one can get two points by defecting when the other cooperates, and zero points if both defect. When only one round of the game is played, most people defect. But when the game is iterated, or repeated for numerous rounds with the same players, cooperation is the norm. When you learn that your partner is a cooperator and not a defector, you become a cooperator yourself.

To test this hypothesis the mathematician and political scientist Robert Axelrod held a contest by inviting people to submit a computer program to play the iterated Prisoner's Dilemma. Pitting the programs against each other for 200 games each, he tallied up the payoff scores and found that the winning program was the simplest one, designed by Anatol Rapoport and called *Tit for Tat.* The program chooses to cooperate on the first round, and then on all subsequent moves it matches the choice of its opponent. Tit for Tat, says evolutionary biologist John Maynard Smith, is an Evolutionary Stable Strategy, or "a strategy such that, if all the members of a population adopt it, no mutant strategy can invade." Tit for Tat is reciprocal altruism—if you'll scratch my back, I'll scratch yours (or the reverse: If you stab me in the back, I'll stab you in the back). But the latter is rarely necessary. Most of the time it pays to cooperate, and most of the time we do.

Is this because we want to be "good" or "moral" people? That may be what it *feels* like now, but current emotions may be proxies for deeper causes. Since our reputations as cooperators must be built over time, we must show consistency from day to day, week to week, and year to year. It would be difficult to fake being a cooperator in order to fool your fellow community members for any length of time. Anthropologist William Irons shows that religion, in addition to providing rules, morals, and enforcement (and numerous other benefits outside the scope of this analysis), furnishes a splendid opportunity to prove loyalty and commitment to the group. If I see you every week in the pews, every month at the confessional, getting circumcised, being bar mitzvahed, not eating meat on Fridays, wearing a yarmulke, singing the psalms of the Lord, not using electricity on the sabbath, facing east to pray, taking the bread and wine as the body and blood of the savior, going to war in the name of God, and even willing to risk death for our group, I know you are someone I can trust. That sort of commitment is hard to fake. If our self-image is that of an honest person, not only are others more likely to perceive us as honest, we are more likely to be honest. We are all fairly good at detecting cheaters and liars, so in order for the cheater or liar to get away with his offense, he has to work very hard at appearing honest. Even if deception is the original intent, in time, with repetition of the ritual, self-deception may take over. Psychics, cult gurus, and other charlatans may very well come to believe in their own outrageous claims for the simple fact that they can deceive their marks better if they themselves believe the lie. Either way, through literally millions of iterations of real-life game-theory events in the course of a lifetime, we learn who are the cooperators and who are the defectors. And through our religion (and, more recently, the state), we come to believe that our actions *really are* moral, just, and right. Our clan *really is* special, perhaps even worth dying for, if our leader or our God so asks.

One of the most common reasons people give for believing in God (see Chapter 4) is that without the existence of a deity there would be no ultimate basis for morality. The source of this belief may be that morality, God, and religion have been so intertwined for so long that there is probably an evolutionary-based epigenetic rule underlying the connection. The Enlightenment concept of human rights—as expressed and fought for in the French and American Revolutions—is relatively new. It is primarily based on the social contract: In order for humans to achieve life, liberty, and happiness they must be free, and their freedoms must be protected by the state through compacts like the Constitution and the Bill of Rights. With proper indoctrination it is possible to get young men so committed to these ideals that they will sacrifice themselves for the larger group. But this has not been easy, so it is probably no accident that beneath the surface of nationalism often lies religion. Our God is better than their God. (Or in the case of the cold war, our God is better than their Godless society.) When Winston Churchill and Franklin Roosevelt together sang "Onward Christian Soldiers" following their meeting cementing the American-British alliance, this was more than ceremonial window dressing. Or consider the words to "The Battle Hymn of the Republic" (especially the final refrain below) written by Julia Ward Howe after a review of the federal troops in Washington (published in the *Atlantic Monthly* in February, 1862):

> Mine eyes have seen the glory of the coming of the Lord;
> He is trampling out the vintage where the grapes of wrath are stor'd
> He hath loos'd the fateful lightning of His terrible swift sword
> His truth is marching on.
>
> • • •
>
> I have read a fiery gospel writ in burnished rows of steel:
> "As ye deal with My contemners, so with you My grace shall deal"
> Let the Hero born of woman crush the serpent with His heel
> Since God is marching on.

The Confederate troops, of course, sang and prayed to the same God to do the same thing to their enemy.

## FROM RELIGION TO GOD

God and religion are many things to many people, and reasons for belief are varied, thoughtful, and momentous. For centuries many a theologian, scholar, and scientist has attempted to explain why people need religion and believe in God. Their efforts have left a legacy of theories and libraries of books on the subject. Edward Tylor and James Frazer viewed religion as animism and magic, whereas Sigmund Freud saw it as an obsessional neurosis. Emile Durkheim said religion is a sacred part of the social structure, while Karl Marx said it is nothing more than another tool of alienation and the opiate of the masses. Mircea Eliade thought religion to be the most sacred part of the human psyche, while E. E. Evans-Pritchard saw religion as society's "construct of the heart," which it needs as much as science's "construct of the mind." Clifford Geertz believed that religion is a cultural system of symbols that act to empower, give meaning, and provide motivation. In evaluating these disparate theories, historian of religion Daniel Pals suggests asking the following questions: "(1) How does it define the subject? (2) What type of theory is it? (3) What is the range of the theory? (4) What evidence does the theory appeal to? (5) What is the relationship between a theorist's personal religious belief (or disbelief) and the explanation he chooses to advance?"

Applying these questions to the theory of religion presented here: (1) *Religion is a social institution that evolved as an integral mechanism of human culture to create and promote myths, to encourage altruism and reciprocal altruism, and to reveal the level of commitment to cooperate and reciprocate among members of the community.* (2) This is a biocultural theory of religion. (3) The range of the theory is limited to deeper, ultimate "why" questions about religion and belief in God. The particulars of any one religion are not the subject of analysis. (4) This is a scientific theory, so the evidence is based on those sciences most allied with the study of myth, religion, and belief in God: archaeology, history, anthropology, sociology, cognitive and social psychology, neurophysiology, behavior genetics, and evolutionary biology. The theory attempts to probe deeply into the core of why people believe in God and religion, but does not focus on specific faiths or customs. (5) See the preface and first chapter of this book for the relationship between my personal religious beliefs and the explanation I have chosen to advance.

This theory of religion has presented a case for how humans evolved from pattern-seeking to storytelling to mythmaking to morality and religion. Where does God fit into this sequence? In short, everywhere. God is a pattern, an explanation for our universe, our world, and ourselves. God is the key actor in the story, "the greatest story ever told" about where we came from, why we are here, and where we are going. God is a myth, one of the most sublime and sacred myths ever constructed by the mythmaking animal. God is the ultimate enforcer of the rules, the final arbiter of moral dilemmas, and the pinnacle object of commitment. And God is the integrant of religion, the most elemental of all components that go into the making of the sacred. God and religion are inseparable. People believe in God because we are pattern-seeking, storytelling, mythmaking, religious, moral animals.

# God and the Ghost Dance

## *The Eternal Return of the Messiah Myth*

*And almost every one, when age,*
*Disease, or sorrows strike him,*
*Inclines to think there is a God,*
*Or something very like Him.*
—Arthur H. Clough, *Dipsychus,* 1850

In Los Angeles there is a radio station catering largely to an African-American audience—KKBT 92.3 FM, also known as "The Beat"—that features an interesting show called *Street Science,* hosted by Dominique Diprima. Since I live near Los Angeles I am a periodic guest, the token skeptic when street science veers into pseudoscience. In 1997 they organized a show about UFOs, and the other guests included Don Ecker, publisher of *UFO Magazine,* and Dwight Schultz, the actor best known as "Barclay" on *Star Trek, The Next Generation.* After the usual banter about blurry photographs and government cover-ups, the host opened up the phone lines. Suffice it to say that after years of doing such shows I am not unaccustomed to interviews in which most of the other guests and call-ins are believers in the topic of discussion. But this time was different. It was not just that these callers believed in UFOs. They believed in a particular type of UFO: a messianic "mothership" circling the Earth that in 1999 will release hundreds of smaller ships that will invade targeted cities.

I did not think too much about the first call or two of this nature, but after half a dozen or so I realized what was going on. Announcing themselves as members of the Nation of Islam (NOI), these callers explained that there is no doubt about the existence of UFOs—their leader, the Minister Louis Farrakhan, had himself been to visit the "Mother Plane" in space, described in the NOI's July 4, 1996, edition of the *Final Call* newspaper as "a human-built planet, a half mile by a half a mile" carrying "1500 smaller baby planes" with bombs "designed for the destruction of this world." The "Mother Plane," the newspaper explained, was visited by Louis Farrakhan himself on September 17, 1985, where he received communication from deceased NOI prophet Elijah Muhammad. For some NOI members, the science fiction film *Independence Day* was nothing short of a documentary. *Final Call*, in fact, explained that the hit film's acronym, *ID4*, "is a biogenetic reference to a genetic inhibitor which ceases certain procedures in evolution and life, according to researchers at M.I.T.," the alma mater, it pointed out, of the "Jewish genius" played by Jeff Goldblum in the film. In fact, according to *Final Call*, the existence of the spacecraft was known to both Ronald Reagan and Mikhail Gorbachev, who met in 1985 to discuss a plan to deal with alien invasion:

> *Col. Colman S. Vonkeviskzky, MMSE, . . . a former Hungarian staff major and a military scientist who said he has been engaged over 45 years in dealing with the United Nations about the UFO phenomenon. In a telephone interview, Col. Vonkevickzky said that former President Ronald Reagan and former Soviet Union President Mikhail Gorbachev spent five hours in confidential talks during the 1985 Geneva Conference, discussing what military strategies to take in the event of alien attack against earth. All the methods used by the government to harm the real "Mother Plane" have failed, just as their efforts to crush the Nation of Islam have failed. . . . Frankly, this government needs help—desperately. I think the best thing they can do is consult with Minister Farrakhan.*

Several callers made reference to the end coming in 1999 when the "Mother Ship" will descend upon Earth, release its smaller ships to topple the "white government," and place the black man back in his rightful position of power. The longer I listened the more incredulous I became. Did these callers *really* believe that the messiah was coming in the form of an alien? Many of them did. But how could Louis Farrakhan, who, regardless of his political beliefs, has always struck me as an intelligent man, believe such nonsense? His message is usually one of this-worldly self-reliance, not otherworldly wishful thinking. Ted Koppel wondered the same thing on the October 15, 1996, episode of *Nightline*:

*Koppel:* Minister Farrakhan, frequently what you say makes eminently good sense and is extremely lucid, and then—and I'm going to read you this quote because we don't have it on videotape—you told congregants just before the Million Man March at a Washington church about the "'Mother Wheel,' a heavily armed spaceship the size of a city which will rain destruction upon white America but save those who embrace the Nation of Islam" [the *New Yorker*]. It sounds like gibberish, but maybe you can explain it.

*Farrakhan:* Well, sir, you can ask President Jimmy Carter if it's gibberish. You can ask some of the astronauts who went up and saw it if it's gibberish. On the front page of the *Washington Post* several years ago, a Japanese pilot was flying across the Bering Strait, and he saw something in his radar that looked like two large aircraft carriers joined together, in terms of size—and an aircraft carrier is 440 yards long—two of them together would be half a mile by a half a mile. This is that wheel that was spoken of in Ezekiel, that has become a reality. It's over the heads of us in North America, and soon you shall see these wheels over the major cities of America. It is above top-secret by the United States government.

As I sat in the studio taking this in, it dawned on me that I was hearing a modern iteration of the nineteenth-century Native American Ghost Dance, where the great spirit would descend to displace the white man and allow the Indians to live freely in their aboriginal home. They would defeat the white soldiers because their bodies would be impervious to bullets. The white man would then disappear, the dead would return to life, and buffalo would once again blanket the plains. I began to explain to the radio audience that such mythological motifs are common in history, especially among oppressed peoples, but Diprima interrupted me, proclaiming: "I respect the Ghost Dance. I believe in the Ghost Dance. I don't want to go down this road. This show is about UFOs, not race."

Is there a connection transversing the century between these two beliefs? There is. In fact, it turns out that there are many Ghost Dance–like myths across time and cultures. The belief in a savior or messiah that, if the proper ritual is performed, will rescue us from our oppression and deliver redemption, fits the classic pattern of myth, in particular, what might be called the *oppression–redemption myth*, or, simply, the *messiah myth*.

## THE GHOST DANCE AS MYTHMAKING

One of my most striking memories of this radio program was the host's comment that she "believes in the Ghost Dance." What did it mean for a twentieth-century African-American woman to believe in a nineteenth-century Native American story? It meant that the story is a myth, an enduring narrative with deep personal meaning and social context. The Ghost Dance in particular, and the messiah myth in general, represent a commingling of eschatological, messianic, and millenarian motifs, with an outer shell of representational rebirth and renewal. The reason this Native American myth was appealing to an African-American woman is because of the common elements of oppression and redemption. For over four centuries both Native Americans and African Americans were conquered, enslaved, and killed in large numbers. In the case of Native Americans, the process involved the dissolution of nearly 500 nations and the elimination of approximately 90 percent of the population (mostly by European diseases for which they had no immunity); in the case of African Americans, the process involved the confiscation of approximately twenty million Africans and the enslavement of the half who survived the journey to the New World. The learned helplessness that comes with such long-term oppression lends itself to the mythos of supernatural intervention.

Suffice it to say that the probability of aliens unfettering African Americans from their perceived oppressors is coequal to the belief on the part of Native Americans that they would become impervious to bullets. Myths, however, even though fictitious, can be a powerful source of fuel to the belief engine that drives our perceptions of the world. In the four centuries from 1492 to 1890, one nation conquered 500 nations, one civilization destroyed 500 civilizations, one culture subsumed 500 cultures. As the sun set on a 13,000-year-old people, the circumstances were ripe for twilight dream time—a slip into the sacred, the numina of the otherworld where messiahs reside and the rebirth of a new age awaits.

## THE 1890 GHOST DANCE

On January 1, 1889, a total eclipse of the sun projected a black disk that streaked across the North American continent. In its path in Nevada was a Paiute Indian named Wovoka ("The Cutter"), known to whites as Jack Wilson. Wovoka was the son of Tavibo, a Paiute from Walker Valley, just south of Virginia City, Nevada. It was in 1870, almost twenty years earlier, that Tavibo and another Paiute named Wodziwob started the first Ghost Dance movement, on the heels of a devastating drought and an epidemic of typhoid and measles that wiped out a tenth of the Paiute population. Tavibo prophesied that a great earthquake would swallow both Indians and whites, but after three days (note the Christian influence) the Indians alone would return, along with fish, game, and plants. Tavibo created a counter-clockwise circular dance around a fire that would bring back the dead, and this evolved (with as many variations as there were groups who adopted it) into the Ghost Dance of 1890. But times were not quite right for the messianic rebirth myth to take off, and the influence of Wodziwob and Tavibo soon faded after 1870. (The Battle of Little Big Horn in 1876, in which General Custer and his soldiers were thoroughly quashed in one of the few major victories for the natives, brought hope that perhaps the tide could be reversed by more earthly means.) Shortly thereafter Tavibo left his family and community, abandoning his teenage son Wovoka to a white rancher named David Wilson, who renamed him Jack. Wovoka/Jack grew up in a deeply Christian family that included daily prayers and readings from the Bible. At the age of twenty he married a Paiute woman, and as a young adult began to pursue the interests of his biological father, including reviving the circular dance that was said to open the Paiute soul to greater spirituality. Between dances he preached an amalgam of Native American Christianity, celebrating faith, Jesus, and the monotheistic God, along with traditional beliefs in the spirits of the mountains, clouds, snow, stars, trees, and antelope.

Wovoka's influence and following grew, and by the 1880s real-world hope for the survival of Native American culture was fading fast. In 1881 their great leader Sitting Bull surrendered after a long and bitter struggle. Where the plains were once covered with the herds of 60 million buffalo, by 1883 a scientific expedition counted a mere 200 head. Finally, any possibility of peaceful coexistence of whites and Indians was greatly inhibited by the government's bureaucratic and military actions, driven by the administration's Indian policy, succinctly summarized by President Benjamin Harrison:

> *First, the anomalous position heretofore occupied by the Indians in this country can no longer be maintained.*
>
> *Second, the logic of events demands the absorption of the Indians into our national life not as Indians but as American citizens.*
>
> *Third, as soon as wise conservation will permit it, the relations of the Indians to the government must rest solely upon the recognition of their individuality [they had to become legal citizens of the United States and not members of an Indian tribe or nation].*
>
> *Fourth, the individual must conform to the white man's ways, peaceably if they will, forcibly if they must.*
>
> *Fifth, compulsory education.*
>
> *Sixth, tribal relationship should be broken.*

Part of the "absorption" process included forcing the Indians to abandon hunting and take up farming. Unfortunately a drought struck the West in the spring and summer of 1890, forcing many of the reservations to go on starvation rations. Some Native Americans tried to return to the ways of their ancestors by "hunting" the cattle rationed to them by the government, but in July this was prohibited by Harrison's new Commissioner of Indian Affairs, Thomas J. Morgan, who declared it a savage holdover from a now-banned primitive way of life. By January 1889, the Indians were essentially finished, and the Ghost Dance was becoming more appealing by the day.

Quite by chance on that first day of January 1889, Wovoka was ill with fever when the solar eclipse approached. As the shadow engulfed him he fell into a hallucination in which he envisioned himself taken to heaven where he could see and speak with God. The Smithsonian anthropologist James Mooney, who documented the Ghost Dance, explains what happened next:

*He saw God, with all the people who had died long ago engaged in their old-time sports and occupations, all happy and forever young. It was a pleasant land and full of game. After showing him all, God told him he must go back and tell his people they must be good and love one another, have no quarreling, and live in peace with the whites; that they must work, and not lie or steal; that they must put away all the old practices that savored of war; that if they faithfully obeyed his instructions they would at last be reunited with their friends in this other world, where there would be no more death or sickness or old age. He was then given the dance which he was commanded to bring back to his people. By performing this dance at intervals, for five consecutive days each time, they would secure this happiness to themselves and hasten the event.*

The umbra of the eclipse receded and the light of day illuminated Wovoka's prophetic mission. The more he spoke, the more his people listened. Indians from districts far and wide came to sit at his feet. Mormons debated whether Wovoka could be the fulfillment of Joseph Smith's prophecy that the Messiah would appear in 1890 (many Mormons believed that Indians were descendants of one of the "Ten Lost Tribes" of the Hebrews). His confidence growing by the week, Wovoka even dictated a letter to President Harrison, explaining that if he would be allowed to deliver God's message to the people of Nevada and the rest of the country he could control the weather, in particular making it rain whenever he wanted (recall this was a drought year). The letter was never delivered.

While Wovoka called himself *a* messiah who was *like* Jesus, he never said he was the Christ. Despite this disclaimer, many (both whites and Indians) referred to him as such, and this started a long process that would cascade into tragedy twelve months later at a creek whose name has become synonymous with the Ghost Dance. As Wovoka's fame grew and Indian delegates from dozens of nations came to listen, white settlers became concerned that something more than a peaceful dance was taking place. The Indian delegates took home with them blessed tokens of Wovoka's power (red ocher, magpie feathers, pine nuts, robes of rabbit fur), and there launched their own Ghost Dance ceremonies. Wovoka instructed them as follows:

*Grandfather said when he die never no cry. no hurt anybody. no fight, good behave always, it will give you satisfaction, this young man, he is a good Father and mother, don't tell no white man.*

*Jesus was on ground, he just like cloud. Everybody is alive agin, I don't know when they will here, may be this fall or in spring.*

*You make dance for six weeks night, and put you foot [food?] in dance to eat for every body and wash in the water. that is all to tell, I am in to you. and you will received a good words from him some time.*

The continual and repetitive motion of the dance, conducted for hours on end, produced profound emotional experiences for the dancers. Some went into a trance, others collapsed and writhed on the ground. It was a spiritual journey from the profane to the sacred. Within months the new religion was taken up by the Utes, Shoshoni, and Washo in Nevada; the Mohave, Cohonino, and Pai in Arizona; and the Arapaho, Cheyenne, Assiniboin, Mandan, Arikara, Caddo, Kichai, Kiowa, Pawnee, Wichita, Comanche, Delaware, Oto, and Sioux in scattered regions throughout the west from California to Oklahoma, and from Texas to Canada. While the ceremony had the common theme of messianic rebirth, the dances varied in detail and were given different names, including "dance in a circle," "the Father's Dance," "dance with clasped hands," and "spirit dance."

*Wovoka, a Paiute Indian and self-proclaimed messiah who spread the salvation of the Ghost Dance.*

Unfortunately for the Native Americans, Indian delegates from the other nations were not the only ones aware of the Ghost Dance. White settlers grew wary, concerned about what the dance might do to stir up trouble, stimulating officials in government to look into the matter. Wovoka's peaceful dance of renewal began to escalate into a war dance when his disciples (as most disciples are wont to do) added their own components to the religion. By the spring of 1890 the Lakota Sioux leader Kicking Bear, not content to give his fate over to the gods, introduced special clothing into the ceremony, declaring "the bullets will not go through these shirts and dresses, so they all have these dresses for war." Wovoka never said anything about war—just the opposite—but Kicking Bear was not about to go peacefully into the closing of the American West.

Someone on the Pine Ridge Reservation told a local resident named Charles L. Hyde about this new development, and on May 29 he wrote to the Secretary of the Interior that he heard the Sioux were plotting a rebellion. Hyde's letter was passed along to the Commissioner of the Bureau of Indian Affairs, who in turn sent copies to the agents of the various Sioux reservations. They were to check it out and report back to Washington. Talk of bulletproof vests and the disappearance of whites did not sound like peaceful absorption to white bureaucrats.

*Ghost dancers (photographed by anthropologist James Mooney).*

Arguably the most famous Indian of the time was Sitting Bull, associated with the massacre of Custer's Seventh Cavalry and later a star in Buffalo Bill Cody's Wild West Show. The government asked Sitting Bull (through an agent named James McLaughlin) to persuade his people to stop the Ghost Dance. He agreed to try if McLaughlin would accompany him to see Wovoka first to determine if there was really anything to fear. McLaughlin declined to go, and on top of that on November 19 he asked permission to withhold all food rations from Indians in Sitting Bull's village. Tensions were mounting. Virtually all activities on the Sioux reservations ceased with the exception of the Ghost Dance. At the Pine Ridge Reservation in South Dakota the newly appointed (and now frightened) agent sent a telegram of alarm to the Bureau of Indian Affairs, dated October 30:

> *Your Department has been informed of the damage resulting from these dances and of the danger attending them of the crazy Indians doing serious damage to others. . . . The only remedy for this matter is the use of military and until this is done you need not expect any progress from these people on the other hand you will be made to realize that they are tearing down more in a day than the Government can build in a month.*

Two weeks later President Harrison ordered the Secretary of War to suppress "any threatened outbreak." Word spread around the country that the Indians were planning one final apocalyptic war against the whites. On October 28, the *Chicago Tribune* ran this headline:

*TO WIPE OUT THE WHITES*
*WHAT THE INDIANS EXPECT OF THE COMING MESSIAH*
*FEARS OF AN OUTBREAK*
*OLD SITTING BULL STIRRING UP THE EXCITED REDSKINS*

*The Chicago Herald*, *Harper's Weekly*, and the *Illustrated American* all sent correspondents to cover the event. The renowned painter of the West, Frederic Remington, was sent to provide images. When facts were lacking, rumors filled the journalists' columns. General Nelson Miles, for example, reported to the Washington press corp that "it is a more comprehensive plot than anything ever inspired by Tecumseh, or even Pontiac." By December thousands of army troops moved onto the Sioux Reservation, including a new Seventh Cavalry. Between the starving, messiah-seeking Indians and the trigger-happy soldiers, something was bound to give. It did on December 15, 1890, when Indian policemen were ordered to arrest Sitting Bull, the man they feared would lead the Indians into war.

The Indian police, now under McLaughlin's command, broke into Sitting Bull's cabin, ordered him to awake, dress, and come with them. There was tragic irony in this situation. Many of the arresting officers had ridden with Sitting Bull and stood by his side through the dog days of the 1870s. But Sitting Bull was an old man now, impotent against the white onslaught, biding his time to the end. As he awaited the saddling of his horse, his remaining loyal followers gathered about him, determined not to let him go. It was a situation that could not end peacefully. At that moment a lieutenant named Bullhead moved in to grab Sitting Bull. A warrior named Catch the Bear pulled out his rifle and shot Bullhead in the side, who, as he twirled and fell shot Sitting Bull in the chest. Gunfire rang out on both sides, and a bullet found its mark in Sitting Bull's head. He died on the spot. Bizarrely, Sitting Bull's horse had been given to him by Buffalo Bill Cody, who had trained the horse to "dance" in his Wild West Show whenever it heard gunfire. As it was being led to its master's cabin gunshots broke out and it began to dance. The horse, said the Indians in a final defiance of the white's prohibition against their new religion, was itself doing the Ghost Dance.

The other Indian leader of great import was the Sioux Chief Big Foot. Following the Sitting Bull debacle, Big Foot was ordered to come to Pine Ridge to help negotiate a peace settlement. On his way he passed near the encampment of Kicking Bear, the Lakota Sioux who introduced the "bulletproof vest" into the Ghost Dance. General Nelson Miles, looking for a fight, saw this as a potentially hostile action on Big Foot's part and moved in to investigate. The commanding major encountered the now sick (with pneumonia) Big Foot and his small, travel-weary band, and ordered them to camp at a creek near the Pine Ridge agency, called Wounded Knee.

On December 28, 1890, 120 men and 230 women and children set up their tepees for the night. Surrounding them were over 500 heavily armed U.S. cavalry troops, Indian scouts working for the Army, and four Hotchkiss artillery cannons. The next morning Big Foot's people were ordered to relinquish all weapons. Some guns were surrendered, but not all. The soldiers went into the tepees to look for more. Meanwhile, Yellow Bird, Big Foot's holy man, launched into the Ghost Dance, reminding his men that the shirts they wore would be impenetrable by the soldiers' bullets. The officers ordered the Indians to strip, hoping to reveal hidden weapons. It was a freezing December winter day. Some of the men refused to obey. The soldiers moved in to frisk them. One spirited young Indian named Black Coyote pulled out a new rifle he had recently purchased, announcing he would not give it up. Two soldiers rushed him from behind and grabbed the rifle. At that moment the Ghost Dancing Yellow Bird threw a handful of dirt into the air, declaring that this was a symbol of the renewal of the Earth promised by their Messiah.

What happened next was much less symbolic. White officers thought it was a signal for the Indians to attack. By chance Black Coyote's gun discharged harmlessly into the air, but it triggered Sioux and soldiers to open fire upon one another. Stray bullets found women, children, and Big Foot in their tepees. Those who managed to make a dash from the camp were cut down by the artillery cannons, firing exploding shells one per second. When it was all over 250 Sioux men, women, and children were dead.

Two weeks after Wounded Knee the last of the Ghost Dancers came out of the Badlands and capitulated to the United States Army. On January 15, 1891, the defiant Kicking Bear gave up his rifle to General Miles at the Pine Ridge agency, and his cause to all eternity. The Ghost Dance was over—it was not the beginning of the end of the Indians, it was the end of the end. The bullets found their mark and the Messiah never came.

## THE ETERNAL RETURN OF THE GHOST DANCE

If we do not "believe in" the Ghost Dance, we can nevertheless understand it as an eternally returning cultural phenomenon of oppressed peoples—one version of the messiah myth. Anthropologist James Mooney certainly understood it this way, in the introduction to his great work on the Ghost Dance:

*The lost paradise is the world's dreamland of youth. What tribe or people has not had its golden age, before Pandora's box was loosed, when women were nymphs and dryads and men were gods and heroes? And when the race lies crushed and groaning beneath an alien yoke, how natural is the dream of a redeemer, an Arthur, who shall return from exile or awake from some long sleep to drive out the usurper and win back for his people what they have lost. The hope becomes a faith and the faith becomes the creed of priests and prophets, until the hero is a god and the dream a religion, looking to some great miracle of nature for its culmination and accomplishment. The doctrines of the Hindu avatar, the Hebrew Messiah, the Christian millennium, and the Hesunanin of the Indian Ghost dance are essentially the same, and have their origin in a hope and longing common to all humanity.*

If the Ghost Dance of 1990's African Americans shares commonalities with that of 1890's Native Americans, and if Mooney is right about the Hindus, Jews, and Christians from centuries past, might this indeed be the result of something deep within our common evolutionary and cultural humanity? Are there other examples of Ghost Dances around the world and across the ages against which we may test this hypothesis? There are.

In his comprehensive anthropological study on the origins of religion, Weston La Barre has carefully documented many such Ghost Dances. In May 1856, for example, during colonial domination of parts of Africa by the English, a South Xhosa girl encountered spirit entities while obtaining water at a nearby stream. She told her uncle, who in turn spoke to the deities who informed him that they would help the Xhosa drive the English from the country. In this version the ritual ceremony that would trigger the English departure was the slaughter of cattle. The girl's uncle, Umhlakaza, ordered his tribesmen to destroy all of their herds as well as the granaries of corn. If this rite was properly carried out, the dead would be resurrected, the old would become young again, illnesses would disappear, herds of fattened cattle would rise from the Earth, and ready-for-harvest millet fields would suddenly appear. It was to be paradise on Earth. What actually happened is that following the mass butchering of some 2,000 head of cattle a famine decimated the Xhosa tribe, nearly driving it into extinction.

A similar story took place in a Maori village in New Zealand at the end of August 1934, when a visionary member of the tribe had a dream in which an angel told him that a Holy Ghost would deliver his people from the whites and return their confiscated lands to them. For days following the dream the Maori fasted, chanted, danced, and waited for the day of deliverance. Then, as some members of millennial cults do today, Maori villagers gave away their belongings (who needs material goods in the next life?). White administrators got wind of the ceremonies and came to investigate. Finding starving children and deprivation-crazed adults, they declared the visionary insane and shipped him off to a mental hospital, thus ending the Maori Ghost Dance.

Searching the globe La Barre finds another example from a Siberian village in July 1904, when an Altai Turk named Chot Chelpan had a vision of a spirit who told him that the land of his people would be returned. Russian Orthodox Church missionaries had discovered that Altai grazing lands were better for farming (for the missionaries, that is!), so they confiscated them from the native peoples. Chot's prayer to the spirits was the Ghost Dance revisited: "Thou art my Burkhan dwelling on high, thou my Oirot descending below, deliver me from the Russians, preserve me from their bullets." Chot instructed his people in an elaborate ceremony that included killing pets, sprinkling milk to the four cardinal compass points, and a number of other rituals, then told the Altains: "Soon their end will come, the land will not accept them, the earth will open up and they will be cast under the earth." Chot's end came in 1905 when he was captured, terminating the Siberian Ghost Dance.

## THE CARGO CULT GHOST DANCE

One of the more curious versions of the Ghost Dance can be found in the Cargo Cults of the South Pacific. According to anthropologist Marvin Harris, Cargo Cults began centuries ago with Pacific islanders scanning the horizon for phantom canoes delivering goods. As the times changed so did the ersatz delivery mechanisms. In the eighteenth century they watched for the sails of sailing vessels, in the nineteenth century they searched for smoke from steamships, and in the twentieth century, particularly after World War II, they scouted for airplanes. The phantom cargo has evolved along with the delivery mechanisms. First it was matches and steel tools, then shoes, sacks of rice, canned meat, knives, rifles, ammunition, and tobacco, and finally it became automobiles, radios, and modern appliances. To the Pacific islanders who only ever saw, heard, or fantasized about the end product of a manufacturing process in some far-off land, the origin of the cargo was a great mystery. With no apparent cause, the airplane arrived with finished products. "The great wealth and power of the whites is a mystery to Melanesians," La Barre concludes. "As far as they could see, whites did no work at all and made no artifacts, and yet got great stores of goods merely by sending out bits of paper, though meanwhile blacks must labor to produce gold and copra. The cargo ships were their link to that mysterious country and the obvious secret of their power."

The Ghost Dance leitmotif intertwined with the Cargo Cults in places like New Guinea where the natives built a thatch-roofed hanger, a bamboo beacon tower, an airstrip manned twenty-four-hours a day by natives wearing simulated uniforms, and even an airplane made out of sticks and leaves. Long dominated by whites who seemed to possess the mysterious powers of the cargo, the natives envisioned the day when their ancestors would return with cargo for them. When that day comes, Harris notes, the natives believed there would be "the downfall of the wicked, justice for the poor, the end of misery and suffering, reunion with the dead, and a whole new divine kingdom."

*A wooden representation of a cargo plane sits next to a cross and a messianic John Frum, said to be a pidgin shortening of the phrase "John from America"—the carrier of cargo goods according to the Melanesian cargo cult.*

Sociologist Peter Worsley, in his classic 1958 study of Cargo Cults in Melanesia, *The Trumpet Shall Sound*, points out that such movements are "by no means peculiar to that part of the world" and that there is in fact evidence that similar phenomena have developed "in most parts of the globe, even in some of the earliest records of civilization." The reason this is important, Worsley adds, is that "the history of apocalyptic religions and of messianism is of special interest to people whose culture has included a central belief in One whom they believed to be *the* Messiah, who died for mankind and with whom they hoped to be reunited in Paradise." He is speaking of Jesus, of course, pointing out that "this messiah, however, has not been the only messiah," and that in addition to multiple Melanesian culture-heroes, spirits, and messiahs, "similar cults have occurred to a greater or lesser extent in other major regions of Oceania," and that "Africa, North and South America, China, Burma, Indonesia and Siberia have also had their share of cults, and the history of Europe provides numerous examples." Does the story of Jesus as the savior fit into this genre of messiah myths?

## JESUS AS MESSIAH MYTH

Cargo Cults are not unlike Messiah Cults, including a first-century one that arose in an eastern province of the Mediterranean surrounding a man who was said to have the power to heal the sick, raise the dead, and preached love, forgiveness, and the worship of the one true God. Roman authorities, fearing that his expanding fellowship might pose a social or political threat, arrested him and had him put to death. Following his execution, his disciples said he rose from the dead, appeared to them to deliver a message, and then ascended to heaven. The messiah was Apollonius of Tyana, who was killed in A.D. 98, six decades after his more famous predecessor, Jesus of Nazareth. As Randel Helms notes: "Readers . . . may be forgiven their error [confusing Apollonius and Jesus] if they will reflect how readily the human imagination embroiders the careers of notable figures of the past with common mythical and fictional embellishments."

The common mythical embellishment of the Jesus legend, like that of the Ghost Dance and Cargo Cults, is the now-familiar oppression–redemption or messiah myth. By the first century A.D. the Jews were engulfed within the Roman empire and feared for their very existence as a people. The regions around and including Nazareth were ruled by King Herod, who endured and responded with violence to numerous Jewish revolts against the Romans. By A.D. 6 Judea was under direct Roman rule. A youthful Jesus, who was now ten years old (probably born in 4 B.C.), must have been painfully aware of the tensions between his people and their oppressors, as well as the biblical promise of a Messiah who would drive out the Romans and reestablish the kingdom of God on Earth. By the time of his three-year ministry from A.D. 27 to 30, Jesus had codified a new theology and an ethical system to sustain his followers until the Second Coming. It was a unique theology, as theologian Burton Mack has noted in his identification of three interconnected ideas that arose in the 30s and 40s following Jesus' death:

> 1. *One was the vague notion of a perfect society conceptualized as a kingdom. The Jesus people latched onto this idea and acted as if the kingdom they imagined was a real possibility despite the Romans. They called it the kingdom of God.*
> 2. *A second idea was that any individual, no matter of what extraction, status, or innate capacity, was fit for this kingdom and could act accordingly if only one would.*
> 3. *. . . the novel notion that a mixture of people was exactly what the kingdom of God should look like.*

Mack also observes that belief in a Messiah that would redeem an oppressed people was certainly not unique to first-century Jews: "This was a notion that many groups had used to imagine a better way to live than suffering under the Romans." Peter Worsley points out that "Christianity itself, of course, as recent interpretations of the Dead Sea scrolls emphasize, originally derived its *élan* from the millenarist traditions of the Essenes and similar sectaries at the beginning of the Christian era. These people looked for the establishment of an actual earthly Kingdom of the Lord which would free the Jews from Roman oppression. Later this doctrine commended itself as a message of hope to the downtrodden of the Roman Empire."

Did Jesus and his followers think he was the Messiah? When Jesus asked his disciples "Who do men say that I am?" he was given the answer: "Thou art the Christ" (Matthew 16:15, 16; *christos*, Greek for *messias*, from *masiah*, Hebrew for *Messiah*). To many early Christians, the Hebrew Bible spoke to them of a returning Messiah: "Behold, the days come, saith the Lord, that I will raise unto David a righteous Branch, and a King shall reign and prosper, and shall execute judgment and justice in the earth." "And there shall come forth a rod out of the stem of Jesse, and a Branch shall grow out of his roots. . . . But with righteousness shall he judge the poor, and reprove with equity for the meek of the earth: and he shall smite the earth with the rod of his mouth, and with the breath of his lips shall he slay the wicked" (Isaiah 11:1, 4). "For unto us a child is born, unto us a son is given: and the government shall be upon his shoulder: and his name shall be called Wonderful, Counsellor, The mighty God, The everlasting Father, The Prince of Peace" (Isaiah 9:6). (See *The Oxford Companion to the Bible*, *The Interpreter's Bible*, and *The New Oxford Annotated Bible* for additional messianic passages and their meanings.)

Such prophecies must have been especially reassuring to a people under the yoke. Indeed, Christianity's founding father, Paul, told the Colossians: "In whom we have redemption through his blood, even the forgiveness of sins" (1:14). To the Hebrews, Paul said: "Neither by the blood of goats and calves, but by his own blood he entered at once into the holy place, having obtained eternal redemption for us" (9:12). Redemption was not only for individuals, but for all of Israel, as Luke (24:19–21) notes: "Concerning Jesus of Nazareth, which was a prophet mighty in deed and word before God and all the people. And how the chief priests and our rulers delivered him up to be condemned to death, and have crucified him. But we trusted that it had been he which should have redeemed Israel." Here we see the foundation of the Christian oppression–redemption myth.

Jesus himself made it clear that ultimate redemption would come within the lifetime of his contemporaries: "Verily I say unto you, There be some standing here, which shall not taste of death, till they see the Son of man coming in his kingdom" (Matthew 16:28). Two thousand years later over one billion people profess Jesus to be the Messiah who not only redeemed Jews from their Roman oppressors, but who will deliver us from ours. (Such is the power of belief systems to rationalize all discrepancies—one being that *this* is the Kingdom of God on Earth.) A particularly striking example comes from the *Los Angeles Times* on the morning of October 8, 1997. In the Sports section, no less, appeared a six-inch by ten-inch advertisement placed by a Bloomington, Indiana–based group called Christ's Soon Return, announcing:

*8 Compelling Reasons Why:*
*Christ Is Coming "Very, Very Soon"*
*How To Be Prepared for History's Greatest Event*

The evidence for the return of the Messiah, we are told, "is overwhelming. It could be any moment." Citing a single "scholar" (who goes unnamed) the advertisement explains that there are no less than

167 "converging clues" that the end is nigh. The ad offers eight: 1. Israel's rebirth. 2. Plummeting morality. 3. Famines, violence and wars. 4. Increasing earthquakes. 5. Explosion of travel and education. 6. Explosion of cults and the occult. 7. The New World Order. 8. Increase in both apostasy and faith. A bonus "clue" is the "Angel Factor," where "As an angel announced Christ's First Coming, there have been recently reported visits from angels saying, 'He is coming very, very soon.' " Why is Christ coming? "Christ will soon come and rescue His people from the approaching 'Great Tribulation'. He will later rule and bring peace on earth—after he judges the world and every person." To avoid our fate and "escape God's judgment, we each must receive His free gift of forgiveness and love." There is no Ghost Dance involved in the redemption (in this case the tyrants from whom we are to be rescued are ourselves, our sinful nature, and a morally moribund world), but one must pray for forgiveness and accept Christ as the Messiah: "Lord Jesus, I believe you are the Son of God and that you died on the cross for my sins to save me from eternal death. I open the door of my life and receive you as my Savior and Lord. I give you my life. Help me to be what you want me to be. Amen."

The ultimate extreme expression of such modern millennial beliefs was the Heaven's Gate calamity of March 27, 1997, when thirty-nine (now forty) members of an end-times cult committed suicide. Instead of the spirit world coming to life and the oppressors disappearing, the members of the Heaven's Gate cult (also called the "Higher Source") would escape the tyranny of their bodies and the immorality of the world by becoming spirits in the next stage of history. Marshall Applewhite and Bonnie Lu Nettles convinced their followers that they were going to TELAH—*The Evolutionary Level Above Humans.*

Just as Wovoka mixed Native American and Christian motifs into a single Ghost Dance myth, the Heaven's Gate belief system was a peculiar mixture of evolution, creationism, reincarnation, and UFOlogy (recall the "spaceship" accompanying comet Hale-Bopp). Before they could attain a spiritual state, members had to enter a gender-free stage—no sexual identity and no sex. In preparation they cropped their hair, abstained from sex, and wore androgynous clothing. According to Applewhite (from his Web page): "They are perfectly beautiful bodies—neither male nor female. They don't have hair that needs to be cut, they don't need to have curlers. They don't need to use makeup. It's a body that exists for the most part, in a nondestructive environment, except when it has to go to a place like planet Earth.

So it's potentially an eternal body—an everlasting body." Oppression on Earth, redemption in TELAH.

## WHY THE MESSIAH MYTH RETURNS

Why, it seems reasonable to ask, would these similar myths recycle through dissimilar cultures and distinct ages? Cultural diffusion may explain some thematic similarities, but a broader hypothesis is that there are a limited number of responses to perceived oppression and the general hardships of the human condition, and the belief in a returning Messiah who will deliver redemption is one of the most common. The specifics vary with varying cultures, but the general theme returns again and again. Why?

History is an exquisite blend of the specific and the general, the unique and the universal. The past is neither one damn thing after another (Heraclitus' river), nor is it the same damn thing over and over (Spengler's life cycles). Rather, it is a series of generally repeating patterns, each one of which retains a unique structure and set of circumstances. History is uniquely cyclical. Wars and battles, witch crazes and social movements, holocausts and genocides, all recycle through history with remarkable periodicity. The reason is that while there are an infinite number of combinations of specific details, there are a limited number of general rules that channel those details into similar grooves. Every historical event is unique, but not randomly so. They are all restricted by the parameters of the system. Such events recycle because the conditions of these parameters periodically come together in parallel fashion.

When social conditions include oppression of a people, there is a good chance that the response will be the belief in a rescuing messiah delivering redemption. The messiah myth, like all myths, may be a fictitious narrative, but it represents something deeply nonfictional about human nature and human history. To this extent it is an important component in answer to the question of how we believe.

Chapter 9

# The Fire That Will Cleanse

## *Millennial Meanings and the End of the World*

*We, while the stars from heaven shall fall,*
*And mountains are on mountains hurled,*
*Shall stand unmoved amidst them all,*
*And smile to see a burning world.*
—Millerite hymn, 1843

On Monday, October 27, 1997, the Dow Jones Industrial Average crashed a record 554.24 points, the biggest one-day drop in history. By the time Ted Koppel's expert guests on *Nightline* reflected on the day's disaster, eastern hemisphere markets were opening to record losses: Korea down 7 percent, Hong Kong down 16 percent, Australia down 10 percent. As Americans scrambled to place their sell orders the next day, the news from Europe was grim: Germany down 5 percent, Belgium down 8 percent, Great Britain down 8 percent. The collapse quickly spread westward: Mexico down 13 percent, Brazil down 15 percent, Venezuela down 12 percent. It looked like the end of the world.

Like all apocalyptic doomsday predictions in history, however, it was not the end. By the close of the next trading day the Dow was up 337 points and investors who bought low that morning were born again. Within five months the Dow was up over 9000, and by the spring of 1998 it summited the Everestian 10,000 with no end in sight. It would appear that reports of the world's death were greatly exaggerated.

Actually, they always are, but for those whose apocalyptic tendencies are measured by biblical benchmarks instead of stock tickers, the stakes are much higher. For example, on a brisk April 29 morning in 1980, Dr. Leland Jensen, a chiropractor and leader of a small religious sect called the Baha'i Under the Provisions of the Covenant, led his devoted followers into fallout shelters in Missoula, Montana, to await the end of the world. Within the first hour, Jensen believed, a full third of the Earth's population would be annihilated in a nuclear holocaust of fire and fallout. Over the course of the next twenty years most of the remaining population would be ravaged by conquest, war, famine, and pestilence. In the year 2000, the Baha'i Universal House of Justice would arise out of the ashes like a phoenix to help establish the thousand-year reign of God's kingdom on Earth. How did Jensen know all this? He had a revelation in the Montana State Prison, while serving a sentence for sexually molesting a fifteen-year-old patient: "I felt a presence only. It talked to me—not in a physical voice but very vividly expressing to me that I was the promised Joshua (prophesied in Zechariah 3)."

This is classic end-times imagery—an apocalyptic revelation, the demise of the world at a millennial marker, the survival of a small chosen group of true believers, the return of the Messiah and peace after massive death and destruction, and even the Four Horsemen of the Apocalypse (conquest, war, famine, and pestilence). The iconography and poetry of the apocalypse is at once beautiful and terrifying, from the horrors wrought on sinners in countless "Last Judgment" paintings (with swirling mixtures of awe-invoking black and red colors) by such masters as Albrecht Dürer, Pieter Brueghel the Elder, and Michelangelo, to the haunting vision in William Butler Yeats' "The Second Coming":

> Turning and turning in the widening gyre
> The falcon cannot hear the falconer;
> Things fall apart; the centre cannot hold;
> Mere anarchy is loosed upon the world,
> The blood-dimmed tide is loosed, and everywhere
> The ceremony of innocence is drowned;
> The best lack all conviction, while the worst
> Are full of passionate intensity.

Today's words and images are more vivid than those of a thousand years ago, with both religious and secular end-times scenarios competing for our attention. Recall The Doors' disquieting vision in their 1967 rock song "The End," visually enhanced a decade later by Francis Ford Coppola in the opening scene of *Apocalypse Now,* as army helicopters torch a Vietnam village the moment Jim Morrison lets loose with a throaty pronouncement "This is the end. . . ." Or Hal Lindsey's 1977 "rapture" scene in the film version of *The Late Great Planet Earth* (narrated by the foreboding baritone voice of Orson Wells), showing Christians snatched from their moving automobiles by a returning Christ. Or more recently in Kevin Costner's apocalyptic films *Waterworld* and *The Postman.* The iconographic theme even finds its way into editorial cartoons, as in a 1996 *New Yorker* contribution featuring a doomsayer with his placard: "THE END IS NEAR http://www.endnear.com."

What are millennial phenomena, how did they develop in the year 1000 and how will they play themselves out in the year 2000, and why do we find them so compelling? Even those who have no particular bent toward ecclesiastical millennialism may find in its secular twin reason enough for legitimate concern. Is the end near? If not, what does the fear of it tell us about ourselves?

## WHAT IS THE MILLENNIUM?

The millennium (literally in Latin *mille* thousand, *annus* year) is a thousand-year block of time that commands our attention because we like to hew the world into tidy categories. Given the average human life span of less than a century, triple-zero increments in a chronology especially stand out. In A.D. 248, for example, Romans celebrated the thousandth anniversary of the mythical founding of the empire by Romulus and Remus with considerable pageantry and fanfare, singing hymns to Apollo and Diana, dancing in the streets and in the hills, and looking forward to being freed from the Barbarian peril. But the millennium takes on deep meaning, and becomes a phenomenon, when associated with the apocalypse (from the Greek *apo* un, *kalypsis* veiling, or revelation)—the catastrophic destruction of evil forces in the world followed by the resurrection of the righteous. And this, in turn, is a form of the larger genre of eschatology, or the study of final events and the end of history.

There are two types of apocalyptic scenarios: religious, where God destroys Satan and sinners and resurrects the virtuous; and secular,

*Three traditional representations of Judgment Day. On the top of the facing page: Albrecht Dürer's 1498* The Opening of the Fifth and Sixth Seals, the Distribution of White Garments Among the Martyrs and the Fall of Stars, *is a depiction of St. John the Divine's vision in Revelation, 6:12–17: "And I beheld when he had opened the sixth seal, and, lo, there was a great earthquake; and the sun became black as sackcloth of hair, and the moon became as blood; And the stars of heaven fell unto the earth, even as a fig tree casteth her untimely figs, when she is shaken of a mighty wind. And the heaven departed as a scroll when it is rolled together; and every mountain and island were moved out of their places. And the kings of the earth, and the great men, and the rich men, and the chief captains, and the mighty men, and every bondman, and every free man, hid themselves in the dens and in the rocks of the mountains; And said to the mountains and rocks, Fall on us, and hide us from the face of him that sitteth on the throne, and from the wrath of the Lamb; For the great day of his wrath is come; and who shall be able to stand." On the bottom of the facing page:* The Last Judgment *by Pieter Brueghel the Elder. Below: Michelangelo's* Last Judgment.

where the destruction of evil comes about by natural or historical forces, and good triumphs over evil. Either way it plays out the same: *destruction followed by redemption,* with the fatalistic twist that The End, or some major break in human history, is inevitable. This apocalyptic millennium is a variation on the *destruction-redemption* and *messiah* myths considered in the previous chapter, as well as the still broader categories of *renewal* and *eschatology* myths reviewed in Chapter 7. These stories of the end resonate deeply in the human breast for the simple reason that we are all aware of the passing of time, that we are locked into that chronology, and that the end of our personal time must come. This is true whether one is religious or not: Everyone who ever lived has died, and so will we, and so will our descendants. Even if all the religious end-times scenarios prove hollow, the Earth itself will be engulfed by the Red Giant the sun will become in another 4.5 billion years. Even if our descendants colonize the galaxy, or other galaxies, the universe will either collapse into a giant black hole, destroying everything in it, or continue expanding until every star in every galaxy runs out of nuclear fuel and is snuffed out like the candles at the end of a liturgical ceremony. Either way, the end is coming. It is literally only a matter of time.

The connection between the millennium and the apocalypse was made most poignantly by St. John the Divine in a vision recounted in Revelation 20:1–10, where he "saw an angel coming down from heaven, holding in his hand the key of the bottomless pit," where he threw Satan "and bound him for a thousand years." According to Revelation, this post-Armageddon event is to be followed by the judgment of sinners and resurrection of the saved, who then "reigned with Christ a thousand years." After this millennium, "Satan will be loosed from his prison and will come out to deceive the nations," join forces with Gog and Magog in one final epic battle—Armageddon—but will be defeated and "thrown into the lake of fire" where he and all of his servants, including all false prophets, "will be tormented day and night for ever and ever."

To some doomsday prophets, this fatidic vision will be played out at the end of human history. But how will we know when we are nearing it? Jesus told his disciples (Luke 21:10–11): "Nation will rise against nation, and kingdom against kingdom; there will be great earthquakes, and in various places famines and pestilences; and there will be terrors and great signs from heaven." Herein lies the problem of interpretation, and one reason why millennial phenomena are so widespread. Since nations and kingdoms have always risen against one another; and great earthquakes, famines, and pestilences are common throughout history; and heavenly signs like comets and eclipses abound in every age, whoever is doing the interpreting sees *themselves* as the chosen generation. For some, the end is always nigh, from Jesus' first-century disciples who took him literally when he said "there shall be some standing here, which shall not taste of death, till they see the Son of man coming in his kingdom" (Matthew 16:28), to Carlulaire de Saint-Jouin-de-Marnes' warning in 964 that "as the century passes, the end of the world approaches," to Ronald Reagan's 1971 admonition that "for the first time ever, everything is in place for the battle of Armageddon and the Second Coming of Christ." Whether it is Satan who brings about the final clash that leads to the terminus of history, or human stupidity through nuclear war, or a chance encounter with an asteroid, the mythic theme of the apocalypse has become a staple of both popular and high culture.

To complicate matters, there are a number of widely divergent beliefs concerning the end-times, especially within Christianity. *Premillennial* Christians, for example, believe that Christ must first return to usher in the millennium. *Postmillennial* Christians, however, believe that Christ will return after humans have already set up God's kingdom on earth. Of course, some individuals have chosen a more moderate middle ground by worrying very little about the precise timing of eschatological events, concentrating simply on the foundational hope of their faith, that is, that *someday* (who knows when?) Jesus will return to judge the living and the dead.

For secular millennialists a similar typology can be constructed. *Premillennial* secularists tend to be pessimistic in their view of humanity and history, where change can only come about after a catastrophe. *Postmillennial* secularists tend to be optimistic and try to work toward a better world *before* disaster strikes. Folklorist Daniel Wojcik, in his compelling history of *The End of the World as We Know It*, makes a similar distinction between *unconditional apocalypticism*, where the end of the world is "imminent and unalterable" and "irredeemable by human effort," and *conditional apocalypticism*, where "within the broad constraints of history's inevitable progression, human beings may forestall worldly catastrophes if they act in accordance with divine will or a superhuman plan." Hal Lindsey offered a prime example of unconditional apocalypticism in *There's a New World Coming*, when he addressed skeptics directly: "To the skeptic who says that Christ is not coming soon, I would ask him to put the book of Revelation in one hand, and the daily newspaper in the other, and then sincerely ask God to show him where we are on His prophetic time-clock." The Montana-based Church Universal and

Triumphant, headed by Elizabeth Claire Prophet, is an example of conditional apocalypticism, where they prepare for the worst by stockpiling foodstuffs and constructing bomb shelters, but pray for the best (so far so good).

Calculating *precisely* when the end will come has generated a mini-publishing industry at the end of this millennium, but it has a long and honored history. Numerous thinkers over the last 2,000 years—from Church Fathers of early Christendom, to theologians of the Middle Ages, to philosophers of Early Modern Europe—concluded that the universe would end exactly 6,000 years after its six-day creation. They based this conclusion on the passage in II Peter 3:8, where "one day is with the Lord as a thousand years." Since God rested on the seventh day, the seventh millennium would begin following the battle of Armageddon. It was a cosmic millennial week.

Jewish tradition, for example, held that there were 2,000 years before the Law (Torah), 2,000 years under the Law, and 2,000 years under the Messiah. The fifth-century Church Father Augustine, in his theological epic, *The City of God,* outlined the six ages of history (plus one still to come) that included: (1) Adam to the Flood, (2) the Flood to Abraham, (3) Abraham to David, (4) David to the Exile, (5) the Exile to Jesus, (6) the (present) Gospel Age, and (7) the (final) Millennium. In a 1525 sermon, the Protestant revolutionary Martin Luther preached that the end would come exactly 6,000 years after Creation, which he dated at 3961 B.C. (pushing the end off until A.D. 2039, leaving plenty of time for his reforms to take effect). Even Christopher Columbus, in his unfinished *Book of Prophecies,* saw himself "predestined to fulfill a number of prophecies in preparation for the coming of the Antichrist and the end of the world." It is entirely possible, in fact, that one of the major motivations for Columbus's voyages of exploration was to fulfill his perceived destiny. The world, he calculated, began in 5343 B.C. and would last 7,000 years (including the final millennium), making the end only a century and a half away, just enough time to save the souls of the newly discovered godless savages of the Indies.

The various dates for the creation computed by countless observers in this scholarly tradition were derived by calculating the ages of the patriarchs, kings, and other biblical peoples, and generally fell in the range between 5500 B.C. (in the Septuagint) and 3761 B.C. (in the still-used Jewish calendar). In 1650, Archbishop James Ussher, in his *Annals of the Old Testament, Deduced from the First Origin of the World,* produced the most comprehensive creation history of his time. In addition to applying Old Testament genealogies, Ussher used astronomical and secular historical data on other societies (especially Babylon), Roman history and the New Testament, and calibrated Hebrew chronology with the Christian calendar. The 6,000-year history of the Earth, he figured, would end precisely 2,000 years after Christ's birth. But when was that? The B.C.–A.D. chronological system was not introduced until the sixth century A.D. According to biblical accounts Jesus was born during the reign of King Herod (recall, Herod talked to the Magi and ordered the slaying of the innocents). Since Herod died in 4 B.C., this is the date often used by theologians for the latest possible birth date of Jesus. Steeped in tradition, Ussher then assumed the creation would have occurred on the first Sunday following the autumnal equinox, which under the old Julian (Roman) calendar would have been Sunday, October 23, 4004 B.C. With a final confident flare, he even deduced the time: high noon! (Ussher figured that since God created the heavens and the Earth and light all on Day One, it must have taken at least half a day to accomplish the feat, so, he concluded, "In the middle of the first day, light was created.")

Since Ussher's book became the most widely read and frequently quoted source for the creation of the world, it was believed by many that the end would come 6,000 years later on October 23, 1996. Since you are reading this book it would appear the end has once again been postponed. But as Stephen Jay Gould points out in *Questioning the Millennium,* something of near-miraculous proportions did happen on that date. Down two games to one in the World Series against the mighty Atlanta Braves, and hopelessly behind 6 to 3 in the eighth inning of the critical fourth game (no one ever comes back from a three-to-one deficit), the New York Yankees pulled off a preternatural comeback to win the game and eventually the series. "So," Gould concludes with a twinkle in his eye, "on the eminently reasonable assumption that God is a Yankee fan (and both a kindly and inscrutable figure as well), He may have used 6000 *Annus Mundi* to send a signal and solicit our earnest preparation before He runs out of reasons for delay and must ring down the truly final curtain on earthly business as usual."

Actually, the date gets pushed off another year because, as Gould notes, there was no year zero in Western mathematics when the B.C.–A.D. system was introduced, so the first year of the first century was 1, not 0. So 6,000 years from 4004 B.C. is actually 1997, not 1996. Did anything of note happen on October 23, 1997? Yes, actually, something did. On that date the United States Energy Department released a statement disclosing that as many as 30,000 nuclear bombs cannot be accounted for in the disassembly process that is part of the latest arms control agreement. According to official records, approximately 70,000 nuclear bombs have been produced since World War II. Of these, 26,735 have been destroyed, 1,741 are awaiting destruction, and 11,000 remain active in the Pentagon's strategic stockpile. Apparently no one knows what happened to the remaining 30,000 bombs. To make matters worse, the Energy Department also admitted that of the 95.5 tons of bomb-grade plutonium produced in the United States since the Second World War, 2.8 tons of it remain unaccounted for—not the end of the world, though perhaps the necessary ingredients for it.

Finally, it should be pointed out that the missing zero at the B.C.–A.D. marker solves the problem of when the next millennium really begins. Since the first century had no zero, it began with the year 1 and did not end until the end of the year 100. This is true for every century since, including ours. Therefore, the twentieth century, and the second millennium, end on December 31, 2000, and the new century and millennium begin the next day, January 1, 2001.

## WHEN PROPHECY FAILS—A.D. 1000

Since myths tend to recur as enduring features of our cultural landscape (primarily due to the fact that there are only so many plot themes in stories, so they are bound to repeat in general outline), we should not be surprised that apocalyptic visions of the end have been reiterated throughout the past two thousand years. The most interesting year, for obvious reasons, is A.D. 1000. Did people then believe the end was nigh? Surprisingly, the matter of what happened at the last triple-zero cleavage in history is not at all clear, primarily owing to the historical black hole known as the Dark Ages. There just is not that much data from which to piece together an adequate picture. We do not know if this absence of evidence is evidence of an absence (of terror), or if the hysteria was expressed in some other fashion not clearly recorded for history. Chroniclers from the sixteenth to the nineteenth centuries, in the finest mode of progressivist history (with a little anti-Catholicism thrown in for good measure), portrayed their ancestors in the year 1000 as irrationally hysterical. The nineteenth-century French historian Jules Michelet, for example, saw evidence of the terror in statues of the period: "See how they implore, with clasped hands, that desired but dreaded moment . . . which is to redeem them from their unspeakable sorrows." In his 1841 classic work, *Extraordinary Popular Delusions and the Madness of Crowds,* Charles Mackay recorded that "during the thousandth year the number of pilgrims increased. Most of them were smitten with terror as with a plague. Every phenomenon of nature filled them with alarm. It was the opinion that thunder was the voice of God, announcing the day of judgment. Fanatic preachers kept up the flame of terror."

The only problem with this account is that there is almost no evidence for it. This apparent nonevent led to an "anti-terror" movement among historians, who pointed out that nothing special happened in 1000 because other dates, such as 909, 950, 1010, and especially 1033 (a thousand years after the crucifixion) were equally apocalyptic. Hillel Schwartz's classic survey of *fins de siécle* from the 990s to the 1990s, *Century's End*, turned up far less panic than the author anticipated. Admitting that he was "feeding on coincidence," Schwartz concluded that any sense of doom at century's ends is a figment of modern historians. Christopher Hitchens amusingly calls this PMS—Pre-Millennial Syndrome. Like its biological counterpart, PMS is cyclical and predictable, but the actual malaise is rarely as bad as what was anticipated. Carnegie-Mellon historian Peter Stearns goes so far as to declare that "the apocalyptic history of the year 1000 turns out to be one of the most successful, large-scale frauds in modern treatments of the past." Well, which is it, fear, feign, or fraud? Is there evidence that demands a verdict?

The latest scholarship on the subject, generated primarily by Richard Landes and Stephen O'Leary from the Center for Millennial Studies at Boston University, is that the dates surrounding and marking the thousand-year anniversary of Christ's nativity and crucifixion (1000 and 1033) "reflect not a variety of equally plausible dates in circulation, but a series of efforts either to speed up the millennium's arrival, to postdate it, or to salvage a coming millennium after its passage." In fact, an eclipse in 968 and the arrival of Halley's comet in 989 were seen by many as signs of the apocalypse. The European-wide famine of 1005–1006 was interpreted as fulfilling one of Jesus' admonitions to his disciples that this would be a sign of the end. In 1033 a mass pilgrimage was made to Jerusalem in preparation for the final judgment. And, especially, the "Peace of God" movements in the 990s and 1030s saw massive throngs of believers gather in open fields to venerate holy relics in the hopes of being healed before the end. So, while the year 1000 did not see mass hysteria, says Landes, "the end is not merely paralyzing terrors; it is also extravagant hope: hope to see an end to the injustice of suffering in this world, hope for a life of ease and delight, hope for the victory of truth and peace." What do we see following the years 1000 and 1033? Landes quotes medieval chronicler Radulfus Glaber's famous passage: "It was as if the whole world had shaken off the dust of the ages and covered itself in a white mantle of churches." Destruction-redemption.

Not all is lost when the end does not come. Hope springs eternal, even for apocalyptic doomsayers. From the Millerites awaiting the end on October 22, 1843 (and again a year later), to the Jehovah's Witnesses who proclaimed that the generation who saw the Great War (World War I) would not pass before the second coming, to the Heaven's Gate followers trip to The Evolutionary Level Above Humans, millennial scenarios do not just mean The End. They also mean The Beginning.

## WHEN PROPHECY FAILS—A.D. 2000

What will happen in the year 2000 and after no one knows, but we can make some predictions based on what people have done in the past when the predicted collapse does not come. It is an interesting study that teaches us a lot about human nature. It turns out we are a remarkably resilient species.

Psychologists who studied Leland Jensen and his Baha'i sect, for example, discovered that when the end of the world came and went, they did not quietly disband and go home. Psychologist Leon Festinger applied his theory of cognitive dissonance to failed prophecy, and argued that the stronger one's commitment to a failing cause, the greater the rationalizations to reduce the dissonance produced by the disappointment. Thus, paradoxically, after the 1980 debacle in the bomb shelters, not only did Jensen and his followers not abandon the cause, they ratcheted up the intensity of future predictions, making no less than twenty between 1979 and 1995! Jensen and his flock employed one or all of the following rationalizations: (1) the prophecy *was* fulfilled—spiritually; (2) the prophecy was fulfilled physically, but not as expected; (3) the date was miscalculated; (4) the date was a loose prediction, not a specific prophecy; (5) the date was a warning, not a prophecy; (6) God changed his mind in order to be merciful; (7) predictions were just a test of members' faith.

The classic case study in millennial resiliency is the 1843 Millerite fiasco, also known as the "Great Disappointment." A one-time deist and farmer from upstate New York, William Miller accepted the 6,000-year theory of creation but rejected Bishop Ussher's specific calculations for the beginning and end. Miller believed he had found errors in Ussher's chronology, concluding that the archbishop was off by 153 years. The end would not come in 1996, but in 1843. Miller published his theory in 1832 and began preaching and acquiring followers in the Boston and New York areas who were impressed that he was even able to pinpoint the date of the end. Using the Jewish year that runs from one vernal equinox to the next, Miller became "fully convinced that sometime between March 21st, 1843, and March 21st, 1844 . . . Christ will come and bring all his saints with him." But when March 21, 1844, came and went without note, a great disappointment set in among many followers. Instead of abandoning the movement, however, the true believers set to task recalculating the Second Coming. It would be the "tenth day of the seventh month of the Jewish sacred year," October 22, 1844. Miller announced at the beginning of the month that "if he does not come within 20 or 25 days, I shall feel twice the disappointment I did this spring." When the new date passed without note, one disciple announced that "our fondest hopes and expectations were blasted, and such a spirit of weeping came over us as I never experienced before. We wept and wept until the day dawned." That disciple was Hiram Edson who, after recovering from the great disappointment, concluded that Miller had misread the Book of Daniel. This was not the end, he said, but only the beginning of God's examination of the names in the Book of Life. To hasten the process, Edson explained, the sabbath should be observed on Saturday, the seventh and last day of the Jewish week, instead of Sunday, and he went on to become a leader of the Seventh-Day Adventist Church.

The Jehovah's Witnesses must hold the record for the most failed dates of doom, including 1874, 1878, 1881, 1910, 1914, 1918, 1920, 1925, and others all the way up to 1975. One of the more novel and audacious rationalizations for failed prophecy came after Armageddon's nonarrival in 1975. In a 1966 book published by the Watchtower Society, *Life Everlasting in Freedom of the Sons of God,* the Witnesses established the date of creation at 4026 B.C., declaring that "six thousand years from man's creation will end in 1975, and the seventh period of a thousand years of human history will begin in the fall of 1975." The Watchtower Society's president, Frederick Franz, at a Toronto, Ontario, rally, *blamed the members themselves.* Because Jesus had stated that no man will know the "day or the hour" of his coming, the Witnesses jinxed the Second Coming: "Do you know why nothing happened in 1975? It was because *you* expected something to happen." Undaunted, they recalibrated again, citing October 2, 1984, as doomsday. Finally, in 1996, the leaders of the church learned the Millerite lesson. In the November 1996 issue of *Awake!*, members discovered that "the generation that saw the events of 1914" would not, after all, be seeing the end of the world. Instead, this oft-quoted line was replaced by a much vaguer "is about to" clause, reducing dissonance indefinitely.

A more recent example occurred on March 31, 1998. This time around it was the prophecy of one Heng-ming Chen, leader of God's Salvation Church presently based in Garland, Texas (a suburb of Dallas), but originating from Taiwan. Chen's original prophecy, published in his guidebook entitled *God's Descending in Clouds (Flying Saucers) on Earth to Save People*, stated: "At 10 A.M. on March 31, 1998, God shall make His appearance in the Holy Land of the Kingdom of God: 3513 Ridgedale Dr., Garland, TX 75041 U.S.A. I guarantee this on my life." What would God look like? Not surprisingly, he would look like Chen, only he would be able to walk through walls, speak numerous languages, and clone himself into as many copies as necessary to greet anyone who came into the home that day. Exactly one year later—March 31, 1999—the chosen few were to travel to a rendezvous point on the shores of Lake Michigan in Gary, Indiana, from where they would board flying saucers to take them to heaven, with a brief stopover at Mars. Chen and several followers went so far as to go to Gary, Indiana, in January 1998 to perform a "purification ceremony" involving rice, fruit, and ceramic dragons, all in the bone-chilling thirty-seven-degree waters of Lake Michigan. In 1997 they traveled through British Columbia and into Alaska, in a quest to find a six-foot-tall, twenty-eight-year old man who looked like Abraham Lincoln, but whom Chen described as the "Jesus of the West." But there was no available account of what they did on the March 31, 1999, date.

Like so many other New Age religions, God's Salvation Church grew out of a cultural milieu fascinated by UFOs. They moved to Texas from Taiwan in the summer of 1997, purchasing over thirty homes for about $500,000 in cash, and moved in about 150 of the faithful. They dressed completely in white, including white cowboy hats and white tennis shoes. Members were told by their leader that he talks to God through a diamond ring on his hand and receives divine messages through golden balls floating in the sky. Why Garland, Texas? Because, the forty-two-year-old "Teacher Chen" (as he is known) explained, it sounds like "God's Land." The other reason, Chen continued, is that in 1999 Asia will succumb to a nuclear holocaust he calls the "Great Tribulation." Proof of the coming disaster, he says, can be seen in the recent storms, fires, and economic problems experienced in Asia. Galactic goings on, not El Niño, are the cause of the severe storms in Asia and America, he explained.

As doomsday grew closer, Chen predicted that God would appear on Channel 18 at 12:01 A.M. on Wednesday, March 25, 1997. When God was a no-show, Chen recanted his prophecy and said that his prediction that God would appear in Garland was "nonsense." With that the media hype was over and the more than one hundred reporters went their separate ways, assuming that a Heaven's Gate replay was unlikely. But Chen's followers remain undaunted. Chin-Hung Chiang, for example, explained that the world actually did end, spiritually: "The world of the spiritual is invisible. It's very difficult to explain what is going on."

Even those who help bring about their own apocalyptic end—Jonestown, Heaven's Gate, and Waco come to mind—there is often a positive spin put on the ultimate outcome. Jim Jones told his flock on November 18, 1978 (as can be heard in the shocking audiotape with screams in the background): "I'm glad it's over. Hurry, hurry my children. . . . No more pain. . . . Death is a million times preferable to ten more days of this life. If you knew what was ahead of you you'd be glad to be stepping over tonight. . . . This is a revolutionary suicide. It's not a self-destructive suicide." Marshall Applewhite and his Heaven's Gate group were the epitome of this brand of apocalyptic spin doctoring. As we all saw in the videotaped final interviews with the suicidal members, they were gleefully looking forward to their passage on a UFO to The Evolutionary Level Above Humans, where there would be no gender, no need for food or sustenance, and "an eternal body—an everlasting body." Heaven's Gate is the passageway to this next level, "a transitional training ground—a proving ground for potential new members of the Kingdom of Heaven." Destruction-redemption.

The most apocalyptic guru of them all was David Koresh, who believed he was God's representative on Earth and as such he would lead his people to the promised land—but only after a great conflagration. As cult expert Richard Abanes concluded from his study of the Branch Davidians: "Koresh would loose fire upon his faithful followers, thereby killing off their old nature and transforming them into flaming entities of divine judgment who would smite the enemy." Whether they set the fire themselves, or the FBI triggered the blaze, or some combination of bureaucratic bumbling and self-fulfilling prophecy caused it, we may never know. But in the Bible of one of Koresh's many wives, Abanes discovered highlighted passages indicating that the Branch Davidians believed the end would come by fire (recall, they named their compound Mt. Carmel). In the margin of the book of Amos 1:2–7, which states "The Lord will roar from Zion . . . and the top of Carmel shall wither . . . I will send fire," Koresh's wife scribbled "The fire that will cleanse."

In this phrase—*the fire that will cleanse*—is the essence of the millennium. But what is the lure?

## THE LURE OF THE MILLENNIUM

Most of us would never be taken in by the likes of Jones, Applewhite, or Koresh, but the attraction of the millennium is not restricted to a handful of religious fanatics and survivalists holed up in compounds and bomb shelters in rural America. In fact, a *U.S. News and World Report* poll conducted November 14 to 16, 1997, found that "66 percent of Americans, including a third of those who admit they never attend church, say they believe that Jesus Christ will return to Earth some day—an increase from the 61 percent who expressed belief in the Second Coming three years ago." Linking the Second Coming to the millennium, an April 1993 poll conducted by Yankelovich Partners for *Time/CNN* found that 20 percent of the respondents answered "yes" to the question, "Do you think that the second coming of Jesus Christ will occur sometime around the year 2000?" One in five is not a trivial figure. And this is only those who hold to a *religious* millennialism. There is a new brand of secular millennial conceptions that envision the end of the world coming by global warming or nuclear war, by genetically engineered viruses or chemical bombs, by overpopulation or mass starvation, or by cosmic collisions or alien encounters. The Four Horsemen of the Apocalypse ride again. Why? What makes the story of the millennium, the apocalypse, the end, so compelling? What is the appeal of these chiliastic movements? To begin to construct an answer, let's step back and look at the larger picture of: (1) humans as pattern-seeking animals; (2) humans as storytelling animals.

**1.**

We evolved to be pattern-seekers—we are the descendants of the most successful pattern finders. But as we have seen, this does not guarantee we will not make errors in our thinking. In fact, it guarantees that we *will* make errors in our thinking, because magical thinking is a spandrel of clear thinking. Recall the two types of thinking errors: *Type 1 Error: Believing a falsehood* and *Type 2 Error: Rejecting a truth.* The belief that the millennium harbors an apocalyptic end of humanity is a Type 1 Error in thinking. It is an error because it is more likely that the apocalyptic images in Revelation were not portents of things to come in the distant future, but commentary on their own times. The Antichrist figure, for example, is believed to allude to the Roman emperor Nero, who killed himself by falling on his sword. The battle of Armageddon described in Ezekiel probably refers to the Scythian invasion of Israel in pre-Christian times. The Bible was written for the people of that time as history, social commentary, and political analysis, not unlike what Nostradamus did in his quatrains written as social and political exegesis on the sixteenth century, rather than prophecy for the twentieth.

2.

We tell stories about the patterns we find in nature. For thousands of years before the advent of writing, myths and religions were sustained by the oral tradition of stories with meaningful patterns—gods and God, supernatural beings and mystical forces, and our place in history, in the world, and in the cosmos. We may live in an enlightenment culture of science and rationality, but we hold on to and cherish our stories nonetheless because it is in our nature to do so. Science does not come naturally. Storytelling does. And our most popular stories are in the forms of myths—stories about life and death, growing up and growing old, rites of passage and marriage, and especially the creation and destruction of the world.

The millennium combines the best in pattern-seeking and storytelling. What could be more dramatic than the pattern of a round number ending in three zeros with a story about the end of time and our redemption to follow? Damian Thompson expressed this nicely in his work on *The End of Time:* "It seems to represent a deep-seated human urge to escape from time, which in the earliest societies was usually met by dreams of a return to a golden past. Apocalypticism offered a radical change of direction, a move *forward* into a world ruled by the saints in which the enemy had been vanquished." Even for those with no particular religious inclinations toward a millennial holocaust, there are plenty of secular versions to go around, starting with what I call the beautiful people myth.

## HEAVEN ON EARTH

Long, long ago, in a century far, far away, there lived beautiful people coexisting with nature in balanced eco-harmony, taking only what they needed, and giving back to Mother Earth what was left. Women and men lived in egalitarian accord and there were no wars and few conflicts. The people were happy, living long and prosperous lives. The men were handsome and muscular, well-coordinated in their hunting expeditions as they successfully brought home the main meals for the family. The tanned, bare-breasted women carried a child in one arm and picked nuts and berries to supplement the hunt. Children frolicked in the nearby stream, dreaming of the day when they too would grow up to fulfill their destiny as beautiful people.

But then came the evil empire—European White Males carrying the disease of imperialism, industrialism, capitalism, scientism, and the other "isms" brought about by human greed, carelessness, and short-term thinking. The environment was exploited, the rivers soiled, the air polluted, and the beautiful people were driven from their land, forced to become slaves, or simply killed.

This tragedy, however, can be reversed if we just go back to living off the land where everyone would grow just enough food for themselves and use only enough to survive. We would then all love one another, as well as our caretaker Mother Earth, just as they did long, long ago, in a century far, far away.

There are actually several myths packed into the beautiful people myth, proffered by no one in particular but compiled from many sources as mythmaking (in the literary sense) for our time. This genre of storytelling, in fact, tucks nicely into the larger framework of golden-age fantasies and has a long and honorable history. The Greeks believed they lived in the Age of Iron, but before them there was the Age of Gold. Jews and Christians, of course, both believe in the golden age before the fall in the Garden. Medieval scholars looked back longingly to the biblical days of Moses and the prophets, while Renaissance scholars pursued a rebirth of classical learning, coming around full circle to the Greeks. Even Newt Gingrich had his own version of the myth when he told the *Boston Globe* on May 20, 1995, that there were "long periods of American history where people didn't get raped, people didn't get murdered, people weren't mugged routinely."

The concept of Heaven on Earth is part of the larger mythic theme represented in Alexander Pope's *Essay on Man:* "The soul, uneasy, and confin'd from home, Rests and expatiates in a life to come." Sometimes that future state of bliss is to be found in another place entirely, as in "the firmament"—an overarching vault resting on pillars at the end of the Earth with windows to view God and the angels above, and from whence the rains come. But as often as not Heaven is a state on Earth, or sometimes even a state of mind. The most famous

Heaven on Earth metaphors, of course, come from the Bible in numerous books. In Isaiah 65:17–18, for example, following God's creation of "new heavens and a new earth," after which "the voice of weeping shall be no more heard in her, nor the voice of crying," the people are to "rejoice" (Isaiah 65:20–23, 25) because:

> *There shall be no more thence an infant of days, nor an old man that hath not filled his days: for the child shall die a hundred years old; but the sinner being a hundred years old shall be accursed.*
>
> *And they shall build houses, and inhabit them; and they shall plant vineyards, and eat the fruit of them.*
>
> *They shall not build, and another inhabit; they shall not plant, and another eat: for as the days of a tree are the days of my people, and mine elect shall long enjoy the work of their hands.*
>
> *They shall not labour in vain, nor bring forth for trouble; for they are the seed of the blessed of the Lord, and their offspring with them.*
>
> *The wolf and the lamb shall feed together, and the lion shall eat straw like the bullock: and dust shall be the serpent's meat. They shall not hurt nor destroy in all my holy mountain, saith the Lord.*

According to *The Interpreter's Bible*, in these Isaiah passages "the meaning is not that the present world will be completely destroyed and a new world created, but rather that the present world will be completely transformed . . . there is no cosmological speculation here." Indeed, in the Hebrew Bible it is not until the book of Daniel—the latest addition to the canon—that one can find reference to humans ascending to heaven. For mortals, heaven generally meant a new Kingdom on Earth, not a place to go where God resides. The shift from earthly paradise to cosmological firmament began in Daniel and was reinforced especially by Jesus who portrayed to his oppressed peoples that redemption was just around the chronological corner. Yet even Jesus made intriguing references to the Kingdom that "has come upon you" (Luke 11:20), that has suffered violence since the time of John the Baptist (Matthew 11:12), and especially in Luke 17:20–21, where he seems to infer that heaven is a state of mind: "And when he was demanded of the Pharisees, when the kingdom of God should come, he answered them and said, The kingdom of God cometh not with observation: Neither shall they say, Lo here! or, lo there! for, behold, the kingdom of God is within you."

## SECULAR HEAVENS

In a biocultural model of religious thought and spirituality, the idea of a Heaven on Earth or a Kingdom of God within should not be restricted to Judaeo-Christian theology, or even to religious traditions of the West and East. Indeed, it is not. The myth of the golden-age past or future can also be secularized, and it has been by modern environmentalists who construct mythical epochs like the one above, where beautiful people have lived or will live in eco-harmony with their environment, which resembles, for all intents and purposes, the heavenly states of world religions.

I first encountered the beautiful people myth as a graduate student in a course co-taught by an anthropologist and a historian in the late 1980s, when both fields were being "deconstructed" by literary critics and social theorists. Anticipating the study of customs, rituals, and beliefs of indigenous preindustrial peoples around the world, I was instead bogged down in books such as Michael Taussig's *The Devil and Commodity Fetishism in South America*, which explicates "Fetishism and Dialectical Deconstruction" or "The Devil and the Cosmogenesis of Capitalism." The anthropologist soon announced that his was a Marxist interpretation of history, seeing the past in terms of class conflict and economic exploitation. The beautiful people lived before capitalism.

Old-line Marxists see communism as the liberating climax of a six-stage evolutionary process that requires the collapse of capitalism. Capitalism is The End. Communism is The New Beginning. Liberal democrats, meanwhile, have their bard in Francis Fukuyama, whose book, *The End of History and the Last Man,* pronounced that the cold war was won by democracy and capitalism. Libertarians' messiah is John Galt, the hero of Ayn Rand's *Atlas Shrugged,* who leads a massive worldwide strike by the men of the mind, forcing civilization to collapse into chaos and anarchy, out of the ashes from which the heroes resurrect the Kingdom of Galt on earth. In the book's final apocalyptic scene, in fact, the heroine, Dagny Taggert, turns to Galt and pronounces: "It's the end." He corrects her: "It's the beginning." The fire that will cleanse.

Radical feminists foresee a day when patriarchy will collapse and men and women will live in egalitarian harmony—the Second Coming is actually a return to an imagined golden age before there were wars, violence, rape, slavery, and the subjugating "isms" that go with male domination. In Riane Eisler's *The Chalice and the Blade,* for example, the author goes back 13,000 years to find history's bogeyman. Before patriarchy there was "a long period of peace and prosperity when our social, technological, and cultural evolution moved upward: many thousands of years when all the basic technologies on which civilization is built were developed in societies that were not male dominant, violent, and hierarchic." As Paleolithic hunting, gathering, and fishing gave way to Neolithic farming, this "partnership model" of equality between the sexes gave way to the "dominator model," and with it came wars, exploitation, slavery, and the like. The solution, says Eisler, is to return to the *equalitarian* partnership model where "not only will material wealth be shared more equitably, but this will also be an economic order in which amassing more and more property as a means of protecting oneself from, as well as controlling others will be seen for what it is: a form of sickness or aberration."

Environmentalist, Marxist, libertarian, and feminist Armageddons are being fought with the belief that the survival of the species is at stake. Either men will lead us into nuclear obliteration, or corporations will sink our environmental lifeboat, or capitalists will spend us into oblivion, or the state will destroy us. But in the end the Antichrist will be defeated, replaced by the Kingdom of Bliss. The fire that will cleanse.

## HOLDING THE CENTER

The fact that such diverse apocalyptic visions can proliferate on the cultural landscape tells us that they are deeply rooted in the human mind. There is something going on here that cries out for an explanation. We saw in the last chapter that sometimes apocalypticism is prevalent among the oppressed, disenfranchised, or marginalized—the Jews oppressed by Romans at the time of Jesus' evangelism and promise of the Kingdom of God on Earth that was soon to come; the 1890 Ghost Dance among the Native Americans who were on the brink of extinction when the prophet Wovoka preached that the Great Spirit would come to destroy the whites and return the buffalo; or the belief among some members of the Nation of Islam (including Farrakhan himself) that a messianic mothership is in orbit around earth that will soon bring deliverance. Hold a people down long enough and learned helplessness arises, leading to feelings of utter futility, which gives rise to fatalism, and that end in apocalypticism, with a hoped-for paradisiacal state to come.

But this is not true for all millennial groups. As the *Time/CNN* poll showed, millions of white, middle class, American Christians believe that the world is soon coming to an end. It would be a long stretch to classify these folks as oppressed, disenfranchised, or marginalized. Likewise, the people who purchased over thirty million copies of Hal Lindsey's *The Late Great Planet Earth* are anything but in a state of learned helplessness or cognitive dissonance. Indeed, some recent polls and studies indicate that religious people, on average, may be both physically and psychologically happier and healthier than non-believers. Apocalypticism requires a different explanation here.

Perhaps it has something to do with the need for justice, where the evils of events like the Holocaust are lessened by the fact that the Lord will mete out retribution in the end. Bad things do happen to good people, but in the end the good will triumph (the fire that will cleanse). This is an ancient theme that even predates the Bible. In the third millennium B.C., for example, the prologue of the Hammurabi law code explains its purpose:

> to cause justice to prevail in the land,
> to destroy the wicked and the evil,
> that the strong might not oppress the weak
> to rise like the sun over the black-headed [people]
> and to light up the land.

The normal ups and downs of life may be more tolerable if you believe that Someone Up There is keeping score and that the tally will be presented to all participants when the game is over (with appropriate rewards and punishments doled out).

More than making things right with the world, millennial visions also help us make sense of the world. Recall that the literal meaning of apocalypse is "unveiling," or "revelation." For some people, a millennial apocalyptic vision, like that of St. John the Divine in the book of Revelation, unveils the secret pattern of life that *must* lie behind the confusing array of history's events. These visions reveal to us the secret pattern set up by God or destiny. Texts like Revelation reveal the hidden scheme of life, thoughtfully and purposefully set up by a God who cares about us and who, perhaps more importantly, is in control. There is a beginning and an end to history, and the events in between fit a larger cosmic design. How much easier it is to suffer the slings and arrows of fate when you know that it is all really part of a deeper, unfolding plan. We may feel like flotsam and jetsam on the vast rivers of history, but the currents are directed toward a final destination in which we play a meaningful role.

Here we see a striking difference between 1000 and 2000. A thousand years ago the world was a relatively simple place where the church was the dominant social structure that provided an inchoate but comprehensive model of the world. Today we face a confusing panoply of competing power structures and explanations virtually impossible to wrap our minds around. If we do not experience an apocalyptic terror, there will at least be some millennial angst, from both religious and secular conceptions of the end. We need restitution and restoration. We want to feel that no matter how chaotic, oppressive, or evil the world is, all will be made right in the end. The millennium as history's end is only acceptable with the proviso that there will be a new beginning. The people in 1000 were given it, and with it they created the Middle Ages. What will we do?

Will the fire cleanse? Will Yeats's anarchy be loosed upon the world and innocence drowned, or will we see ourselves through this historical fissure and arise to create the next epoch, whatever it may be? Perhaps this time the falcon will hear the falconer, the centre will hold back the blood-dimmed tide, the best, and even sometimes the worst will retain conviction. And may we all be full of passionate intensity in anticipation of our future, whatever it holds.

Chapter 10

# Glorious Contingency

## *Gould's Dangerous Idea and the Search for Meaning in an Age of Science*

*Through no fault of our own, and by dint of no cosmic plan or conscious purpose, we have become, by the grace of a glorious evolutionary accident called intelligence, the stewards of life's continuity on earth. We have not asked for that role, but we cannot abjure it. We may not be suited to it, but here we are.*

—Stephen Jay Gould, *A Glorious Accident,* 1997

In one of his final public addresses before his death, recorded live at the celebration of the fiftieth anniversary of the United Nations in the Cathedral of St. John the Divine in New York City, the astronomer Carl Sagan waxed poetic about our place in the universe and its profound implication for the relationship of science and religion:

> *One of science's alleged crimes is revealing that our favorite, most reassuring stories about our place in the universe and how we came to be are delusional. Instead, what science reveals is a universe much older and much vaster than the tidy, anthropocentric proscenium of our ancestors. We have found from modern astronomy that we live on a tiny hunk of rock and metal third from the sun, that circles a humdrum star in the obscure outskirts of an ordinary galaxy, which contains some four hundred billion other stars, which is one of about a hundred billion other galaxies that make up the universe, and according to some current views, a universe that is one among an immense number, perhaps an infinite number, of other universes. In this perspective the idea that our planet is at the center of the universe, much less that human purpose is central to the existence of the universe, is pathetic.*

In his 1977 book, *The First Three Minutes*, the physicist Steven Weinberg speculated on the human need for centrality, but he was even more direct in his assessment of where we actually fit in the cosmic scheme of things:

> *It is almost irresistible for humans to believe that we have some special relation to the universe, that human life is not just a more-or-less farcical outcome of a chain of accidents reaching back to the first three minutes, but that we were somehow built in from the beginning. It is even harder to realize that this present universe has evolved from an unspeakably unfamiliar early condition, and faces a future extinction of endless cold or intolerable heat. The more the universe seems comprehensible, the more it also seems pointless.*

Was our existence foreordained from the beginning, or are we nothing more than a "farce," a fluke product of a "chain of accidents?" Modern astronomers and physicists may be the theologians of science, but these questions date back at least to the ancient Greek historians and philosophers who, twenty-five hundred years ago, identified a central tension in the nature of change, as to what *must* be versus what *may* be—that which happens *necessarily* versus what happens *contingently*. Is our existence a *necessity*—that is, are things such that it could not have been otherwise? Or is our existence a *contingency*—something that need not have been? Must we choose between contingency and necessity? Is there not an interactive middle ground that more adequately describes the history of the universe, the world, and life? There is.

One of the most common reasons people give for believing in God is that the universe, the world, and life appears to be designed—in other words, it looks necessary, not contingent. If the universe, the world, and life were not necessary, however, it would imply that there is no designer. And without a designer there is no necessary meaning to life other than what we humans impose upon it. If life is contingent, then we might not have been: Rewind the tape of life and play it again and we would not be here. This is what makes contingency such a "dangerous" idea. Most people find the prospects of this worldview existentially devastating. In fact, contingency can be both liberating and empowering.

## IF THE TAPE WERE PLAYED TWICE

I first discovered the notion of contingency in 1987 when I entered a doctoral program in history at Claremont Graduate School. In preparation for a course in the philosophy of history, I turned to the *Syntopicon* (109 "great ideas") of the *Great Books of the Western World* and read what the great minds of history said about *fate and chance, universal and particular,* and especially *necessity and contingency.* Here were the grand and timeless debates about history and the nature of change. To my surprise and disappointment, however, not only did we not discuss what these great minds said about these great ideas, we did not even study these great ideas. Instead we explored the possibility that it was not possible to know any ideas, or understand any authors, great or not. Later I realized I was caught squarely in the middle of the postmodern, deconstructionist movement. I abandoned hope for the future of the philosophy of history.

Two years later, however, my flame of optimism was rekindled by the publication of a book that would help launch a resurgence in thinking about the nature of history. But it was not written by a historian. It was written by a paleontologist. Stephen Jay Gould's *Wonderful Life: The Burgess Shale and the Nature of History*, has become something of a watershed for those who study contingency and complexity, especially applied to organisms, societies, and history, and discussions of it can be found in many works. Walter Fontana and Leo

Buss, for example, ask in the title of their chapter "What Would Be Conserved If 'The Tape Were Played Twice'?" This is a direct reference to Gould's suggestion in *Wonderful Life* that if the tape of life were rewound to the time of the organisms found in the Canadian outcrop known as the Burgess Shale, dated to about 530 million years ago, and replayed with a few contingencies tweaked here and there, humans would most likely never have evolved. So powerful are the effects of contingency that a small change in the early stages of a sequence can produce large effects in the later stages. Edward Lorenz calls this the *butterfly effect* and by now the metaphor is well known: A butterfly flaps its wings in Brazil, producing a storm in Texas. The uncertainty of our past and unpredictability of our future created by contingency is what makes this such a challenging idea to historians and scientists, whose models and laws call for a search for unifying generalities, not capricious happenstances.

Gould's dangerous idea, therefore, did not go unnoticed. Stuart Kauffman, one of the pioneers of complexity in explaining the self-organization of complex systems, references Gould and *Wonderful Life* and asks about the Cambrian explosion of life: "Was it Darwinian chance and selection alone . . . or did principles of self-organization mingle with chance and necessity"? Mathematicians Jack Cohen and Ian Stewart published a feature story on "Chaos, Contingency, and Convergence" in *Nonlinear Science Today,* centered around *Wonderful Life. Wired* magazine's Kevin Kelly devotes several pages to Gould's contingency. Philosophers also got in on the discussion. Murdo William McRae published a critique entitled "Stephen Jay Gould and the Contingent Nature of History." And, most exhaustively, Daniel Dennett devoted a Brobdingnagian chapter to Gould and this idea in his book *Darwin's Dangerous Idea.*

Most of these authors have criticisms of Gould's theory, and some are valid. Fontana and Buss contend that plenty would be conserved if the tape were rerun again. Kauffman argues for necessitating laws of self-organization that defy contingency. Cohen and Stewart point out: "Nowhere in *Wonderful Life* does Gould give an adequate treatment of the possible existence of evolutionary mechanisms, convergences, universal constants, that might constrain the effects of contingency." Kelly has actually run Gould's thought experiment in a sandbox with contrary results: "First thing you notice as you repeat the experiment over and over again, as I have, is that the landscape formations are a very limited subset of all possible forms." McRae concludes: "Gould's argument for contingency ultimately returns to the notions of progress and predictability it set out to challenge." And Dennett calls Gould "the boy who cried wolf," a "failed revolutionary," and a "refuter of Orthodox Darwinism."

## THE MISMEASURE OF CONTINGENCY

One of the surprising things about all of these criticisms is that they appear to have missed or misunderstood the meaning of contingency and what Gould believes is its relationship to necessitating laws of nature. The reason for these misunderstandings is twofold. The first is the *problem of meaning*—contingency does not mean random, chance, or accident. The second is *the problem of emphasis*—contingency does not exclude necessity. Identifying and solving these problems can not only show us what is right about Gould's dangerous idea, but also helps us understand how to find meaning in a contingent universe.

### The Problem of Meaning

Many of those who oppose the idea of a predominantly contingent universe have misread *contingency* for *accidental* or *random.* Jack Cohen and Ian Stewart, for example, have stated explicitly that, "The survivors, who produced us, did so by contingency, by sheerest accident;" "Gould [argues] that contingency—randomness—plays a major role in the results of evolution . . .", and Gould "sees the evolution of humanity as being accidental, purely contingent." Yet Gould states quite clearly in *Wonderful Life*:

> I am not speaking of randomness, *but of the central principle of all history*—contingency. *A historical explanation does not rest on direct deductions from laws of nature, but on* an unpredictable sequence of antecedent states, *where any major change in any step of the sequence would have altered the final result. This final result is therefore dependent, or contingent, upon everything that came before—the unerasable and determining signature of history. [Emphasis added.]*

As Gould notes, contingency is *an unpredictable sequence of antecedent states,* not randomness, chanciness, or accident.

Daniel Dennett likewise takes Gould to task in a chapter entitled "Tinker to Evers to Chance," a play on words linking Gould's love of baseball—the three names represent the most famous double-play combination in baseball history—to chance, which Dennett identifies with contingency. But contingency does not mean chance, nor does it mean random, despite Dennett's conclusion: "The fact that the Burgess fauna were decimated in a mass extinction is in any case less important to Gould than another conclusion he wants to draw about their fate: their decimation, he claims, was *random.*" True, mass extinctions may *seem* random, as when an asteroid hits the Earth. But by contingency Gould means a conjuncture of preceding states that determine subsequent outcomes. Just as astronomers knew exactly when and where Comet Shoemaker-Levy 9 was going to strike Jupiter in July of 1995 (and nailed the timing and location precisely), astronomers from (say) Mars, observing Earth 65 million years ago could have calculated the collision with the Yucatán peninsula with pinpoint accuracy. But the effects of those impacts could not have been adequately computed (and in the case of the Jupiter hit were not), because of the number of contingencies involved.

The eventual rise of *Homo sapiens,* is even more contingent with *millions* of antecedent states in our past. Each event in the sequence has a cause, and thus is determined, but the eventual outcome is unpredictable because of contingency, not randomness or chance. The Burgess extinction may have been determined, but the sequence of events leading up to it, and those following, all the way to humans, were contingent. On this point Dennett says he is confused about what Gould means by "we" when he says we would not be here again if we reran the tape:

> *There is a sliding scale on which Gould neglects to locate his claim about rewinding the tape. If by "us" he meant something very particular—Steve Gould and Dan Dennett, let's say—then we wouldn't need the hypothesis of mass extinction to persuade us how lucky we are to be alive. . . . If, at the other extreme, by "us" Gould meant something very general, such as "air-breathing, land-inhabiting vertebrates," he would probably be wrong.*

Dennett's confusion seems, well, confusing. By "we" Gould means the species *Homo sapiens,* no more, no less, and he has stated so on numerous occasions, including in *Wonderful Life*: "Replay the tape a million times from a Burgess beginning, and I doubt that anything like *Homo sapiens* would ever evolve again."

One might claim that these misunderstandings are caused by the fact that Gould has not offered a formal definition of contingency. That is true, so one must read him broad and deep. But it is there in dozens of examples and several informal definitions. In his essay "The Panda's Thumb," Gould shows that the thumb—actually the radial sesamoid bone of the panda's wrist—is not a predictable design of nature's necessitating laws of form, but an improvised contraption constructed from the history of what came before. In "The Panda's Thumb of Technology," Gould argues that the evolution of the QWERTY typewriter keyboard (denoting the first six letters from the left on the top letter row) supports his theory of contingency: "To understand the survival (and domination to this day) of drastically sub-optimal QWERTY, we must recognize two other commonplaces of history, as applicable to life in geological time as to technology over decades—contingency and incumbency." He then defines contingency as "the chancy result of a long string of unpredictable antecedents, rather than as a necessary outcome of nature's laws. Such contingent events often depend crucially upon choices from a distant past that seemed tiny and trivial at the time. Minor perturbations early in the game can nudge a process into a new pathway, with cascading consequences that produce an outcome vastly different from any alternative."

This process is sometimes called *path dependency,* where systems get slotted into channels, and the QWERTY example is illuminating. Regular users of computers are locked by history into the QWERTY keyboard, designed for nineteenth-century typewriters whose key striking mechanisms were too slow for human finger speed. Even though more than 70 percent of English words can be produced with the letters DHIATENSOR, a quick glance at the keyboard will show that most of these letters are not in a strong striking position (home row struck by the strong first two fingers of each hand). All the vowels in QWERTY, in fact, are removed from the strongest striking positions, leaving only 32 percent of the typing on the home row. Only about 100 words can be typed exclusively on the home row, while the weaker left hand is required to type over 3,000 different words alone not using the right hand at all. Another check of the keyboard reveals the alphabetic sequence (minus the vowels) DFGHJKL. It appears that the original key arrangement was just a straight alphabetical sequence, which made sense in early experiments before testing was done to determine a faster alignment. The vowels were removed to slow the typist down, to prevent key jamming. This problem was eventually remedied, but by then QWERTY was so entrenched in the system (through manuals, teaching techniques, and other social necessities) that it became virtually impossible to change. Unless the major typewriter and computer companies, along with typing schools, teachers and publishers of typewriter manuals, and a majority of typists all decide to change simultaneously, we are stuck with the QWERTY system indefinitely.

Gould's biological version of this process is what he calls the *Panda Principle:* "The complex and curious pathways of history guarantee that most organisms and ecosystems cannot be designed optimally." Extending this principle to technology we might call it the *QWERTY Principle:* "Historical events that come together in an unplanned way create inevitable historical outcomes."

## The Problem of Emphasis

In the philosophy of history journal *Clio,* Murdo William McRae writes: "In spite of all his dedication to contingency and its attendant questioning of progress and predictability, Gould equivocates often enough to cast doubt upon the depth of his revolutionary convictions. . . . At times he insists that altering any antecedent event, no matter how supposedly insignificant, diverts the course of history; at other times he suggests that such antecedents must be significant ones." The reason for the apparent "equivocation" is that Gould knows contingency *interacts* with necessity, but in his writings he sometimes emphasizes the former over the latter to make a particular point. Again, Gould does not offer a formal definition of necessity, yet it is there in his writings. After he first defined what he meant by contingency, in 1987, he immediately noted that "incumbency also reinforces the stability of a pathway once the little quirks of early flexibility push a sequence into a firm channel. Stasis is the norm for complex systems; change, when it happens at all, is usually rapid and episodic." And in *Wonderful Life* Gould asks and answers the question of emphasis:

> *Am I really arguing that nothing about life's history could be predicted, or might follow directly from general laws of nature? Of course not; the question that we face is one of scale, or level of focus. Life exhibits a structure obedient to physical principles. We do not live amidst a chaos of historical circumstance unaffected by anything accessible to the "scientific method" as traditionally conceived. I suspect that the origin of life on earth was virtually inevitable, given the chemical composition of early oceans and atmospheres, and the physical principles of self-organizing systems.*

Daniel Dennett goes much farther, accusing Gould of attempting to refute the quintessential driving mechanism of evolution itself, natural selection: "Can it be that Gould thinks his thesis of radical contingency would refute the core Darwinian idea that evolution is an algorithmic process? That is my tentative conclusion." It is hard to imagine how Dennett came up with this notion since it is not to be found in Gould's writings. The problem, it would seem, stems from the fact that when one wants to *emphasize* a previously neglected facet of nature, it might appear that something is being displaced. I asked Gould about Dennett's charge and he responded as follows:

> *My argument in* Wonderful Life *is that there is a domain of law and a domain of contingency, and our struggle is to find the line between them. The reason why the domain of contingency is so vast, and much vaster than most people thought, is not because there isn't a lawlike domain. It is because we are primarily interested in ourselves and we have posited various universal laws of nature. It is because . . . we want to see ourselves as results of lawlike predictability and sensible products of the universe in that sense.*

To distance his pure Darwinism from Gould's contingently modified version, Dennett makes an intriguing distinction between two types of metaphorical building devices: *skyhooks,* or "miraculous lifters, unsupported and insupportable," and *cranes,* "no less excellent as lifters, and they have the decided advantage of being real." Skyhooks are for wishful-thinking whimps who can't handle the cold, hard reality of natural selection's crane: "A *skyhook* is a 'mind-first' force or power or process, an exception to the principle that all design, and apparent design, is ultimately the result of mindless, motiveless mechanicity. A *crane*, in contrast, is a subprocess or special feature of a design process that can be demonstrated to permit the local speeding up of the basic, slow process of natural selection, and that can be demonstrated to be itself the predictable (or retrospectively explicable) product of the basic process." Dennett accuses Gould of trying to sneak in a skyhook while he and his brave brethren—the unalloyed Darwinians—face the crane maker with brutal honesty. In fact, Dennett spends no less than fifty typeset pages trying to convince his readers that Gould is a skyhooker. Me thinks the gentleman doth protest too much. In my opinion, Dennett, and some others who adhere to a strict Darwinian adaptationist program, may be trying to find in nature a nonexisting pattern that shows us—*Homo sapiens*—as the nearly inevitable result of evolution. Dennett's crane of relentless natural selection is, for him, a skyhook—"a 'mind-first' force or power or process" that, run over and over, would produce us again and again. It is something akin to an evolutionary theology, a secular cosmogony that finds us as the pinnacle of progressive cerebral evolution.

## CONTINGENT-NECESSITY

The issue of contingency and necessity remains one of the great issues of our time because it touches on such deeply meaningful issues as free will and determinism, fate and destiny, and our place in the cosmos and in history. No one captured this better than Karl Marx, who opened the second paragraph of *The Eighteenth Brumaire* with these now classic lines: "Men make their own history, but they do not make it just as they please; they do not make it under circumstances chosen by themselves, but under circumstances directly found, given and transmitted from the past. The tradition of all the dead generations weighs like a nightmare on the brain of the living."

For the next century historians sought out those transmitted circumstances in the form of historical "laws," culminating in 1942 with the publication of Carl Hempel's influential paper entitled "The Function of General Laws in History," in which he concluded: "There is no difference between history and the natural sciences: both can give an account of their subject matter only in terms of general concepts, and history can 'grasp the unique individuality' of its objects of study no more and no less than can physics or chemistry." Hempel was wrong about general laws, but right about history and the natural sciences; not, however, in the direction one might think. History is not governed by Hempel's laws (which he describes as "universal conditional forms"), but neither are the physical and biological worlds to the extent we have been led to believe. Scientists are coming to realize that the Newtonian clockwork universe is filled with contingencies, catastrophes, and chaos, making precise predictions of all but the simplest physical systems virtually impossible. As noted, we could predict precisely when and where Comet Shoemaker-Levy 9 would hit Jupiter, but we could muster at best only a wild guess as to the effects of the impacts on the Jovian world. The guess was completely wrong. Why? Contingency.

There is irony in Hempel's quest for general laws in history. For decades historians chased scientists in quest of universal laws, but gave up and returned to narratives filled with capricious, contingent, and unpredictable elements that make up the past. Meanwhile, a handful of scientists, instead of chasing the elusive universal form, began to write the equivalent of scientific narratives of systems' histories, integrating historical contingencies with nature's necessities, as Gould observes: "This essential tension between the influence of individuals and the power of predictable forces has been well appreciated by historians, but remains foreign to the thoughts and procedures of most scientists." Indeed, contingency is not Gould's idea at all. Twenty-five hundred years ago Aristotle explained "that which cannot be otherwise is necessarily as it is," yet "an event might just as easily not happen as happen," and this is contingency. In a sense, science has been one long struggle to tame the contingent beast by finding necessitating laws that govern nature. Contingency becomes dangerous in Gould's hands because he is a scientist, demonstrating how even a subject as predictable and subservient to natural law as planets and their moons, when examined closely, reveal so much uniqueness and individuality that while "we anticipated greater regularity . . . the surfaces of planets and moons cannot be predicted from a few general rules. To understand planetary surfaces, we must learn the particular history of each body as an individual object—the story of its collisions and catastrophes, more than its steady accumulations; in other words, its unpredictable single jolts more than its daily operations under nature's laws." Simply put, history matters.

Historians and philosophers have been cognizant for millennia of this basic tension between what may not be at all and what cannot be otherwise, between the particular and the universal, between history and nature, between contingency and necessity. But such synonyms can only take us so far (and may lead to problems of meaning and emphasis). Precise definitions are needed to formulate a model of change. Thus in this analysis *contingency* will be taken to mean *a conjuncture of events occurring without design,* and *necessity* to mean *constraining circumstances compelling a certain course of action.* Contingencies are the sometimes small, apparently insignificant, and usually unexpected events of life—the kingdom hangs in the balance awaiting the horseshoe nail. Necessities are the large and powerful laws of nature and trends of history—once the kingdom has collapsed, the arrival of 100,000 horseshoe nails will not help a bit. Leaving either contingency or necessity out of the formula, however, is to ignore an important component in the development of historical sequences. The past is constructed by both components, and therefore it might be useful to combine the two into one term that expresses this interrelationship—*contingent-necessity*—taken to mean *a conjuncture of events compelling a certain course of action by constraining prior conditions.*

Contingency and necessity, long seen to be opposites on a continuum, are not mutually exclusive models of nature from which we must choose. Rather, they are descriptions of change that vary in the amount of their influence in the historical sequence. No one denies that such forces as politics, economics, religion, demographics, and geography impact individuals falling within their purview. Contingencies, however, exercise power sometimes in spite of these forces. At the same time they reshape new and future paths to be taken—think of cassette tapes winning out over eight-tracks, or VHS tapes defeating Beta. It is not that the victor is absolutely superior (and that the invisible hand of the free market always selects the best product), but that quirky events may give one a market edge over the other, and once we start down that path it may be difficult to leap the ever-deepening trough—the *QWERTY Principle* in action.

There is in this system a rich matrix of interactions among contingencies and necessities, varying over time, in what I call the *model of contingent-necessity,* which states: *In the development of any historical sequence the role of contingencies in the construction of necessities is accentuated in the early stages and attenuated in the later.*

There are corollaries that encompass six aspects of the model, including:

> *Corollary 1: The earlier in the development of any historical sequence, the more chaotic the actions of the individual elements of that sequence are; and the less predictable are future actions and necessities.* In other words, chaos reigns early, making long-term prediction all but impossible—think of the initial stages in the development of a storm and how poor meteorologists are at predicting when, where, and how strong the weather pattern will be.
>
> *Corollary 2: The later in the development of any historical sequence, the more ordered the actions of the individual elements of that sequence are; and the more predictable are future actions and necessities.* In other words, order reigns late, increasing predictive power—think of the late stages in a weather system and how accurate meteorologists are at pinpointing when, where, and how strong the storm will be.
>
> *Corollary 3: The actions of the individual elements of any historical sequence are generally postdictable but not specifically predictable, as regulated by Corollaries 1 and 2.* In other words, in all stages in a sequence, early and late, it is much easier to look back to reconstruct how and why it unfolded as it did, but always difficult to say what is going to happen next—think of the fall of the Berlin Wall in August of 1989 and the subsequent collapse of the Soviet Union, neither of which was anticipated by even the most seasoned politicians and authoritative political scientists.
>
> *Corollary 4: Change in historical sequences from chaotic to ordered is common, gradual, followed by relative stasis, and tends to occur at points where poorly established necessities give way to*

*dominant ones, so that a contingency will have little effect in altering the direction of the sequence.* In other words, historical pathways are cut gradually and deeply, stabilizing the system so that order dominates over chaos—think of how most countries usually are stable, secure, and resist change of all sorts.

*Corollary 5: Change in historical sequences from ordered to chaotic is rare, sudden, followed by relative nonstasis, and tends to occur at points where previously well-established necessities have been challenged by others so that a contingency may push the sequence in one direction or the other.* In other words, when historical pathways change, they do so quickly and only under conditions where the system becomes unbalanced—think of the sociopolitical conditions of August 1914, when the assassination of the Austrian Archduke Franz Ferdinand triggered the outbreak of World War I.

*Corollary 6: Between origin and bifurcation, sequences self-organize through the interaction of contingencies and necessities in a feedback loop driven by the rate of information exchange.* In other words, the hewing of a historical channel is driven by a feedback mechanism between the forces within the system and the forces without—think of mass hysterias and witch hunts that feed on themselves, with the exchange of information among accusers, informants, victims, and bystanders driving the system faster and deeper until it collapses.

At the beginning of a historical sequence, actions of the individual elements (atoms, molecules, organisms, people) are chaotic, unpredictable, and have a powerful influence on the future development of that sequence. But as the sequence slowly but ineluctably evolves, and the pathways become more worn, the chaotic system self-organizes into an orderly one. The individual elements sort themselves, and are sorted into their allotted positions, as dictated by what came before, with the conjuncture of events compelling a certain course of action by constraining prior conditions—contingent-necessity.

In the language of contingent-necessity, a bifurcation, or "trigger of change," is *any stimulus that causes a shift from the dominance of necessity and order to the dominance of contingency and chaos in a historical sequence,* such as inventions, discoveries, economic and political revolutions, war, famine and disease, immigrations and emigrations, and so on. A trigger of change, however, will not cause a shift at just any point in the sequence. Corollary 5 states that it will be most effective when well-established necessities have been challenged by others so that a contingency may push the sequence in one direction or the other. This trigger point is *any point in a historical sequence where previously well-established necessities have been challenged by others so that a trigger of change (contingency) may push the sequence in one direction or the other.* Similarly, the butterfly effect, or the *trigger effect*—described in Corollaries 1 and 2—*is the cascading consequences of a contingent trigger of change in a historical sequence.* The power of the trigger depends on when in the chronological sequence it enters. The flap of the butterfly's wings in Brazil may indeed set off a tornado in Texas, but only when the system has started anew or is precariously hanging in the balance. Once the storm is well under way, the flap of a million butterfly wings would not alter the outcome for the tornado-leery Texans. The potency of the sequence grows over time.

Corollary 6 describes feedback systems whose outputs are connected to their inputs in such a manner that there is constant change in response to both, like microphone feedback in a P.A. system. The mechanism that drives the feedback loop is the rate of information exchange, as in the stock market that booms and busts in response to a flurry of buying or selling, or social movements such as witch crazes that self-organize, grow, reach a peak, and then collapse, all described by Corollaries 1 to 6.

Chaos theory and the model of contingent-necessity describe change in the same manner, as the Nobel laureate Ilya Prigogine notes when observing that in chaos the "mixture of necessity and chance constitutes the history of the system." Similarly, necessity and contingency are the shaping forces for historical sequences—humans making their own history but not just as they please. According to Prigogine, all systems, including historical ones, contain subsystems that are "fluctuating." As long as the fluctuations remain modest and constant, relative stasis in the system is the norm. If the fluctuation becomes powerful enough that it upsets the preexisting organization and balance, a major change or revolution may occur, at which point the system may become chaotic. Necessity takes a system down a certain path until it reaches a bifurcation point. At this time contingency plays an exaggerated role in nudging the system down a new path, which in time develops its own powerful necessities such that contingency is attenuated until the next bifurcation. It is the alloy of contingency and necessity that guides and controls the presence or absence of these bifurcations, and elsewhere I have provided numerous historical examples.

## GLORIOUS CONTINGENCY: A LITTLE TWIG CALLED *HOMO SAPIENS*

The model of contingent-necessity and its corollaries is a formalization of Gould's dangerous idea. In an essay entitled "Fungal Forgery," Gould applied the model to a complex insect-flower system to show how it could have evolved, but in a very unpredictable manner in its early stages: "Fungal pseudoflowers are late necessities, and they give us no reason to suppose that the complex contingent prerequisite for this sensible story—the evolution of the insect-flower system—has any similar predictability." Before turning from this fascinating particular to broader generalities about contingency, Gould offered his usual caveat: "I do not, of course, deny that the history of life includes predictable events and recurrent patterns. I do, however, suspect that most predictable aspects of life lie at too 'high' a level of generality to validate what really stirs and troubles our souls—the hope that we might ratify as a necessary event the evolutionary origin of a little twig called *Homo sapiens.*" But beyond such anthropomorphic concerns, Gould shows why necessities may not always dominate:

> *As an interesting consequence of Shermer's model, we may ask why life as a whole doesn't finally settle down to globally predictable unrolling, whatever the massive contingency of initial stages. Shermer points, correctly I think, to the importance of infrequent and highly disturbing events (such as mass extinction for faunas or punctuated equilibria for lineages) in derailing the stasis or predictable unrolling of systems otherwise stabilized. The theoretical importance of rare, and sometimes cataclysmic, events—as the preservers and reinvigorators of global contingency—may best be appreciated in the light of such historical models.*

Gould then returns to his familiar metaphor of the tape, applying the model to the entire history of life:

> *But if I could rerun the tape of life from the origin of unicellular organisms, what odds would you give me on the reevolution of this complex and contingent insect-flower system . . . ? Would we see anything like either insects or flowers in the rerun? Would terrestrial life originate at all? Would we get mobile creatures that*

*we could call animals? Fine-scale predictability only arises when you are already 99 percent of the way toward a particular result—and the establishment of this 99 percent lies firmly in the domain of unrepeatable contingency.*

The contingent evolution of insect-flower systems, however, is not what makes contingency dangerous. It is that contingent little twig called *Homo sapiens* that tasks us. We want to be special. We want our place in the cosmos to be central. We want evolution—even Godless evolution—to have been directed toward us so that we stand at the pinnacle of nature's ladder of progress. Rewind that tape of life and we want to believe that we (*Homo sapiens*) would appear again and again. Would we? Most likely not. There are simply too many contingent steps along the way, too many trigger points where the sequence could have bifurcated down some other equally plausible path. Alfred Russel Wallace, the codiscoverer of natural selection, toward the end of his life realized this in his book, *Man's Place in the Universe*: "The ultimate development of man has, therefore roughly speaking, depended on something like a million distinct modifications, each of a special type and dependent on some precedent changes in the organic and inorganic environments, or in both. The chances against such an enormously long series of definite modifications having occurred twice over . . . are almost infinite." And Wallace did not know what we know about human evolution: His "million distinct modifications" is probably off by orders of magnitude. We now know that human evolution goes back millions of years, and that is just for the lineage leading to us. What if we rewound the tape to include the evolution of all primates, or all mammals, or all life on Earth? *Trillions* of distinct modifications over the last three billion years since life began would need to proceed along similar lines to produce our little twig a second time.

Is the cosmos itself so contingent? If we rewound the tape back to the beginning of the universe would there be another Big Bang, another universe just like ours? No one knows, but if recent cosmological models pan out it would appear that there are a near infinite number of bubble universes all with slightly different laws of nature. Chances are another universe like ours would reappear, which means that galaxies like ours with stars like ours would form again and again. Recent evidence also leads us to believe that planetary formation is a commonplace event in the galaxy. It is still a little soon to be drawing any definite conclusions, but with enough stars (roughly 400 billion in our galaxy alone), chances are there will be other Earth-like planets, maybe hundreds of thousands of them, the right distance from the home star to give rise to life. It would appear that physical systems are more governed by necessity, while living systems are more governed by contingency.

But this is oversimplifying matters. The actual evolution of life on a planet is really governed by contingent-necessity, and since we cannot remove living organisms from their physical environment, these relative estimates of potential "other Earths" depend on when in the sequence the tape begins again. Moreover, since no one really cares about whether cockroaches would reappear, let's cut to the chase and ask whether a primate species with a big enough brain to have consciousness, symbolic language, religion, awareness of its own mortality, and a developed enough system of thought to ask this very question would evolve again. We cannot run the experiment, of course, but we do not need to, because history has done it for us. The fossil record, while still fragmented and desultory, is complete enough now to show us that over the past thirty million years we can conservatively estimate that hundreds of primate species have lived out their lives in the nooks and crannies of rain forests around the world; over the past ten million years dozens of great ape species have forged specialized niches on the planet; and over the last six million years, since the hominid split from such great apes as gorillas, chimps, and orangutans occurred, dozens of bipedal, tool-using hominid species have struggled for survival.

If these hominids were so necessitated by the laws of evolutionary progress, why is it that only a handful of those myriad pongids and hominids have survived? If braininess is such an inevitable product of necessitating trends of nature, then why has only *one* hominid species managed to survive long enough to ask the question? What happened to those big-brained hominids *Homo habilis, Homo rudolfensis, Homo ergaster, Homo erectus, Homo heidelbergensis,* and *Homo neanderthalensis?* If big brains are so great, why did all but one of their owners go extinct (including the Neanderthals, whose brains were slightly larger than our own)? And before them, what happened to the bipedal, tool-using *Australopithecines: anamensis, afarensis, africanus, aethiopicus, robustus, boisei,* and, most recently, *garhi?* Discovery after discovery coming out of Africa reveals our ancestors to be puny, small-brained creatures walking upright, using tools, and eating meat, allegedly the ingredients that go into making big brains. If necessitating evolutionary progress were so potent, then why aren't there a dozen modern humanlike species that should have arisen out of these *Australopithecine* ancestors? Historical experiment after experiment reveals the same answer: We are a fluke of nature, a quirk of evolution, a *glorious contingency.*

## THE FULL IMPACT OF CONTINGENCY

It is not surprising that the idea of glorious contingency does not have a wide following among the religious. But what is unexpected is that many scientists still cling to a more sophisticated notion of progress as "trends," where humans—or sentience, cognition, big brains, or some other form of advanced mentation—sit atop the phylogenetic bush because evolution "moves" in this direction. In more extreme versions, such as in Freeman Dyson's *Infinite in All Directions* or Frank Tipler's *The Physics of Immortality,* it seems as if the universe "knew" we were coming, as argued in the strong anthropic principle. Even more modest progressivists manage to find a special place for humans on an evolutionary pedestal. Evolution does not "know" we are coming, but run that tape of life again and a species very like us would once again sit atop the heap. Philosopher of science Michael Ruse calls such evolutionism the "secular religion of progress." Surveying the writings of some of today's leading evolutionary biologists, and reading "the message between as well as on the lines," Ruse concludes: "If one came away thinking that evolution is progressive and that natural selection is the power behind the throne, one would be thinking no more than what one had been told." The full impact of contingency is that even this belief in progress is wrong. There is no evolutionary trend toward us.

As Gould shows in his 1996 book, *Full House,* these "apparent trends can be generated as by-products, or side consequences, of expansions and contractions in the amount of variation within a system, and not by anything directly moving anywhere." Gould claims that things like .400 hitting in baseball are not "things" at all, in the Platonic sense of fixed "essences." They are artifacts of trends, which disappear when the overall structure of the system changes over time. No one has hit .400 in baseball since Ted Williams did it in 1941 (for every ten times at bat he got four hits), and this unsolved mystery continues to generate arguments about why it hasn't happened since. The mystery is now solved, says Gould. It is not because players were better then (what he calls the Genesis myth: "There were giants on the earth in those days"—or as Williams himself put it: "The ball isn't dead, the hitters are, from the neck up"), or because players today have tougher schedules, night games, and cross-country travel. (Rod Carew says night games are easier on the eyes and travel by jet beats a train any day.) It is because the overall level of play—by everyone from Tony Gwynn and Eddie Murray to Backup Bob and Dugout Doug—has inexorably marched ever upward toward a hypothetical outer wall of human performance. Paradoxically, .400 hitting has disappeared because today's players are *better,* not worse. But *all of them* are better, making the *crème de la crème* stand out from the mediocre far less than before. The best players may be absolutely better (better training, equipment, diet) than players fifty years ago, but they are *relatively* worse compared to the average level of play. It was easier for Ted Williams to "hit 'em where they ain't" fifty years ago than it is for Wade Boggs today, because every position in the field is manned by players whose average level of play is much better than before. Consider these numbers: Only seven other players have hit .400 since 1900, and three of those in one year (1922). Add Williams in 1941 and the list is complete at eight, out of tens of thousands who have played. And the difference between .400 and George Brett's .390 in 1980, for example, based on his 175 hits in 449 at-bats, is *five hits!* That computes to only one hit in every thirty-two games. How many times did Brett face top relievers in late innings, or defensive alignments (based on computer analyses of his hitting style) that Williams and Cobb never faced? Surely at least once every thirty-two games. Williams's feat of 1941 would not be discussed today except for three hits (the difference between .406 and .399 in his 185 hits out of 456 at-bats). Would Williams have been deprived of one hit per fifty-four games by today's players? Most assuredly.

So what? For Gould, the disappearance of .400 hitting is just one of many examples of how systems change over time and how our bias of progress and complexity has led us to misunderstand historical change. "All of these mistaken beliefs arise out of the same analytical flaw in our reasoning—our Platonic tendency to reduce a broad spectrum to a single, pinpointed essence. This way of thinking allows us to confirm our most ingrained biases—that humans are the supreme being on this planet; that all things are inherently driven to become more complex; and that almost any subject can be expressed and understood in terms of an average." In baseball there is a bell-curve variation from worst to best players; what has happened in the past century is that while the league average has remained the same, the "spread" has shrunk as the entire system has marched closer toward that outer limit. It is this spread that matters, not the single point on it. As an example of the latter Gould relates his personal battle with abdominal mesothelioma, a rare and usually fatal form of cancer for which he was given eight months to live. That was in 1982. What happened? The "eight months" was a median that did not describe the *variation within the entire system* (the spread) which, fortunately for Gould, has a long right tail on which he is located.

As in baseball and disease prognosis, evolution can be illustrated by a bell curve of organisms from simple cells to complex mammals of today. But what else could evolution have done?, Gould asks. In the spread of life, there is a left wall of simplicity—any simpler and it would not be alive. For life to evolve it could only have gotten more complex—evolution reflects "an increase in total variation by expansion away from a lower limit, or 'left wall,' of simplest conceivable form." It's the same thing with size: "Size increase is really random evolution away from small size, not directed evolution toward large size."

Why is this idea revolutionary? Because change is a result of the whole system (the "full house") expanding, not a progressive march of an average "toward" something. Evolution is not "going" anywhere in a teleological sense. It is massively contingent, and we are but a minor twig on the richly branching bush of life. "The vaunted progress of life is really *random motion away from simple beginnings, not directed impetus toward inherently advantageous complexity.*" With that the full impact of the Darwinian revolution is felt. We are not even special in the impersonal world of materialistic evolution. Where, then, shall we turn?

## CONTINGENCY AND FREEDOM

In numerous places Dennett accuses Gould of "radical contingency," particularly with regard to its significance for human freedom: "If we can just have contingency—radical contingency—this will give the *mind* some elbow room, so it can *act*, and be *responsible* for its own destiny, instead of being the mere effect of a mindless cascade of mechanical processes! This conclusion, I suggest, is Gould's ultimate destination."

Nowhere that I know of has Gould modified contingency with "radical" (i.e., to the exclusion of necessity, or to the degree that necessity becomes irrelevant, which is what most philosophers mean by *radical* contingency). Yet I partly agree with Dennett. Whether it is Gould's ultimate destination or not, it *is* the ultimate implication of contingency. But contingency is not in contrast with the algorithm of natural selection—Dennett's "mindless cascade of mechanical processes." Contingency *interacts* with the necessitating force of natural selection. Natural selection is both constrained by contingencies and, in turn, confines them—for example, genetic mutations, chromosomal aberrations, and asteroid-triggered mass extinctions. Natural selection is also constrained by other necessitating forces such as geography, climate, and self-organizing complexity. Natural selection may be Darwin's dangerous idea, but it is not the only one. (Contingency would also seem to undermine critics' charges that Gould's Marxist beliefs have shaped his evolutionary theories: contingency not only subverts evolutionary determinism, it negates socioeconomic determinism, the very foundation of Marxist ideology, because, as Gould himself notes, "when we realize that the actual outcome did not have to be, that any alteration in any step along the way would have unleashed a cascade down a different channel, we grasp the causal power of individual events. Contingency is the affirmation of control by immediate events over destiny.")

Contingency helps us think about human meaning and freedom within a scientific perspective. Although all contingencies are caused—and thus determinism lives in the model of contingent-necessity—the number of contingent causes, and the complexity of their interactions with necessities, make the predetermination of human action essentially impossible; but because of this, the determination of human action on history becomes possible. An analogy between the physical and behavioral sciences is helpful: The movement of atoms in space, like the movement of people in the environment, is caused, but their collisions (atomic) and encounters (human) happen by contingent-necessity. Contingency leads to collisions and encounters; necessity governs speed and direction. An effect, dependent upon the activity of one or more causes, may seem to be produced by accident but is really the result of contingent-necessity, or *a conjuncture of events compelling a certain course of action by constraining prior conditions*. The words *compelling* and *constraining* were chosen to convey powerful influence but not omnipotence. Since we cannot possibly understand the innumerable and interactive causes of our actions, and since we will never know the initial conditions of our own personal histories, we *feel* free. And why not? No cause or set of causes we select to examine as the determiners of human action can be complete, thus they cannot be considered as determining causes, only influencing ones. There will always be other causes left unexamined. Human freedom arises out of this ignorance of causes, and the model of contingent-necessity explains why. And because of the trigger effect of contingency, and its cascading consequences, we are also free to change our history. Therefore: *Human freedom is action taken with an ignorance of causes within a conjuncture of events, that compels and is compelled to a certain course of action by constraining prior conditions.*

## IT'S A WONDERFUL LIFE

Though the majority of Gould's focus has been on paleontological contingencies, his exemplar for human historical systems is the 1946 holiday film classic by Frank Capra—*It's a Wonderful Life.* Jimmy Stewart plays George Bailey, a small-town building and loan proprietor who, after decades of hard, honest work feels his life has been a failure because he sees nothing of the results of his efforts, and his youthful dreams of seeing and changing the world have seemingly been lost to age and responsibility. Further, some of his friends have managed to break away from the small town to make more money. Where others have ventured out to see the world, George only fantasized about it. His own brother is a decorated war hero, who saved the lives of many men in battle. But George has done seemingly little. His life seems stalled and stagnant, and when financial and familial pressures finally build beyond control on Christmas Eve, George decides to take his life by leaping into the rapids of a nearly frozen river. Fortunately he is interrupted by his guardian angel—Clarence Oddbody—who, knowing George's humanitarian disposition, jumps in the river before him, triggering George to follow him in to save his life. In recovery, George unloads his problems on Clarence, and then exclaims that he wishes he were never born. Clarence grants him his wish, taking George out of the historical picture and rerunning the story of what his little town of Bedford Falls would have been like without him.

Suddenly things are not what they used to be, and the changes are mostly slanted toward the negative. The people George helped financially are instead poor and wretched, the buildings he constructed are nonexistent, his wife is a lonely unmarried librarian, his children unborn, and the town is renamed "Pottersville," after the treacherous banker whose miserly ways prevented those George had helped from ever owning their own homes. His brother, whom George saved in childhood, is not there to save other lives in that specific battle, with the contingent consequences that the lives the brother saved are now also gone. As Clarence guides George through his now unfamiliar surroundings, he is dismayed and shocked. The history of his town is quite different without the influence of George Bailey. He never realized just how many people were dependent upon his seemingly routine existence. "Strange, isn't it?," queries Clarence to George at the appropriate moment of enlightenment. "Each man's life touches so many other lives, and when he isn't around he leaves an awful hole, doesn't he?"

In the end, of course, Clarence restores the historical sequence to its original condition, with George's contingent influence intact, and makes a reassuring pronouncement to him: "You see, George, you really had a wonderful life." In this sense, then, we are all individuals of power and importance. Whether we like it or not, whether we know it or not, every encounter and every action, can and does make some degree of difference, ranging from virtually negligent to powerfully diverting. A seemingly innocuous decision, carefully placed in time and circumstance, may affect uncounted others in multitudinous ways.

Because of the trigger effect and contingent-necessities, and the fact that at any point in the system it could be early as well as late (since we do not know when our personal system will end), one never knows which actions will or will not make a difference. Only the historian looking back is privileged to so judge. It is this lack of foresight and prognostication that makes the potential for the power of contingency and individuality so puissant. Since we do not know for certain which actions will matter and which will not, it is as rational as not to assume the former than the latter. It may be nothing but wishful thinking to desire one's place in history to be contingently significant, but since we do not know, why not act as if it does?

## FINDING MEANING IN A CONTINGENT UNIVERSE

I am often asked by believers why I abandoned Christianity and how I found meaning in the apparently meaningless universe presented by science. The implication is that the scientific worldview is an existentially depressing one. Without God, I am bluntly told, what's the point? If this is all there is, there is no use. To the contrary. For me, quite the opposite is true. The conjuncture of losing my religion, finding science, and discovering glorious contingency was remarkably empowering and liberating. It gave me a sense of joy and freedom. Freedom to think for myself. Freedom to take responsibility for my own actions. Freedom to construct my own meanings and my own destinies. With the knowledge that this may be all there is, and that I can trigger my own cascading changes, I was free to live life to its fullest.

This is not to say that those who are religious cannot share in these freedoms. But for me, and not just for me, a world without monsters, ghosts, demons, and gods unfetters the mind to soar to new heights, to think unthinkable thoughts, to imagine the unimaginable, to contemplate infinity and eternity knowing that no one is looking back. The universe takes on a whole new meaning when you know that your place in it was not foreordained, that it was not designed for us—indeed, that it was not designed at all. If we are nothing more than star stuff and biomass, how special life becomes. If the tape were played again and again without the appearance of our species, how extraordinary becomes our existence, and, correspondingly, how cherished. To share in the sublimity of knowledge generated by other human minds, and perhaps even to make a tiny contribution toward that body of knowledge that will be passed down through the ages—part of the cumulative wisdom of a single species on a tiny planet orbiting an ordinary star on the remote edge of a not-so-unusual galaxy, itself a member of a cluster of galaxies billions of light years from nowhere, is sublime beyond words.

Since we are such a visual primate, perhaps images can help capture the feeling. The Hubble Telescope Deep Field photograph on the following page reveals as never before the rich density of galaxies in our neck of the universe, is as grand a statement about the sacred as any medieval cathedral. How vast is the cosmos. How contingent is our place. Yet out of this apparent insignificance emerges a glorious contingency—the recognition that we did not have to be, but here we are. In fact, compare this slice of the cosmos to two of the most hallowed and sacrosanct structures on Earth—both medieval in age but on opposite sides of the planet, literally and figuratively—Machu Picchu and Chartres Cathedral. Machu Picchu captures the numina through an interlocking relationship between nature and humanity that generated in me an almost mystical connection across space and time with the ancients who had once lived and loved atop this 8,000-foot precipice. This is the "lost city" in so many ways. When I stood inside Chartres Cathedral with my soulmate, lit candles, and we promised each other our eternal love, it was a more sacred moment than any I have experienced. Skeptics and scientists cannot experience the numinous? Nonsense. You do not need a spiritual power to experience the spiritual. You do not need to be mystical to appreciate the mystery. Standing beneath a canopy of galaxies, atop a pillar of reworked stone, or inside a transept of holy light, my unencumbered soul was free to love without constraint, free to use my senses to enjoy all the pleasures and endure all the pains that come with such freedom. I was enfranchised for life, emancipated from the bonds of restricting tradition, and unyoked from the rules written for another time in another place for another people. I was now free to try to live up to that exalted moniker—*Homo sapiens*—wise man.

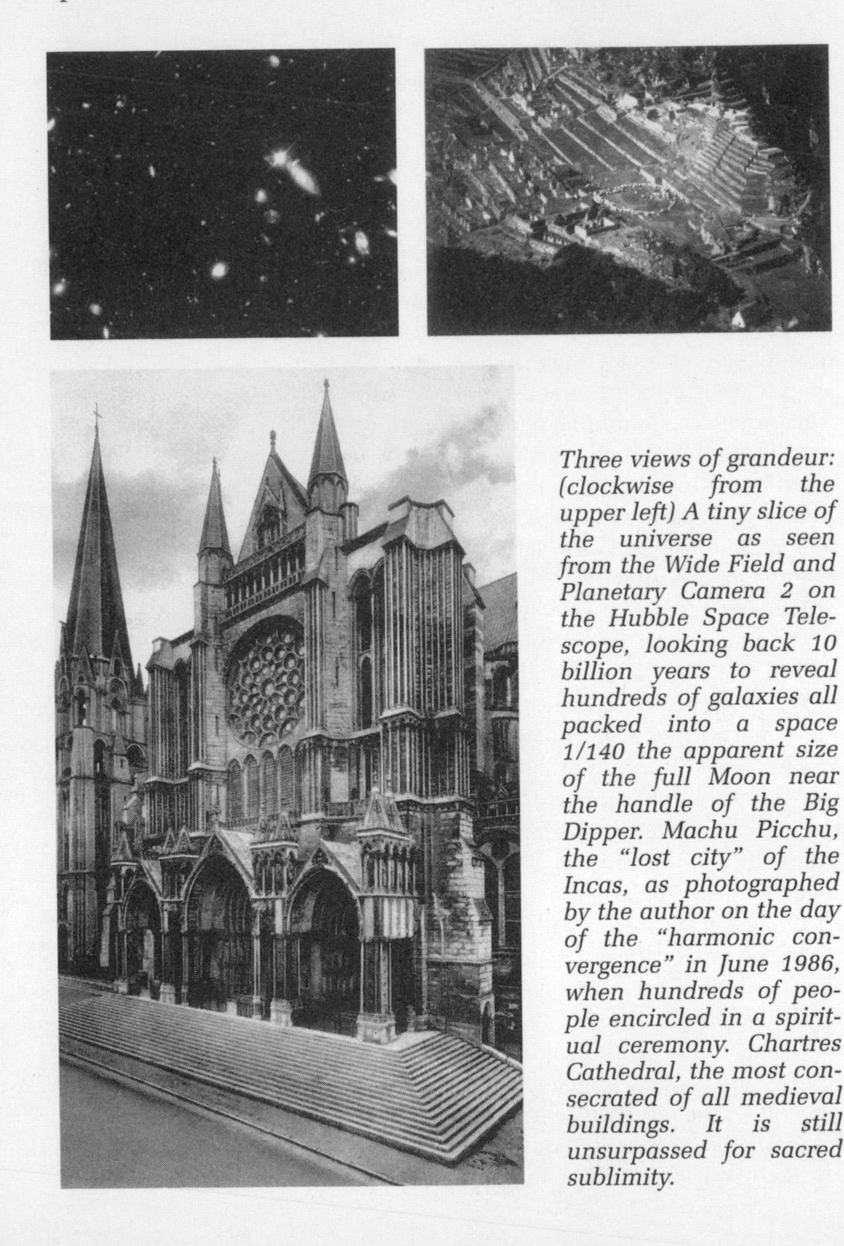

*Three views of grandeur: (clockwise from the upper left) A tiny slice of the universe as seen from the Wide Field and Planetary Camera 2 on the Hubble Space Telescope, looking back 10 billion years to reveal hundreds of galaxies all packed into a space 1/140 the apparent size of the full Moon near the handle of the Big Dipper. Machu Picchu, the "lost city" of the Incas, as photographed by the author on the day of the "harmonic convergence" in June 1986, when hundreds of people encircled in a spiritual ceremony. Chartres Cathedral, the most consecrated of all medieval buildings. It is still unsurpassed for sacred sublimity.*

# AFTERWORD TO THE SECOND EDITION

## God on the Brain

About a decade ago when I began research on the question of why people believe in God, I asked a colleague in a religious studies program at Occidental College (where I was teaching at the time) to recommend the latest pathbreaking scientific work in this area. "William James's 1890 *Varieties of Religious Experience,*" he responded sardonically, explaining that in his opinion the field was largely moribund.

That's an exaggeration, of course, but his point was that with the exception of a handful of psychologists teaching at theological seminaries, mainstream social and cognitive scientists had largely ignored the question. This has changed dramatically in the past decade, as the renewed debate on the relationship of science and religion has exploded onto the cultural landscape and scientists from a variety of fields have jumped into the fray. Much of this research, along with my own, appeared in the first edition of this book. In this new chapter for the revised second edition I would like to review, critique, and comment on the research that has come out since the original edition of *How We Believe.*

### THE NEUROPHYSIOLOGY OF GOD

I begin with a book with the intriguing title *Why God Won't Go Away*, by Andrew Newberg and Eugene D'Aquili, both medical doctors, with Newberg holding joint appointments in radiology and religious studies at the University of Pennsylvania, and D'Aquili, now deceased, a professor of psychiatry at Penn. God won't go away, the authors argue, because the religious impulse is rooted in the biology of the brain. When Buddhist monks meditate and Franciscan nuns pray, for example, their brain scans (these scientists used the single photon emission computed tomography, or SPECT) indicate strikingly low activity in the posterior superior parietal lobe, a bundle of neurons the authors have dubbed the OAA, or Orientation Association Area, whose job it is to orient the body in physical space (people with damage to this area have a hard time negotiating their way around a house). When the OAA is booted up and running smoothly there is a sharp distinction between self and non-self. When OAA is in sleep mode—as in deep meditation and prayer—that division breaks down, leading to a blurring of the lines between reality and fantasy. Is this what happens to monks who feel a sense of oneness with the universe, or with nuns who feel the presence of God?

Yes, say the authors, who believe they have "uncovered solid evidence that the mystical experiences of their subjects—the altered states of mind they described as the absorption of the self into something larger—were not the result of emotional mistakes or simple wishful thinking, but were associated instead with a series of observable neurological events. . . ." It is an odd distinction to make, which the authors do throughout the book. "A skeptic might suggest that a biological origin to all spiritual longings and experiences, including the universal human yearning to connect with something divine, could be explained as a delusion caused by the chemical misfirings of a bundle of nerve cells."

Indeed, I am one such skeptic, but I fail to see the difference (outside a minor linguistic distinction) between a delusion and a decrease in OAA activity. In this case, delusion is simply a descriptive term for what happens when the OAA shuts down and the brain loses the ability to distinguish self from non-self. But it is still all in the brain. Unless, of course, you believe that these neurologically triggered mystical experiences actually serve as a conduit to a real spiritual world where God (or what the authors call the "Absolute Unitary Being") resides. That is, in fact, what they believe: ". . . our research has left us no choice but to conclude that the mystics may be on to something, that the mind's machinery of transcendence may in fact be a window through which we can glimpse the ultimate realness of something that is truly divine." Thankfully they are honest enough to admit that this conclusion "is a terrifically unscientific idea" and that to accept it "we must second-guess all our assumptions about material reality." In the end they do just that.

The strength of *Why God Won't Go Away* lies in the original research conducted by the authors, and the brain correlates of mystical states they have identified, that together go a long way toward explicating the experiences of religious mystics. For the billions of believers who have never had a mystical experience, however, explanations for their faith are more likely grounded in the psychology and sociology of belief where, for example, the number one predictor of anyone's religious faith is that of their parents, modified by siblings and peer groups, mentors, education, age, cultural experiences, and other variables (see Chapter 4). This is not a critique, since the authors focused their attention on the neurological correlates of belief only, but the book does unravel when they seek an evolutionary origin for religion.

As compelling as such evolutionary explanations are—and surely this must be where the ultimate reason for belief lies (see Chapter 7)—much of the authors' case depends on explanation in the just-so storytelling mode. (Critics of sociobiology will find much fodder for their cannons here.) We are told, for example, that religion alleviated the "existential gloom" facing our paleolithic ancestors who were "taken off their game by the soul-sapping notion that no matter how hard they struggled, how skillfully they hunted, how fiercely they battled, or how creatively they thought, death was always waiting, and that their lives added up to nothing in the end. The promises of religion protected early humans from such self-defeating fatalism, and allowed them to struggle tirelessly but optimistically for survival." That's interesting. Prove it.

The authors also fall into the trap of thinking of human evolution as almost entirely centered around men on the hunt, a paradigm abandoned decades ago in favor of more sophisticated models of social evolution that stress the importance of relationships, hierarchy, dominance, cooperation, reciprocal altruism, and various forms of social exchange. It is out of this paradigm, in conjunction with psychosocial models, that a fuller explanation for why God won't go away is to be found (again, see Chapter 7).

In related research, a story that broke as I was writing this chapter came out in the pages of the journal *Nature*. Swiss neuroscientists Olaf Blanke, Margitta Seeck, Stephanie Ortigue, and Theodor Landis, from the University Hospitals of Geneva and Lausanne, through electrical brain stimulation of a forty-three-year old woman who was suffering from severe epileptic seizures, discovered a part of the brain that can induce Out-of-Body Experiences. Out-of-Body Experiences (OBEs) are typically associated with Near-Death Experiences (NDEs), and have a long tradition of harboring religious and spiritual overtones, as if the experience itself was a conduit to a transcendent state or spiritual dimension. The scientists repeatedly generated OBEs in this woman through electrical stimulation of her brain's right angular gyrus, part of the temporal lobe that is thought to play a role in the way that the brain analyzes sensory information and monitors the difference between self and non-self (as in the Orientation Association Area, or OAA, described in the research above). Blanke and associates believe that when the angular gyrus misfires it can produce the sense of floating outside of the body: "Stimulation at this site also elicited

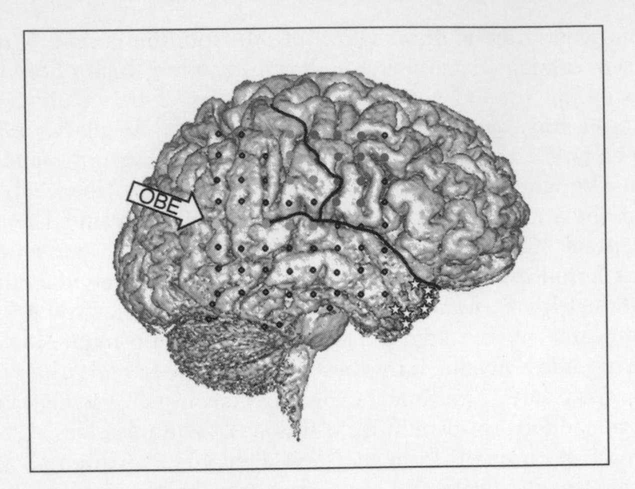


Figure 1. *Three-dimensional surface reconstruction of the right hemisphere of the brain from magnetic-resonance imaging.*

illusory transformations of the patient's arms and legs (complex somatosensory responses) and whole-body displacements (vestibular responses), indicating that out-of-body experiences may reflect a failure by the brain to integrate complex somatosensory and vestibular information." Figure 1 shows the area of the brain electrically stimulated to produce OBEs.

In initial mild stimulations, the patient reported that she was "sinking into the bed" or "falling from a height." More intense stimulation led her to report "I see myself lying in bed, from above, but I only see my legs and lower trunk." Two additional stimulations induced "an instantaneous feeling of 'lightness' and 'floating' about two meters above the bed, close to the ceiling." They then asked the patient to stare at her outstretched legs while they stimulated her brain. This led her to seeing her legs "becoming shorter." When they had her first bend her legs and then applied the electrical stimulation, "she reported that her legs appeared to be moving quickly towards her face, and [she] took evasive action." The same thing happened with her arms when the experiment was duplicated. Blanke and associates concluded: "These observations indicate that OBEs and complex somatosensory illusions can be artificially induced by electrical stimulation of the cortex. The association of these phenomena and their anatomical selectivity suggest that they have a common origin in body-related processing, an idea that is supported by the restriction of these visual experiences to the patient's own body. . . . It is possible that the experience of dissociation of self from the body is a result of failure to integrate complex somatosensory and vestibular information."

This is an exceptionally important study that goes a long way toward providing a normal explanation for what has long been considered to be paranormal phenomena, associated not only with near-death experiences, but with remote viewing, alien abductions, auditory and visual hallucinations (particularly, for our purposes here, those affiliated with religious epiphanies), and other mental ephemera and psychological anomalies. This study should stimulate other neuroscientists to explore adjacent regions of the brain to see if they can replicate other such phenomena, such as alien abductions and visual and auditory hallucinations. Although caution is called for because the subject pool was only one, all our brains are wired in a similar manner so there is little reason to think that stimulation of this brain region in other patients will not corroborate the finding. In fact, last December the British medical journal *Lancet* published a Dutch study in which 344 cardiac patients were resuscitated from clinical death. About 12 percent reported Near-Death Experiences where they saw the light at the end of a tunnel. Some even reported speaking to dead relatives. If these studies are corroborated it means yet another blow against those who believe that the mind and spirit are somehow separate from the brain, from pure neural activity. In reality, all experience is mediated by the brain, and these studies are another step in the long historical tradition where mysterious phenomena are subsumed under the blanket of science and naturalism.

Paranormal beliefs in general, in fact, may be related to brain chemistry. The July 2002 issue of *New Scientist* magazine reported the proceedings from a meeting of the Federation of European Neuroscience Societies in Paris, in which a study was presented showing that people with high levels of dopamine are more likely to find significance in coincidences and pick out meaning and patterns where there are none. The research was conducted by neurologist Peter Brugger of the University Hospital in Zurich, Switzerland. Brugger's subjects consisted of twenty self-professed believers and twenty self-confessed skeptics. His methodology was to briefly flash images on a screen to see if there was a difference between believers and skeptics on what they thought they observed. In one experiment real faces and scrambled faces were shown. In another experiment real and scrambled words were flashed. Brugger found that believers were much more likely than skeptics to see a word or face when there was not one. Skeptics were more likely to miss real faces and words when they appeared on the screen. The dopamine variable was added when

Brugger gave his subjects L-dopa, normally used to relieve the symptoms of Parkinson's disease by increasing levels of dopamine in the brain. (Dopamine is also involved in the brain's reward and motivation system and has some relevance for the treatment of drug addiction.) Although both groups made more mistakes under the influence of L-dopa, skeptics became more likely to interpret scrambled words or faces as the real thing. This finding suggests that paranormal thoughts and beliefs may be associated with high levels of dopamine in the brain. The significant effect is that L-dopa makes skeptics less skeptical. By contrast, and surprisingly, L-dopa did not seem to increase the tendency of believers to see coincidences or relationships between the words and images. Brugger concluded that this could mean that there is a plateau effect for believers, with more dopamine having relatively little effect above their belief threshold.

## GODS, ANGELS, AND ESP: WHAT PEOPLE BELIEVE ABOUT THE SUPERNATURAL

It is possible that studies like these will begin to explain at least some of the reasons why people believe in paranormal, supernatural, and spiritual entities. Just how many believe in such ephemera? According to the Scripps Howard News Service, in a study conducted in collaboration with Ohio University, Americans overwhelmingly believe in the angels that heralded the birth of Jesus 2000 years ago and think they still walk the Earth today. In a survey of 1,127 adults, one out of every five Americans believes he or she has seen an angel or knows someone who has. Of those 1,127 adults polled, 77 percent answered "Yes" to the question: "Do you believe angels, that is, some kind of heavenly beings who visit Earth, in fact exist?" Another 73 percent believe angels still "come into the world even in these modern days." Belief in angelic beings cuts across almost all ranges of education, income, and lifestyle. Women and young people are slightly more likely to believe than are men or older Americans, but a majority of almost every demographic group are angel believers.

Consider the following account given by Catherine Forbes, seventy-two, of Derby, Kansas: "Yes, I absolutely believe in angels. I met one." The circumstances of this experience are telling. After the death of her husband, Forbes decided to take a trip to Jerusalem with a friend in 1953. On their way through the Dallas airport they got lost and became anxious. "All of a sudden, the nicest voice I ever heard said, 'May I help you?' I turned around and saw a clean-cut young man, just the most handsome, beautiful man. He picked up my luggage and showed me where to go and which people I was to be traveling with. I turned around to thank him, and he had absolutely disappeared." Although there was no flash of light, the helpful young man had disappeared from sight in an apparently impossible fashion, she said. "I know some people will think I'm off my rocker, but I know what I saw."

There is no doubt that Forbes's experience was a real one. The question, however, is: was the source of her experience inside or outside the brain? The scientific evidence shows that such experiences are brain-generated, mediated by past experiences (in the form of memories) and the context in which they occur (in this case the airport during an episode of extreme grief). There is no need to call forth supernatural explanations when natural ones will do.

Yet, the angel belief poll was emblematic of the larger trend in beliefs in spiritual and paranormal phenomena. The Gallup News Service, for example, reported on June 8, 2001, the results from a survey they conducted on paranormal and spiritual beliefs. "The results suggest a significant increase in belief in a number of these experiences over the past decade, including in particular such Halloween-related issues as haunted houses, ghosts, and witches. Only one of the experiences tested has seen a drop in belief since 1990: devil possession. Overall, half or more of Americans believe in two of the issues: psychic or spiritual healing, and extrasensory perception (ESP), and a third or more believe in such things as haunted houses, possession by the devil, ghosts, telepathy, extraterrestrial beings having visited earth, and clairvoyance."

There were interesting differences in beliefs by various subpopulations. For example:

> Age: *Younger Americans—those 18–29—are much more likely than those who are older to believe in haunted houses, in witches, in ghosts, that extraterrestrials have visited Earth, and in clairvoyance. There is little significant difference in belief in the other items by age group. Those thirty and older are somewhat more likely to believe in possession by the devil than are the younger group (perhaps as a result of seeing The Exorcist?).*
> Gender: *Women are slightly more likely than men to believe in ghosts and that people can communicate with the dead. Men, on the other hand, are more likely than women to believe in only one of the dimensions tested: that extraterrestrials have visited Earth at some point in the past.*
> Education: *Americans with the highest levels of education are more likely than others to believe in the power of the mind to heal*

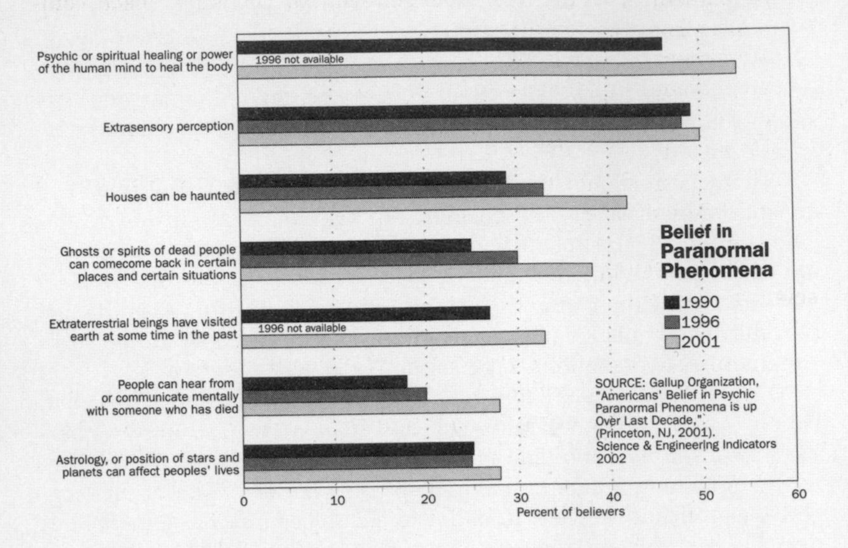


Figure 2. *Changing belief in paranormal phenomena.*

> *the body. On the other hand, belief in three of the phenomena tested goes up as the educational level of the respondent goes down: possession by the devil, astrology, and haunted houses.* Importance of Religion: *Perhaps not surprisingly, the major difference in belief in these phenomena by importance of religion focuses on the devil: 55 percent of those who say religion is very important in their daily lives say they believe in devil possession, compared to just 14 percent of those who say religion is "not very" important to them. Religion is also correlated with belief in extraterrestrials: Those for whom religion is very important are less likely to say they believe in beings from other worlds that may have visited this planet than are those who are less religious.*

The change in belief percentages over the past decade are seen in Figure 2.

The National Science Foundation found similar percentages of belief in pseudoscience as the Gallup Pollsters did in belief in the paranormal. In April 2002, the NSF published their biennial report on the state of science understanding and public attitudes toward science, which included a section on the relationships between science and pseudoscience. The results were alarming:

30 percent of adult Americans believe that UFOs are space vehicles from other civilizations
60 percent believe in ESP
40 percent think that astrology is scientific
32 percent believe in lucky numbers
70 percent accept magnetic therapy as scientific
88 percent agree that alternative medicine is a viable means of treating illness

The NSF survey summarized the overall findings on pseudoscience this way:

> *Belief in pseudoscience, including astrology, extrasensory perception (ESP), and alien abductions, is relatively widespread and growing. For example, in response to the 2001 NSF survey, a sizable minority (41 percent) of the public said that astrology was at least somewhat scientific, and a solid majority (60 percent) agreed with the statement "some people possess psychic powers or ESP." Gallup polls show substantial gains in almost every category of pseudoscience during the past decade. Such beliefs may sometimes be fueled by the media's miscommunication of science and the scientific process.*

As for alternative or complementary medicine, the NSF report highlighted their findings as such:

> *Alternative medicine, defined here as any treatment that has not been proven effective using scientific methods, has been gaining in popularity. One study documented a 50 percent increase in expenditures for alternative therapies and a 25 percent increase in the use of alternative therapies between 1990 and 1997. Also, more than two-thirds of those responding to the NSF survey said that magnetic therapy was at least somewhat scientific, although no scientific evidence exists to support claims about its effectiveness in treating pain or any other ailment.*

Of the various alternative modalities, the survey reported magnets as the most popular. "Among those who reported using energy healing, the most frequently cited technique involved the use of magnets. In 2001, NSF survey respondents were asked whether or not they had heard of magnetic therapy, and if they had, whether they thought that it was very scientific, sort of scientific, or not at all scientific. A substantial majority of survey respondents (77 percent) had heard of magnetic therapy. Among all who had heard of this treatment, 14 percent said it was very scientific and another 54 percent said it was sort of scientific. Only 25 percent of those surveyed answered correctly, that is, that it is not at all scientific."

Education by itself is no paranormal prophylactic. Although belief in ESP decreased from 65 percent among high school graduates to 60 percent among college graduates, and belief in magnetic therapy dropped from 71 percent among high school graduates to 55 percent among college graduates, that still leaves over half fully endorsing such claims! And for embracing alternative medicine, the percentages actually increase, from 89 percent for high school grads to 92 percent for college grads.

On a positive note the survey revealed that "for the first time, a majority (53 percent) of NSF survey respondents answered 'true' to the statement 'human beings, as we know them today, developed from earlier species of animals,' bringing the United States more in line with other industrialized countries in response to this question." The report also noted, however, that the teaching of creationism still finds majority support, in that "although a majority (60 percent) of people surveyed in a Gallup poll were opposed to the Kansas State Board of Education's decision to delete evolution from the state's science standards (a decision that was later reversed), more than two-thirds favored teaching both evolution and creationism in U.S. public school classrooms."

We can glean a deeper cause of this contradiction in another statistic: 70 percent of Americans still do not understand the scientific process, defined in the study as grasping three concepts: probability, the experimental method, and hypothesis testing. One solution is more and better science education, as indicated by the fact that 53 percent of Americans with a high level of science education (nine or more high school and college science/math courses) understand the scientific process, compared to 38 percent with a middle level (six to eight such courses) science education, and 17 percent with a low level (less than five such courses). The NSF report concluded:

> *Although more than 50 percent of NSF survey respondents in 2001 had some understanding of probability, and more than 40 percent were familiar with how an experiment is conducted, only one-third could adequately explain what it means to study something scientifically. Understanding how ideas are investigated and analyzed is a sure sign of scientific literacy. Such critical thinking skills can also prove advantageous in making well-informed choices at the ballot box and in other daily living activities.*

## GOD AND EVOLUTION

The key here is teaching how science works, not just what science has discovered. An article published in *Skeptic*, Volume 9, Number 3, presents the results of a study that found no correlation between science knowledge (facts about the world) and paranormal beliefs. The authors, W. Richard Walker, Steven J. Hoekstra, and Rodney J. Vogl, concluded: "Students that scored well on these [science knowledge] tests were no more or less skeptical of pseudoscientific claims than students that scored very poorly. Apparently, the students were not able to apply their scientific knowledge to evaluate these pseudoscientific claims. We suggest that this inability stems in part from the way that science is traditionally presented to students: Students are taught what to think but not how to think."

In no area of human knowledge is this observation more true, and critical thinking in such desperate demand, than when scientific claims appear to conflict with religious tenets. Here I am thinking of the creation-evolution controversy, which continues to inflame many religious Americans as they try to come to grips with the findings of modern science. (Indeed, the day I finished writing this chapter, the Cobb County, Georgia, board of education voted to include a sticker in all public high school biology textbooks indicating that evolution is just one theory among many to explain the development of life, and that creationism and Intelligent Design Theory should also be included in the curriculum of biology classes.) In March of 2001 the Gallup News Service reported the results of a survey that found 45 percent of Americans agree with the statement "God created human beings pretty much in their present form at one time within the last 10,000 years or so," while 37 percent preferred a blended belief that "Human beings have developed over millions of years from less advanced forms of life, but God guided this process," and a paltry 12 percent accepted the standard scientific theory that "Human beings have developed over millions of years from less advanced forms of life, but God had no part in this process."

In a forced choice between the "theory of creationism" and the "theory of evolution," 57 percent chose creationism against only 33 percent for evolution (10 percent said they were "unsure"). Only 33 percent of Americans think that the theory of evolution is "well supported by evidence," while slightly more (39 percent) believe that it is not well supported, and that it is "just one of many theories." One reason for these disturbing results can be seen in the additional finding that only 34 percent of Americans consider themselves to be "very informed" about evolution. Clearly the 66 percent who do not consider themselves well informed have not withheld their judgment on the theory's veracity.

This fact was brought to light for me in the overwhelming response to my February 2002 *Scientific American* column on evolution and Intelligent Design creationism. Where I typically receive a couple of dozen letters a month, for this one no less than 134 were submitted (117 men, 4 women, 13 unknowns—a ratio equivalent to the magazine's gender split).

When I first started writing for *Scientific American* I found reading critical letters mildly disconcerting, until I hit upon the idea that these are a form of data to be mined for additional information on what people believe and why—an informal vox populi. Conducting a content analysis of all 134 letters, I discovered a pattern within the cacophonous chaos that gave me additional insight into why people reject the theory of evolution. Initially I read through them all quickly, coding them into about two dozen one-line categories that summed up the reader's point. I then coalesced these into six taxonomic classes, and reread all the letters carefully, placing each into one or more of the six (many readers made more than one point), giving a total of 163 ratings from which the following percentages were derived:

Excerpts from the letters illuminate each taxon. (Although most were friendly and reasonable, one fellow opined that my column "could have been written in 1939 by a Nazi," while another said that "Michael Shermer must not only be a sceptic but also stupid in the 3rd degree the way he talks about 'Intelligent Design.'") I was initially surprised to discover that only 7 percent agreed with me about the

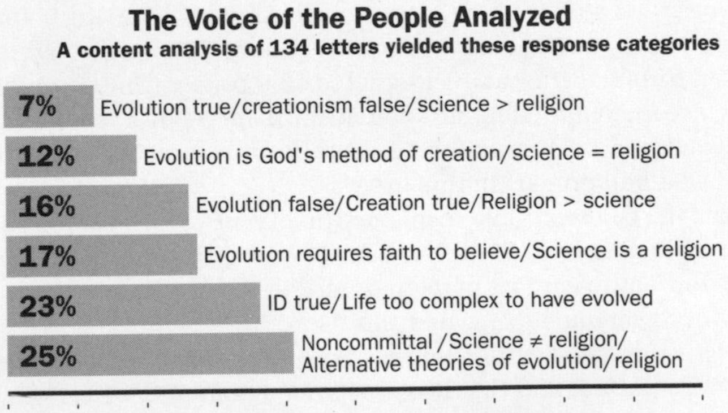

**The Voice of the People Analyzed**
**A content analysis of 134 letters yielded these response categories**

7% Evolution true/creationism false/science > religion
12% Evolution is God's method of creation/science = religion
16% Evolution false/Creation true/Religion > science
17% Evolution requires faith to believe/Science is a religion
23% ID true/Life too complex to have evolved
25% Noncommittal /Science ≠ religion/ Alternative theories of evolution/religion

Figure 3. *Why people do not accept evolution.*

veracity of evolution (and the emptiness of creationism), with one reader going so far as to claim, "The defenders of science behave too well. No amount of evolution education will counter the deliberate, sly, selective ignorance of creation 'science.'" Nearly double that number argued that evolution is God's method of creating life, such as one correspondent who agreed "that evolution is right—but still I see GOD in the will and cunning intention in the genetic system of all living organisms and in the system and order present in the laws of nature. Seeing all the diversity in the methods of camoflage in animals and plants for an example, I know that there is a will behind it." Another reader sees creation and evolution "as complementary to each other. Put simply, since all parts of the universe follow intelligible law as educed through human intelligence, and such a law is a principle or cause, it follows that the universe as a whole must be the effect of the operation of a singular all-encompassing Principle."

Critics of evolution in the third taxon hauled out an old canard every evolutionary biologist has heard: "I want to point out that evolution is only a theory." And: "To my knowledge evolution is just a theory that has never been put to the test successfully and is far from being conclusive." That evolution requires faith to believe (the fourth category) found many adherents among readers, such as these: "There are so many vast chasms that evolutionists paint over with broad strokes, they act as if their faith is fact as often as a creationist." Or: "On my view, a key part of being a rational skeptic is consistent dedication to the standards and methods of critical thinking and logic. In his zeal to defend his faith in evolutionary theory, Dr. Shermer violates those standards." My favorite letter in this class echoed a standard refrain we hear often at *Skeptic* magazine about inadequate or misplaced skepticism (with a cc list that included "Pres. George W. Bush, V.P. Dick Cheney, and The Members of The US Congress, American Academy of Science; Dr. Dean Edell, America's Doctor; Dr. Laura, America's Jewish Mother"!): "I applaud your SKEPTICISM when it comes to Creationism and Astrology and 'Psychic Phenomena'; but how can you be so THICK HEADED when it comes to the GLARING WEAKNESSES of Darwinian Evolution??? Honestly, you come across as both a 'brain-washed apologist' and a 'High School Cheerleader' for Darwinian Evolution." Charles Darwin, he's our man. If he can't do it no one can.

The penultimate taxon was that Intelligent Design creationism must be true because life is simply too complex to be explained by evolution. For example: "ID theorists also see a variety of factors, constants, and relationships in the construction of the universe which are so keenly well-adjusted to the existence of matter and life that they find it impossible to deny the implication of intelligent purpose in those factors. Materialists see the same thing and wave their hands vaguely and mutter mystical phrases about 'Anthropic Principles.' What the materialist calls the anthropic principle, the IDer calls the Designer."

Intriguingly, the greatest number of responses fell into a noncommittal position where readers expounded on the relationship of science and religion, often presenting their own theories of evolution and creation as alternatives to the models under discussion. For example: "Evolution is not a theory. It is an analytic approach. There are three elements of science: operation, observation, and model. An observation is the result of applying an operation, and a model is chosen for its utility in explaining, predicting, and controlling observations, balanced against the cost of using it." And: "There is nothing that scientists have ever discovered, or could ever discover, that can prove or disprove the existence of God. The Bible is a tool for the illumination of the heart, not the revelation of observable facts. Thus there is no conflict between the Bible and science—there is even an amazing synergy between the two—when each is kept in its proper place."

It has been my experience that correspondents in this final classification, like questioners in the Q & A sessions of lectures I present at colleges and universities, are less interested in my opinion and more intent on launching their own ideas into the cultural ether. With no subject is this more apparent than for evolution; it is here we face the ultimate question of genesis and exodus: where did we come from and where are we going?

## FAITH, RELIGION, AND THE SOUL: WHY RELIGION MATTERS

Since the initial publication of *How We Believe*, books on religion, particularly on the relationship of science and religion and on the origin and purpose of religion, have tumbled off the presses in droves. There is money to be made in the religion publishing business, not the least of which is due to the fact that there are so many believers. Oxford University Press's newly released second edition of its *World Christian Encyclopedia* reports that of the earth's 6.1 billion humans fully 5.1 billion of them, or 84 percent, declare themselves believers who belong to some form of organized religion. Christians dominate at just a shade under two billion adherents (Catholics count for half of those), with Muslims at 1.1 billion, Hindus at 811 million, Buddhists at 359 million, and ethnoreligionists (animists and others in Asia and Africa primarily) accounting for most of the remaining 265 million.

But as the editors note, such overall numbers tell us little. There are, in fact, 10,000 distinct religions of ten general varieties (in decreasing size and inclusiveness of cosmoreligion, macroreligion, megareligion, and so forth), each one of which can be further subdivided and classified. For example, Christians (classified as members of a cosmoreligion because it is open to all) may be found among 33,820 different denominations.

The variety of non-Christian religions is also stunning, with worldwide distribution outstripping Christian religions despite the tireless efforts of evangelists to convert as many souls to Christ as possible. (One irritation with this encyclopedia is its Christian bias—its senior editor is Reverend David B. Barrett, who heads the Global Evangelization Movement, making one wonder if all this data is being collected to calibrate how long it will take to reduce this rich diversity to one cosmo-macro-mega Christian religion.) One table, for example, tracks the number of Christians (69,000) and non-Christians (147,000) by which the world will increase over the next 24 hours. A diagram reveals the global convert/defector ratio, adjusted for births and deaths, indicating that the sphere of evangelism continues to expand into non-Christian belief space.

A visual companion to the encyclopedia is Oxford's *New Historical Atlas of Religion in America*, which is packed with 260 color maps and charts printed on thick glossy paper to enhance the fine detail and shades of geographical differences between and among the various religious sects that inhabit the landscape. This new edition of religious historian Edwin Gaustad's 1962 classic includes the arrival of religious colonialists to the New World over the past four decades, including Hindus, Jains, Sikhs, Buddhists, and especially Muslims, who have enjoyed a fourfold increase in America. Likewise, the number of Baha'is has risen nearly proportional to the membership drop in many mainstream religions, such as Episcopalians, Methodists, and Presbyterians. By contrast, "Southern Baptists" might better be labeled "All Over America Baptists," as their ranks have swollen well into the northern territories. Likewise, the "Bible Belt" is now wider than a weightlifter's leather girdle. Their ranks have even penetrated the formerly impenetrable Mormon-dominated Utah; but, in turn, in three-quarters of the counties west of the Rockies, the Mormon church ranks in the top three religious denominations. Most revealing are the historical maps and charts that track the changing demographics of American religion. Conservative pundits who proclaim that we need to return to the good old days when America was a Christian nation better look closely at Figure 4.16, showing that church membership as a percentage of the U.S. population over the past century and a half has increased from 25 percent to 65 percent. If America is going to hell in an immoral handbasket, it is happening when church membership is at an all-time high, and a greater percentage of Americans (90–95 percent) proclaim belief in a God than ever before.

Why do so many people believe and belong? One answer is that it is good for us. Studies show that religious people live longer and healthier lives, recover from illness and disease faster, and report higher levels of happiness. While most of these effects are probably due to lifestyle, diet, and exercise (e.g., religious people drink and smoke less), there *is* something about having family, friends, and a community that enhances life and longevity. An interesting book entitled *Aging with Grace* explores this thesis through a remarkable study of 678 nuns ranging in age from 75 to 104, lovingly told by Dr. David Snowdon, once a Catholic altar boy and now a distinguished epidemiologist and the director of the Nun Study at the Sanders-Brown Center on Aging at the University of Kentucky. As the book of Proverbs proclaims, "A merry heart doeth good like a medicine, but a broken spirit drieth the bones." It turns out that a powerful predictor of which nuns would live the longest was the positive emotional content contained in their youthful writings, even when the analysis was controlled for age, education, and linguistic ability. The lowest emotional group averaged 86.6 years old at death, the highest emotional group averaged 93.5 years old at death. Snowdon also argues that profound faith, along with prayer and contemplation, "have a positive influence on long-term health and may even speed the healing process," but then oddly concludes: "We do not need a study to affirm their importance to the quality of life." Oh yes we do, if we want to make this a scientific claim. In fact, prayer and healing is now a hot field of study in medicine, but to date the studies have been severely flawed, failing to control for intervening variables and lacking consistent findings across comparative studies. I have no doubt that Snowdon is right about the importance of community and close relationships, but you don't need God or religion for that. All humans benefit from any type of social commitment because we are a social primate species.

There is another side to this story that recent research is illuminating, and that is that religious beliefs are not always a source of comfort during illness. In fact, in some cases, they may actually increase the risk of dying. A study conducted at Duke University Medical Center and Bowling Green State University, whose results were published in the August 13, 2002, issue of *Archives of Internal Medicine*, found that of the nearly 600 older hospital patients (95 percent of whom were Christian) negative feelings evoked by religious beliefs sometimes predicted mortality. Some of the key variables that increased the risk of death were feelings of being "abandoned or punished" by God, "believing the devil caused the illness," or "feeling abandoned by one's faith community." "The study reminds us that religion . . . can, at times, be a source of problems in itself," the lead author, Kenneth Pargament, concluded. Additional findings included: patients who reported feeling alienated from God or who blamed the devil had a 19 to 28 percent increased risk of dying during the following two years, although (and surprisingly) there was no association of gender, race, diagnosis, brain function, independence, depression, or quality of life with mortality. Duke University's Dr. Harold Koenig noted that anger and frustration were normal grief responses when people discovered health problems. Those who were religious and were able to reconnect with God and their spiritual feelings could use those resources for support. But those who continued to experience conflict could be making their health worse. "Those people are in trouble and doctors need to know about it. Doctors need to be assessing their patients for these kinds of feelings."

What we're really after here in our search for scientific answers to the question of why people believe in God is the undergirding beneath the panoply of religious faiths. For Michael Barnes, a professor of religious studies at the University of Dayton, the commonality is to be found in the thinking process itself. In his cleverly argued book, *Stages of Thought*, Barnes uses Piaget's stage theory of development to argue that cultures, like individuals, develop in stages from easier cognitive skills to harder ones, and that not only religion and faith, but science and reason have followed this general pattern. Both religion and science evolved from simple to complex because complex cognitive thinking first requires simple cognitive technologies such as writing and formal logic, as well as simple social institutions that reinforce those skills that can then be built into formal religions and sciences. Barnes is certainly correct about science and technology, because they are cumulative and complex and depend significantly on what came before. I'm not so sure about religion. Is monotheism really more cognitively challenging than polytheism, itself more complex than animism? Might it not be the opposite, where the world is so much easier to explain with one God than many, and many gods simpler than the spirit-haunted world of so-called primitive peoples? A stronger case for Barnes's cognitive model can be made within religions and especially for theology, which has turned the question of God's existence into a quagmire of syllogisms and contorted logic (see Chapter 5).

On one level it is that very stage of advanced cognitive development that Huston Smith rails against in his book *Why Religion*

*Matters*, a passionate personal manifesto for why society must return to its more fundamental roots of basic spirituality. While not completely disparaging science (his oncologist *did* save his life), Smith claims that it has trapped us in a tunnel whose floor is scientism, whose walls are liberal democracy, higher education, and a morally sterile legal system, and whose ceiling is a cowardly media. It is a closed system that excludes old-time religion. To get it back we must exit the tunnel and embrace the sacred. "The sacred world is the truer, more veridical world, in part because it includes the mundane world." Barnes would describe Smith's mundane world as an early stage of cognitive development, and that does appear to be the level at which Smith thinks religion should operate. Religion matters, he says, because "there is within us—in even the blithest, most lighthearted among us—a fundamental disease. It acts like an unquenchable thirst that renders the vast majority of us incapable of ever coming to full peace." Maybe for thee, but not for me. And that's the problem with Smith's book. It is, by its nature, personal and anecdotal, and so ultimately can tell us nothing more about why God and religion persist for anyone beyond Smith and those he copiously quotes in support.

What can inform us about these persistent questions? Science. Although it has its limitations, science is the best method ever devised for answering questions about our world and ourselves. Therefore, Pascal Boyer's *Religion Explained: The Evolutionary Origins of Religious Thought* is a penetratingly insightful scientific analysis of religion because as an anthropologist he understands that any explanation must take into account the rich diversity of religious practices and beliefs around the world, and as a scientist he knows that any explanatory model must account for this diversity. Boyer is at his ethnographic best in describing the countless peculiar religious rituals he and his anthropological brethren have recorded, and especially in identifying the shortcomings of virtually every explanation for religion ever offered. You name it, Boyer has an exception to it. To that end, anyone offering a theory of religion should read this book before transducing thought to ink. As a consequence, however, Boyer himself fails to provide a satisfactory explanation because he knows that religion is not a single entity that is the result of a single cause. "There cannot be a magic bullet to explain the existence and common features of religion, as the phenomenon is the result of *aggregate relevance*—that is, of successful activation of a whole variety of mental systems." Here the book bogs down in the jargon-laden field of cognitive science, as the author struggles to unite an array of disparate findings, but comes up empty handed. "Is there some religious center in the brain, some special cortical area, some special neural network that handles God-related thoughts? Not really . . . religious persons are not different from nonreligious ones in essential cognitive functions." Then what is the origin of religious faith and belief? For Boyer they "seem to be simple by-products of the way concepts and inferences are doing their work for religion in much the same way as for other domains." In other words, religion requires no special explanation, an answer many will find unsatisfactory.

Whatever its origin, what does the future hold for religion? One avenue for the ever-burgeoning religious landscape is cyberspace, the subject of the aptly titled *Give Me that Online Religion* by Brenda Brasher. It is a delightful romp through spiritual cyborgs, virtual monks, and the new world of cyberspirituality. Global prayer-chains, e-prayer wheels, cybercast seders, and neo-pagan cyber-rituals are all practiced from home, finally making Martin Luther's proclamation of "every man his own priest" a virtual reality. Even mainstream religions have gone online, offering adherents and potential converts a smorgasbord of doctrines to download (except Scientology, whose lawyers pounced on an ex-member who was posting their religious documents like Torquemada on a relapsed heretic). Much of this book will leave you LOL (for the computer illiterate that's laughing out loud), my favorite being Brasher's discussion of the more than 800,000 web "shrines" devoted to Princess Diana and other celebrities. "Scanning fan sites, it is easy to believe that the spiritual discipline of *imitato Christus* has been replaced by *imitato* Keanu Reeves." For those who do not wish to risk choosing the wrong God to achieve immortality, read about the transhumanists, who believe that some day we will be able to download our minds from our protein brains that survive only about a century, to silicon-chip brains that can last hundreds of centuries, by which time they can be downloaded into something more permanent still, ad infinitum to infinity. Is this in any sense possible? The transhumanists think it is, but since the technology is not yet available cryonics is a temporary solution, a quick fix if you will. Recall the brouhaha that developed shortly after baseball legend Ted Williams died, when his son whisked the body away to Phoenix, Arizona, where it was cryonically frozen at minus 320 degrees. The hope is that one day "Teddy Ballgame" would be resurrected to play again. If Williams's body were reanimated one day, would it still be the cranky perfectionist who was the last to hit .400? In other words, even if future cryonics scientists could bring him back to life, would it still be "him"? Is the "soul" of Ted Williams also in deep freeze along with his brain and body?

Duke University philosopher Owen Flanagan would probably answer "yes," if by soul we mean the pattern of Ted Williams's memories, personality, and personhood, and if the freezing process did not destroy the neural network in the brain where such entities are stored. But as for some ethereal entity that continues past physical death (whether buried, cremated, or frozen), Flanagan would offer an emphatic "no." In his latest book, *The Problem of the Soul*, a courageous and daring look into the heart of what it means to be human, Flanagan builds a bridge between two irreconcilable views of the mind: the humanistic/theological versus the scientific/naturalistic. The former includes a place within our brains for nonphysical mind, free will, and a soul, but fails to offer any tangible proof that such things even exist. The latter is grounded in solid empirical data but fails to show how humans as evolved animals can lead moral and meaningful lives. Flanagan's purpose is to reconcile the two, and he has done so successfully in this crisply reasoned work. "Can we do without the cluster of concepts that are central to the humanistic image in its present form—the soul and its suite—and still retain some or most of what these concepts were designed to do?" Flanagan's answer is an emphatic "yes." To that I add "amen."

It may simply be that I resonate well with Flanagan because I am a nonbelieving, nontheistic, naturalistic scientist. After a lifetime spent reading the obfuscating works of philosophers and theologians twisting logic into pretzelian contortions to prove such unprovable concepts as God, the soul, and free will, I want to stand up and cheer when I read passages such as this one from Flanagan's opening salvo: "There is no point beating around the bush. Supernatural concepts have no philosophical warrant. Furthermore, it is not that such concepts are displaced only if we accept, from the start, a naturalistic or scientific visions of things. There simply are no good arguments—theological, philosophical, humanistic, or scientific—for beliefs in divine beings, miracles, or heavenly afterlives."

How then, without such ephemera, can we find meaning in this meaningless cosmos? By broadening the scope of science. Flanagan convincingly demonstrates that the scientific quest to understand our place in the cosmos and our relation to other beings, including and especially our own species, itself generates both awe and reverence—feelings that were previously the exclusive domain of religion: "There is benevolence and compassion expressed by a feeling of connection to all creatures, indeed even to the awesome inanimate cosmos." This connection comes through knowing something about creatures and the cosmos, and Flanagan spends most of the book discussing the nature of what it means to be human, how brains can create minds (that are not separate from neurons), why free will is not necessarily incompatible with the deterministic assumption behind making free moral choices, how natural selves exist and retain most of the benefits of supernatural selves (souls) with the exception of immortality, and how ethical principles can be derived (and consequent moral behaviors generated) through a purely naturalistic worldview. Here the reading slows a little as Flanagan reviews all the major competing views before delivering his verdict on them along with his alternatives (for example, it takes fifty pages to dispense with the soul and another fifty pages to rebuild it through a natural system). But the effort pays off, as when he delivers this brilliant denouement showing how it is not the answers of science that provide transcendence, it is the quest: "It is becoming, worthy, and noble. It is the most we can aim for given the kind of creature we are, and happily it is enough. If you think this is not so, if you want more, if you wish that your life had prospects for transcendent meaning, for more than the personal satisfaction and contentment you can achieve while you are alive, and more than what you will have contributed to the well-being of this world after you die, then you are still in the grip of illusions. Trust me, you can't get more. But what you can get, if you live well, is enough."

It is enough for Flanagan. And it is enough for me and the (roughly) 60 percent of practicing scientists who, according to a 1996 survey by Ed Larson, have no belief in God or an afterlife. But will it ever be enough for the masses? Can we convince hundreds of millions of people—even billions of souls—that the scientific worldview is good enough? The realist in me remains pessimistic. But the idealist in me wants more—a worldview where science is presented as a humanistic and humane enterprise. Science is constructive, not destructive. A few structures (like the soul) may be demolished to make room for the new edifice, but many of the contents of the old building will be preserved in the new. That is the cumulative and uplifting nature of science.

## IN SEARCH OF SPIRITUAL MEANING

There is a humorous scene in Woody Allen's *Hannah and Her Sisters*, when his unfulfilled and neurotically Jewish character fails to find meaning in alternate religious expressions after visiting a Catholic church and returning home with a loaf of white bread, a jar of mayonnaise, and a crucifix. The reason, of course, is that the trappings and facade of a religion will not get you to that deeper place where so many desire to go. This is the deeper side of the psychology of religion, the exploration of which I found illuminating in Martha Sherrill's narrative account entitled *The Buddha from Brooklyn*. There are no grand theories here, no sweeping pronouncements about "what it all means," but it is a compelling case study in the search for spiritual meaning in an age of materialism. This is the story of Catharine Burroughs, born and raised in Brooklyn as Alyce Louise Zeoli, who was severely abused as a child but found redemption first as a psychic and spiritual counselor in suburban Maryland, then as a Tibetan tulku, or reborn lama (a type of living Buddha—thus accounting for the book's alliterative title) when she was told by a visiting Tibetan religious leader (His Holiness Penor Rinpoche, who sits just beneath the Dalai Lama himself) that she was the reincarnation of a sixteenth-century Tibetan saint. This conjuncture of events led to her becoming the first American woman to ever achieve such high religious status in this faith.

Now holding the honorific title of Jetsunma Ahkön Norbu Lhamo, or just Jetsunma, Zeoli/Burroughs (name changes are common in this story) founded a Tibetan Buddhist center in Maryland in 1986 and quickly developed a cast of loyal followers, which we meet one by one in detail through the sensitive and searching eyes of Sherrill, who is herself seeking spiritual balance. The history of how Burroughs turned to the mystical in response to her tragic upbringing (including cigarette burns on her body and beatings with a radiator brush), however, is not where the power of this story is to be found. The downtrodden bootstrapping themselves into happiness is vintage Americana and not especially interesting outside of the particulars of how it was done.

Where Sherrill's insight is most valuable is in introducing us not only to the tenets of Tibetan Buddhism, but in exploring the fascinating ways it has been modernized for the 1990s. Imagine a Buddha who wears makeup, paints her nails red, and shops at mall stores while also believing that we will all reincarnate "countless times, as bugs and animals, even descend into the ghost realms and hell realms, before you achieved liberation from the endless hamster wheel of death and rebirth." One of the reasons the characters in this story continue reinventing themselves is that in Tibetan Buddhism "the student progresses toward enlightenment by practicing intense introspection and retraining the mind, learning to see the world differently. The student is taught, sometimes rather painfully, to abandon the notion of self (it is a delusion anyway) and to go in search of his or her own Buddha nature."

Tibetan Buddhism, however, is not just focused inwardly; in fact, true enlightenment comes through "being of benefit to all sentient beings," as Sherrill explains. "Everything in Tibetan Buddhism is about sentient beings—and ending the suffering of sentient beings. You say sentient beings instead of 'human beings' because you don't want to exclude anybody, and sentient is a way of describing all life-forms that are conscious, sensate—all people, all animals, all bugs and fish, including the invisible realms, the ghost realms and the hell realms." This is not a religion for the spiritually faint of heart. "There are eighteen different hells in Tibetan Buddhism, and there are countless beings there, too, all hoping to be released."

Are people released from their private hells in this religion? This is the subtext of Sherrill's narrative as she explores the many ways people deal with the slings and arrows of modern life, and the answer is a highly qualified one. Some do, some don't. Some leave too early, some stay too long. Some are undercommitted, some are overcommitted. The story of Betsy Elgin (aka Elizabeth, aka Alana) is an especially troubling one. Attractive, mid-thirties, happily married with children (but with the usual doubts about the meaning of it all), Elgin first encountered Zeoli (or should it be Alana met Jetsunma?) when the latter was doing psychic readings for twenty bucks a pop. Occasional meetings became regular rituals, time with the Buddha took precedence over time with the family, and before long, she recalls, "I remember laying in bed thinking, Here I am in my perfect town house with my perfect little kids and my perfect little husband and everything . . . but why do I feel so empty?"

To find out she consulted a channeled entity named Santu, who told her to divorce her husband, which she promptly did, moving into an apartment with her two daughters. In the sociological study of cults this is what is known as detachment—the individual is removed from her traditional reference sources and isolated into the new group that now controls her life. Elgin later confessed as much: "I was needy and compulsively fixated on her and our friendship. I was trying to make that replace what I had given up. I felt a need to have something." It still wasn't enough, as Sherrill explained: "She went through a period of promiscuity, becoming sexually involved with several men at the center and others outside. For nearly two years 'I was either at the temple,' she said, 'or out on a date. It was a bit crazy.' Her daughters were left alone at night. Eventually the older one moved out to live with her father. Her younger girl, just sixteen, was 'close to the edge.'" Unfortunately, this is not an isolated incident, and Sherrill considers the sometimes subtle differences between a cult and a religion.

Still, a much more common story is like that of Sherrill herself, a successful journalist at a prestigious publication (*The Washington Post*), who discovers in her self-searching that "aside from the well-trod pleasures of the quotidian—holidays at the beach, dance parties—you could still feel a greater need for something else entirely. You could feel a hunger and emptiness. You could be tormented by unanswered questions. Modern life leaves many people feeling insignificant and a bit lost. If you were living a spiritual life—and believed you were helping to end suffering—that could make you feel quite potent."

Indeed, and here we begin to approach that mystery of mysteries of why people believe. Sherrill does not give us the answer, but she does offer one explanation that carries an important qualifier: "There is nobility in sacrifice—any sacrifice. And as much as I didn't want to admit this, there is in fact a sort of ladder that people seem to ascend in order to be liberated from self-concern and see themselves as part of something larger. And sometimes people do ridiculous things to get there."

The rub, of course, is in finding that larger something without losing yourself along the way. It is the journey of a lifetime, a voyage we all must take if we want to find deeper spiritual meaning.

# APPENDIX I

# What Does It Mean to Study Religion Scientifically? Or, How Social Scientists "Do" Science

Chapter 4, "Why People Believe in God," involves a considerable amount of statistical analysis, so what follows is a brief explanation of what it means to study religion scientifically, and how social scientists "do" science. Human behavior, including religious behavior, is so complex that we must use statistics and probabilities to understand cause-and-effect relationships. Because humans are pattern-seeking animals, we cannot rely on intuitive guesses, as this might result in a scholarly version of "seek and ye shall find"—we may end up discovering what we are looking for in the data.

A good example of this can be seen in the powerful influence of Karl Marx's theory that social class is the most powerful force in human thought and behavior. For the past century historians intuitively interpreted historical events in this light. Starting with the assumption that social class dominates history, it is very easy to seek and find historical examples of it. In a 1992 biography of Charles Darwin by social historians Adrian Desmond and James Moore, for example, the authors struggled mightily to find evidence to prove their hypothesis that the Darwinian Revolution was class driven, with the upper class in opposition and the lower class in support. The biggest problem for the authors, of course, was that Darwin himself, from womb to tomb, was solidly embedded in the monied aristocracy of nineteenth-century England, thus their book's subtitle is *The Life of a Tormented Evolutionist* (tormented by the anxiety produced in leading this lower-class revolution). It was not until social scientists began to *test* this hypothesis that it became clear that socioeconomic status—SES, as it is called—is only one of many variables influencing human action, and more often than not it is a minor variable. In his 1996 book, *Born to Rebel,* for example, Frank Sulloway showed—contrary to what historians have believed for over a century—that proponents and opponents in the Darwinian Revolution, as well as the Copernican Revolution, the Protestant Reformation, the French Revolution, the civil rights movement, and many others, were *not* divided by social class. Whether someone was upper class, middle class, or lower class had little to no influence on his or her thinking. But anecdotes and narrative writing—the tools of the historian—will not reveal this because there is no mechanism to tell you if you are not simply seeking and finding anecdotes and quotes to support your hypothesis. The *only* way to know what is really going on is to conduct a formal test to find out which variables are significant and which are not.

One of the most common statistics used by social scientists is the *correlation coefficient,* represented by *r,* which has a range from .00 to ±1.0—from no relationship to a perfect relationship. The relationship of height and weight, for example, shows a high correlation whereas, say, height and I.Q. shows a low correlation. A negative correlation, signified by a minus sign in front of the value of *r,* represents associated values in opposite directions, as in golf skill and golf scores—as the first goes up, the second goes down. In the social sciences most correlations fall in the .00 to .50 range. The correlation of religious interests between identical twins raised apart, for example, is $r = .49$, a very significant figure that indicates a strong genetic component to religiosity—fully half of the variation between people on their religious interests can be accounted for by their genes. [*Note:* Normally psychologists square the $r$ to obtain the percentage of variance explained by genetic factors (e.g., an $r$ of .71 generates an $r^2$ of .50, or 50 percent of the variance accounted for), but in the case with twins this is not necessary because, as Arthur Jensen explained: "Most psychologists have learned to treat correlations as the square root of variance explained. But it is incorrect to take the square of twins or other kinship correlations to determine the proportion of variance attributable to genetic or environmental effects. The unsquared correlation itself is correctly interpreted as a proportion."]

The study reported in Chapter 4 was conducted by MIT social scientist Frank Sulloway and me. We employed a *multiple regression* analysis of our religiosity data, which is a statistical tool employed when more than one variable is a significant predictor of another. For example, we found that education negatively correlates with religiosity—as education goes up, religiosity goes down. But we also discovered that age and parental conflict are associated with a decrease in religiosity, while gender, parents' religiosity, and being raised to be religious, are all associated with an increase in religiosity. Since there are multiple causes of religiosity, we are obliged to conduct a multiple regression analysis to tease apart how these numerous variables operate separately and together.

The *p* value represents the *probability* that a given correlation could result by chance. A correlation is considered statistically significant if there is only a probability of 1 in 20 (or $p < .05$), of its being due to chance. A *p* value of .01 means that there is only 1 chance in 100 that the correlation happened by chance. A *p* of .001 is 1 in 1,000, and a *p* of .0001 means the likelihood of the correlation being due to chance is 1 in 10,000. Most of the correlations we found in our study were significant at the $p < .0001$ level or better. We report our sample size represented by *N*. Although our total sample size was 2,707, the *N* will vary from statistic to statistic, since not everyone answered every question on the survey. In general, our sample size was larger than most encountered in social science research (many of which rely on introductory psychology courses for their subject pools), especially in religious surveys. For example, Kenneth Pargament, in his comprehensive 1997 *The Psychology of Religion and Coping,* provides an extensive appendix listing 261 studies, including their sample sizes. The average *N* was 324.

Finally, to relate all this to a practical matter of concern to us all, Sulloway has shown that even seemingly small correlations of, say, .10, are "equivalent to improving your chances of surviving a potentially fatal disease, assuming that you take an effective medication, from 45 percent to 55 percent. This improvement represents an increase in survival of 22 percent over the base rate (55/45 = 1.22)." Sulloway's point is that if our lives were threatened by a deadly disease, most of us would gladly take a drug that would give us a 22 percent probability of improvement. A correlation of .30 represents a near doubling of survival probability from 35 percent to 65 percent, while a correlation of .50 is equivalent to a tripling of the probability of survival, from 25 percent to 75 percent. In this manner, small correlations often represent real and powerful effects. In other words, significant correlations matter, even if they appear small.

In the scientific study of religion—mainly involving three fields of the psychology of religion, the sociology of religion, and, to a lesser degree, the anthropology of religion—correlations, multiple regressions, and probabilities are the tools wielded by social scientists who want to better understand the nature of *Homo religiosus.* Strangely, however, these fields are almost wholly neglected by social scientists studying psychology, sociology, and anthropology in general—strange, considering how deeply important and vastly universal is the religious impulse. You will find only a couple of textbooks in these areas, and rarer still is the college course covering the scientific study of religion (with the possible exception of some seminaries and theological departments). I attribute this paradoxical dearth to the hands-off nature of religion in general—religion is something to be followed, God is someone to be worshipped. To focus the narrow and intense beam of scientific light into this often dark and murky corner of the human condition can be blinding at first. As I have discovered in conducting this empirical study, to most folks there is something mildly unsettling about being asked personal and penetrating questions about their most deeply held and cherished religious beliefs. Still, if you want to understand the human condition the study of religious belief cannot be neglected, and since science is the best method yet devised for uncovering cause-and-effect relationships, we must apply that method whenever and wherever possible.

# APPENDIX II

# Why People Believe in God—The Data and Statistics

The source of the general survey sample was Survey Sampling, Inc., in Fairfield, Connecticut, the same organization that provides random samples of Americans for many of the most notable political, social, and cultural surveys conducted by social scientists and the media. Before the mailing we tested numerous versions of the survey on approximately a thousand people, refining the questions so that the answers accurately reflected what we hoped to measure. Based on the feedback from these test surveys, we believe that the instrument we used to collect the data provides an accurate reflection of what Americans believe about God, some of the most important influencing variables on their belief, and why they believe. The statistics were run on BMDP and SYSTAT. The graphics in this appendix were initially produced by Frank Sulloway using SYSTAT. These graphics were subsequently redesigned by *Skeptic* magazine art director Pat Linse using Adobe Illustrator. The page numbers that follow are linked to the discussion of this survey in Chapter 4.

## END NOTES CORRELATIONS, DATA, AND STATISTICS

**Page 76.**

The correlation between religious conviction and belief in God, in the skeptics survey, is $r = .46$ ($N = 1650$, $t = 20.82$, $p < .0001$).

**Page 76.**

The correlation between religious conviction and belief in God, in the general survey, is $r = .63$ ($N = 960$, $t = 25.07$, $p < .0001$).

**Page 76.**

The difference between the two correlations, $r = .46$ and $r = .63$, is significant ($z = 4.82$, $p < .0001$).

**Pages 77–78.**

What follows is a list of the reasons skeptics say they (a) believe in God, (b) do not believe in God, and (c) why they think other people believe in God, in order of the number of responses given in the written portion of the survey. Answers have been grouped under a summary response that represents a paraphrasing of the originals.

### Why Skeptics Believe in God

1. Good design/natural/beauty/perfection/complexity of the world or universe (29.2 percent).

2. It is comforting, relieving, consoling, gives meaning and purpose to life (21.3%).

3. Experience of God in everyday life/God is in us (14.4 percent).

4. Just because/faith/need to believe in something (11.4 percent).

5. Without God there would be no morality (6.4 percent).

6. The Bible says so (5.5 percent).

7. The universe is God (4.0 percent).

8. Raised to believe in God (3.0 percent).

9. God has a plan for the world, history, destiny, and us (3.0 percent).

10. To account for good and avenge evil in the world (.10 percent).

*Cumulative total: 99.1 percent.* Other answers included "God answers prayers."

### Why Skeptics Think *Other People* Believe in God

1. It is comforting, relieving, consoling, gives meaning and purpose to life (21.5 percent).

2. Need to believe in an afterlife/fear of death and the unknown (17.8 percent).

3. Lack of exposure to science/lack of education/ignorance (13.5 percent).

4. Raised to believe in God (11.5 percent).

5. Good design of the world/natural beauty/perfection/ complexity (8.8 percent).

6. Culture is religious (7.2 percent).

7. Social/need for community (5 percent).

8. Brainwashed (4.5 percent).

9. Genetics/evolution (4.1 percent).

10. Just because/faith/need to believe in something (2.1 percent).

*Cumulative total: 96.0 percent.* Other answers included "I don't know," "religion is a meme virus," "to account for good and avenge evil in the world," and "schizophrenic/mad/nuts."

### Why Skeptics Do Not Believe in God

1. There is no proof for God's existence (37.9 percent).

2. There is no need to believe in God (13.2 percent).

3. It is absurd to believe in God (12.1 percent).

4. God is unknowable (8.3 percent).

5. Science provides all the answers we need (8.3 percent).

6. The Problem of Evil: pain, suffering, children dying, wars, holocausts, genocides, etc. (7.0 percent).

7. God is a product of the mind and culture (4.0 percent).

8. God is just another explanation for uncertainties and the unknown (3.1 percent).

9. God and religion are just a means of social control (2.4 percent).

10. Religion is bad for society, history, religious wars, religious crimes, etc. (2.1 percent).

*Cumulative total: 99.4 percent.* Other answers included "God is a product of primitive beliefs transferred to us," and "the burden of proof is on believers to prove God, not on us to disprove God."

**Page 79.**

The correlation between religiosity and being raised religiously is $r = .17$ ($N = 2084$, $t = 6.06$, $p < .0001$).

**Page 79.**

The correlation between religiosity and gender is $r = .17$ ($N = 2084$, $t = 8.55$, $p < .0001$).

**Page 79.**

The correlation between religiosity and parents' religiosity is $r = .12$ ($N = 2084$, $t = 4.03$, $p < .0001$).

**Page 79.**

The negative correlation between religiosity and education (as one goes up the other goes down) is $r = -.17$ ($N = 2084$, $t = -8.05$, $p < .0001$).

**Page 79.**

The negative correlation between religiosity and age (as one goes up the other goes down) is $r = -.08$ ($N = 2084$, $t = -3.68$, $p < .0001$).

**Page 79.**

The negative correlation between religiosity and parental conflict is $r = -.10$ ($N = 2084$, $t = -4.86$, $p < .0001$).

**Page 79.**

For the interaction between parental religiosity and parental conflict, as they relate to religious doubt, the partial correlation is $r = .08$ ($N = 926$, $t = 2.36$, $p < .01$); controlled for the two main effects.

**Page 80.**

For the interaction between attending church when growing up and parental conflict, as they relate to attending church now is $r = -.15$ ($N = 2595$, $t = -7.63$, $p < .0001$); controlled for the two main effects.

**Page 80.**

The negative correlation between interest in science and religiosity is $r = -.26$ ($N = 2341$, $t = -13.33$, $p < .0001$).

**Page 80.**

The correlation between interest in science and education is $r = .13$ ($N = 745$, $t = -3.70$, $p < .0001$), between science and gender is $r = -.24$ ($N = 745$, $t = -6.88$, $p < .0001$), between science and conscientious is $r = .09$ ($N = 745$, $t = 2.66$, $p < .0001$), and between science and openness to experience is $r = .21$ ($N = 745$, $t = 6.10$, $p < .0001$).

**Page 80.**

The negative correlation between being raised religiously and interest in science is $r = -.26$ ($N = 2341$, $t = -13.33$, $p < .0001$).

**Page 80.**

The correlation between religiosity and age of serious doubt is $r = .15$ ($N = 464$, $t = 3.19$, $p < .0001$).

**Page 80.**

The negative correlation between religiosity and political liberalism is $r = -.40$ ($N = 916$, $t = -13.34$, $p < .0001$).

**Page 82.**

The negative correlation between openness to experience and religiosity is $r = -.14$ ($N = 736$, $t = -3.75$, $p < .0001$).

**Page 82.**

The correlation between openness and religious doubt is $r = .18$ ($N = 744$, $t = 4.88$, $p < .0001$).

**Page 82.**

The negative correlation between openness and change (diminution) in religiosity is $r = -.09$ ($N = 719$, $t = -2.40$, $p < .01$).

**Page 82.**

The negative correlation between openness and the rate of church attendance is $r = -.11$ ($N = 557$, $t = -2.55$, $p < .01$).

**Page 82.**

The correlation between birth order and openness is $r = .11$ ($N = 526$, $t = 2.55$, $p < .01$).

**Page 82.**

The correlation between openness and political liberalism is $r = .28$ ($N = 705$, $t = 1.77$, $p < .0001$). The correlation between tender-mindedness and religiosity is $r = .12$ ($N = 435$, $t = 2.40$, $p < .05$). The partial correlation between birth order and liberalism is $r_p = .09$ ($N = 554$, $t = 2.10$, $p < .04$, where $r_p$ is a partial correlation). Finally, the correlation between birth order and tender-mindedness is ($r_p = .14$, controlled for the variables in the model).

**Page 83.**

The correlation between gender and "rational" reasons for belief in God (apparently intelligent design of the world, without God there is no basis for morality, the existence of evil, pain and suffering, and scientific explanations of the world) is $r = -.16$ ($N = 2085$, $t = -7.49$, $p < .0001$).

**Page 83.**

The correlation between gender and "emotional" reasons for belief in God (emotional comfort, faith, and desire for meaning and purpose in life) is $r = .19$ ($N = 2054$, $t = 8.92$, $p < .0001$).

**Page 83.**

The correlation between education and rational reasons for God's existence is $r = .14$ ($N = 2085$, $t = 6.36$, $p < .0001$).

**Page 83.**

The negative correlation between education and emotional reasons for God's existence is $r = -.16$ ($N = 2054$, $t = -7.13$, $p < .0001$).

**Page 84.**

The correlation between openness and rational reasons for God's existence is $r = .11$ ($N = 714$, $t = 2.84$, $p < .005$).

**Page 84.**

The negative correlation between openness and emotional reasons for belief in God is $r = -.12$ ($N = 710$, $t = -3.16$, $p < .002$).

**Page 84.**

The negative correlation between preferring rational reasons for God's existence and being raised religiously is $r = -.10$ ($N = 2085$, $df = 9/2075$, $t = -3.30$, $p < .001$).

**Page 84.**

The negative correlation between preferring rational reasons for God's existence and parents' religiosity is $r = -.07$ ($N = 2085$, $t = -2.48$, $p < .01$).

**Page 84.**

To the two questions (a) *In your own words, why do you believe in God, or why don't you believe in God?* and (b) *In your own words, why do you think most other people believe in God?*, the reasons people say they believe in God, and why they think other people believe in God, in order of the number of responses given in the written portion of the survey, are presented here. To control for possible experimenter bias in placing subjectively written answers into discrete categories, I had a research assistant score the same answers and found an interrater agreement—that is, consistency between raters in scoring—above 90 percent, more than acceptable in social science research. For example, for the more-frequent answer for why people believe in God—"Good design/natural beauty/perfection/complexity of the world or universe"—the interrater agreement was 91.4 percent; for the number-one answer for why other people believe in God—"It is comforting, relieving, consoling, gives meaning and purpose to life"—the interrater agreement was 98.4 percent.

### Why People Believe in God

1. Good design/natural beauty/perfection/complexity of the world or universe (28.6 percent).

2. Experience of God in everyday life/God is in us (20.6 percent).

3. It is comforting, relieving, consoling, gives meaning and purpose to life (10.3 percent).

4. The Bible says so (9.8 percent).

5. Just because/faith/need to believe in something (8.2 percent).

6. Raised to believe in God (7.2 percent).

7. God answers prayers (6.4 percent).

8. Without God there would be no morality (4.0 percent).

9. God has a plan for the world, history, destiny, and us (3.8 percent).

10. To account for good and avenge evil in the world (1.0 percent).

*Cumulative total: 99.8 percent.* Other answers that did not seem to fit into any category (and too few in number to justify a category of their own) included "because God is most powerful," "because He loves us," "the Jews survived," "near-death experiences," and "no other explanation."

### Why People Think Other People Believe in God

1. It is comforting, relieving, consoling, gives meaning and purpose to life (26.3 percent).

2. Raised to believe in God (22.4 percent).

3. Experience of God in everyday life/God is in us (16.2 percent).

4. Just because/faith/need to believe in something (13.0 percent).

5. Fear of death/unknown (9.1 percent).

6. Good design/natural beauty/perfection/complexity of the world or universe (6.0 percent).

7. The Bible says so (5.0 percent).

8. Without God there would be no morality (3.5 percent).

9. To account for good and avenge evil in the world (1.5 percent).

10. God answers prayers (1.0 percent).

*Cumulative total: 94.0 percent.* This question doubled the number of categories into which answers could reasonably be classified (compared to the first question), including (in order of quantity): "I don't know," "Jesus/God saved them," "the culture is religious," "fear of death," "conversion experience," "need to believe in an afterlife," "need for community," and a few unique answers including "stupidity," "they are alive," "because of the narcotic effect," and "it is the right thing to do."

## HOW WE BELIEVE AND DISBELIEVE

### The Variables That Shape Our Religiosity and Belief in God

These graphs present the most important variables that shape our religiosity and belief in God, including:

1. Belief in God in relation to being raised religiously;
2. Religiosity in relation to being raised religiously;
3. Religiosity in relation to education;
4. Belief in God in relation to age;
5. Religiosity in relation to age;
6. Belief in God in relation to parental conflict and being raised religiously or nonreligiously;
7. Religiosity in relation to parental conflict;
8. Being raised religious or nonreligiously.

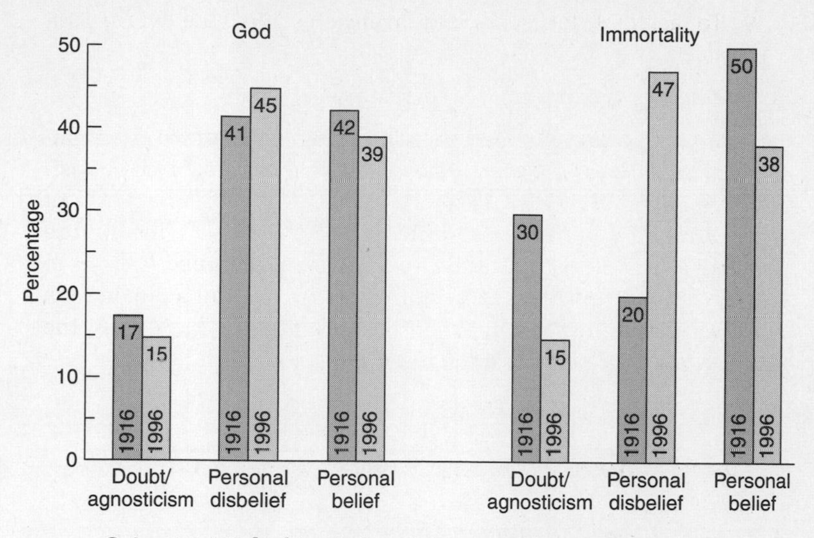


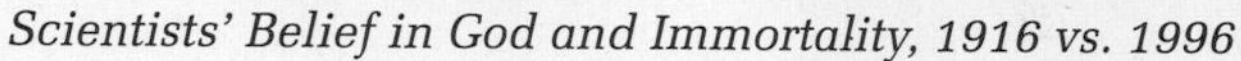
*Scientists' Belief in God and Immortality, 1916 vs. 1996*

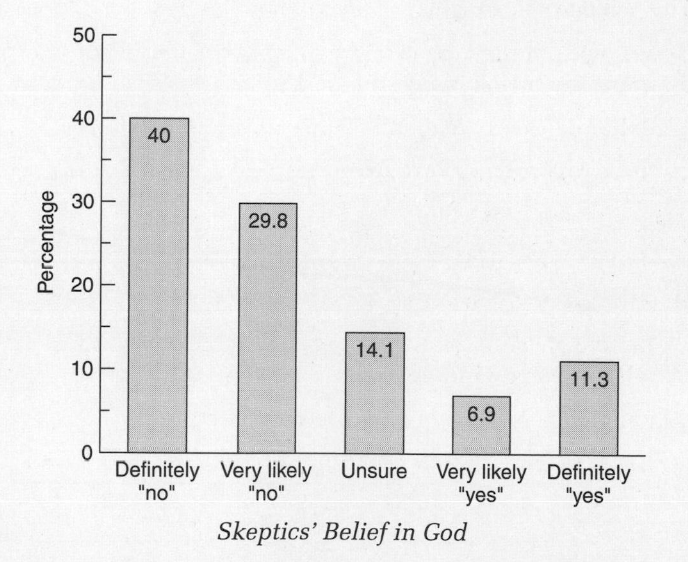


*Skeptics' Belief in God*

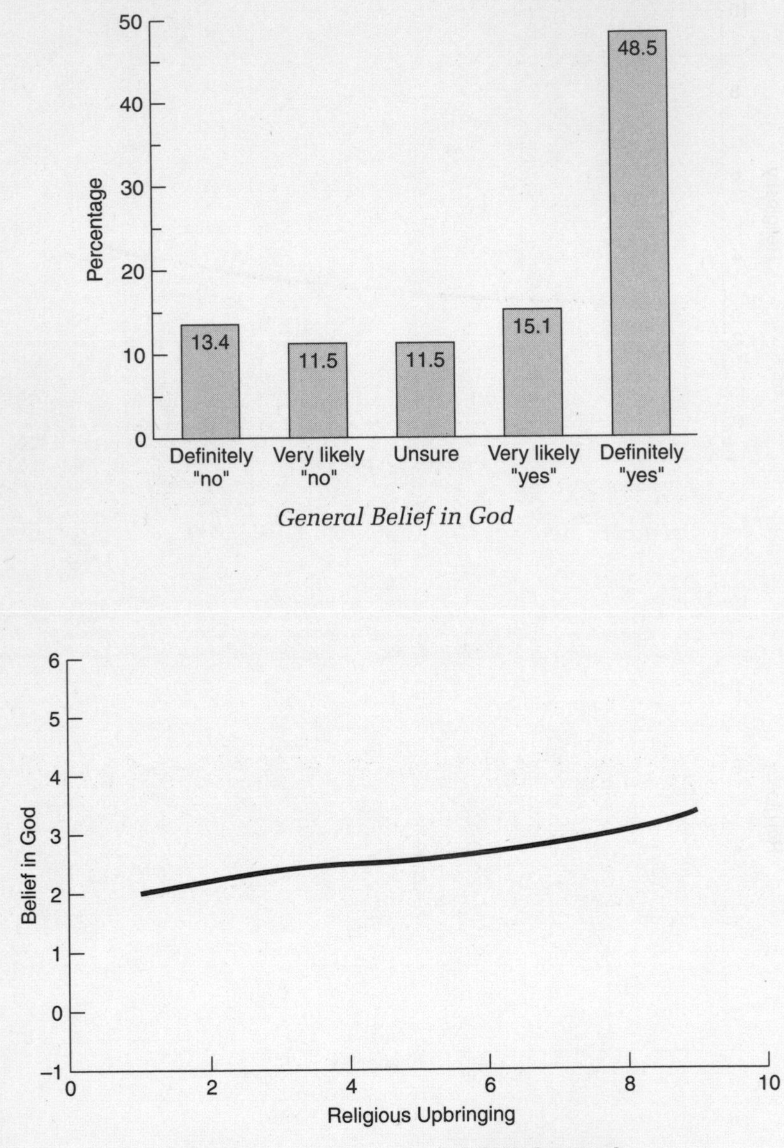


*General Belief in God*

*Religious Upbringing and Belief in God*

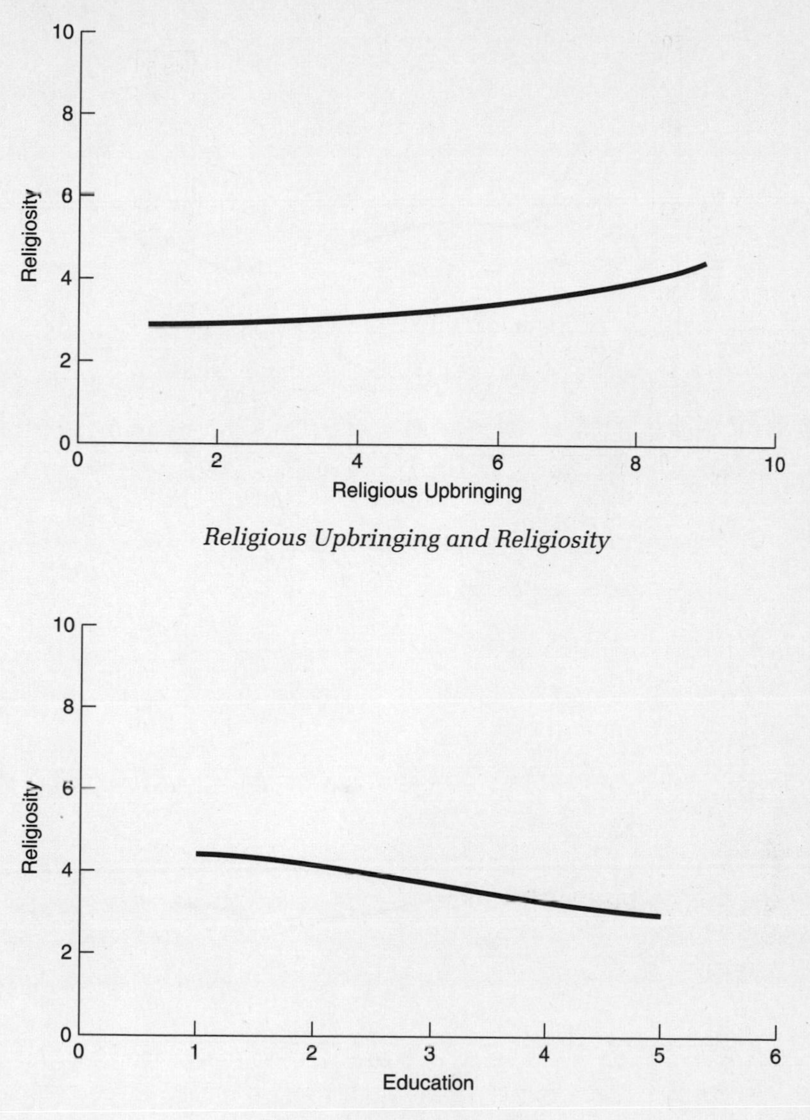


*Religious Upbringing and Religiosity*

*Education and Religiosity*

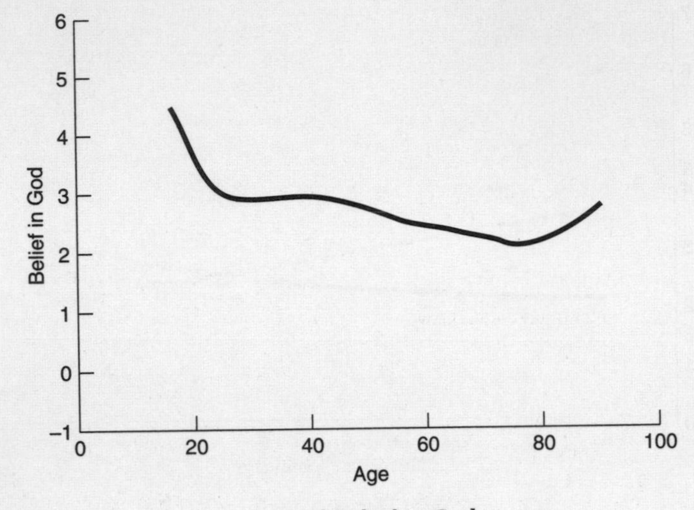


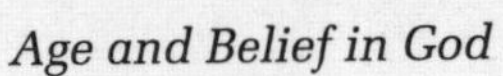
*Age and Belief in God*

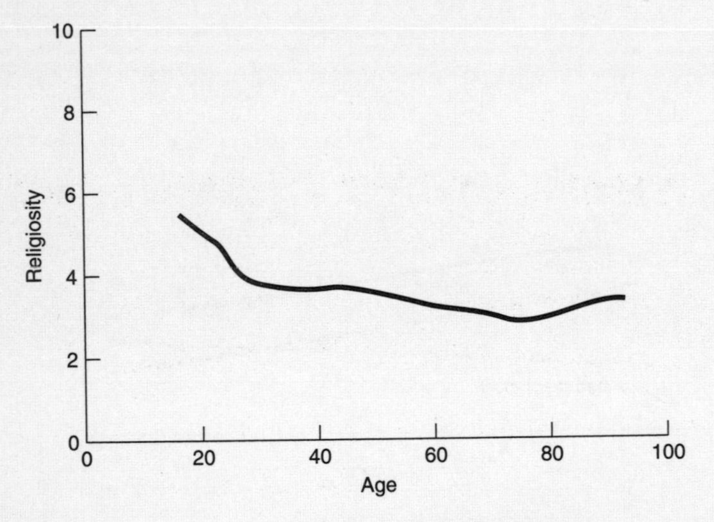


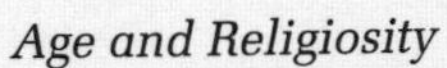
*Age and Religiosity*

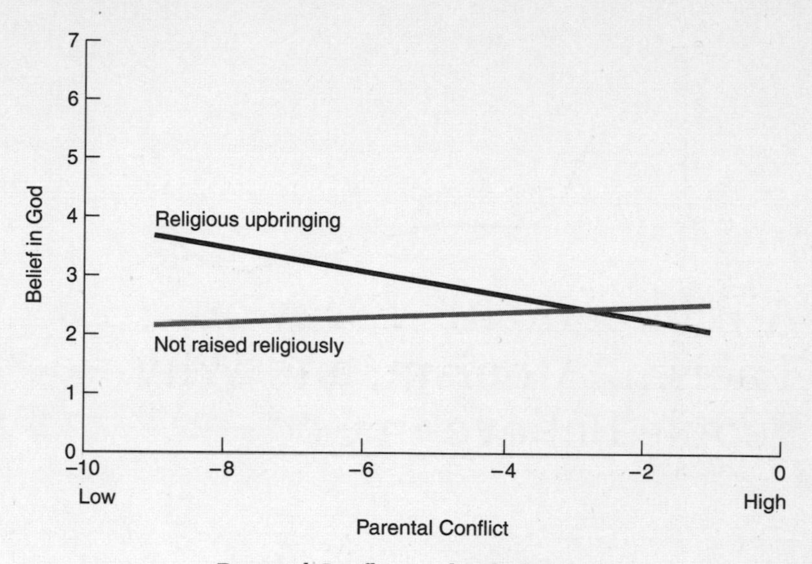


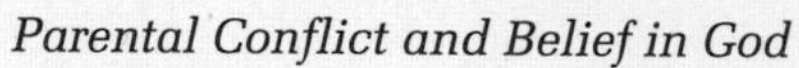
*Parental Conflict and Belief in God*

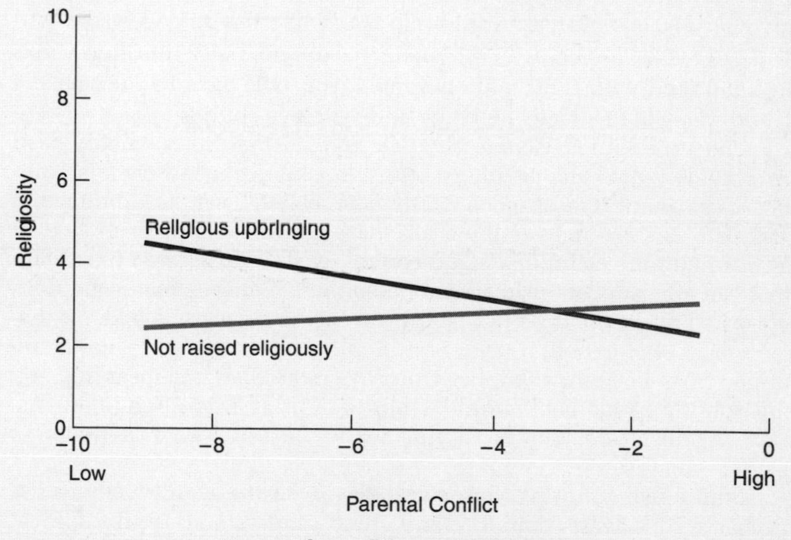


*Parental Conflict and Religiosity*

# A Bibliographic Essay on Theism, Atheism, and Why People Believe in God

I have been careful to document all of my sources in the text. For further reading in this area, I recommend going to the original sources, which I list in the Notes that follow this short essay. Not included in my analysis, however, are additional books that the serious reader will want to read. This list is by no means complete, but with these books in hand, and with their bibliographies, you will have before you the grand scope of theological writing and religious studies.

The philosopher George Smith's book, *Atheism,* is widely read and highly touted among atheists. In it he claims that atheism is nothing more than "the absence of theistic belief," yet he admits that "atheism is sometimes defined as 'the belief that there is no God of any kind, or the claim that a god cannot exist'." He goes on to explain that "an atheist is not primarily a person who believes that a god does not exist; rather, he does not believe in the existence of a god." If theism is "belief-in-god," says Smith, atheism is simply "no-belief-in-god." Yet he admits that the word has more than one meaning, and the subtitle of his book would seem to gainsay this disclaimer: *The Case* Against *God.* Shouldn't this read *The Case Against* Belief *in God?*

Smith also confirms the pejorative use of the word: "Atheism is probably the least popular—and least understood—philosophical position in America today. It is often approached with fear and mistrust, as if one were about to investigate a doctrine that advocates a wide assortment of evils—from immortality, pessimism and communism to outright nihilism." Smith cites the influential twentieth-century philosopher, A. E. Taylor, who, in his 1947 book,

*Does God Exist?*, blames atheism for the two World Wars: "The world has directly to thank [atheism] for the worst evils of 'modern war'." With atheists never amounting to more than a couple of percentage points of any population, it is hard to imagine who had instigated and fought those wars. In my opinion Smith's book on atheism is the best available, yet reading it only reinforced my conviction that the agnostic/nontheist position defended in Chapter 1 is the most reasonable one with our present understanding of the universe and our current social conditions.

Michael Martin, who has probably thought and written about atheism as much as anyone in history, in his 1990 magnum opus, *Atheism: A Philosophical Justification,* agrees with Smith that "atheists have been attacked for flaws in their moral character: it has often been alleged that they cannot be honest and truthful. For example, in 1724 Richard Bentley, an English Christian apologist, maintained that 'no atheist as such can be a true friend, an affectionate relation, or a loyal subject'." Martin also cites the Evidence Amendment Act of 1869, where "atheists in England were considered incompetent to give evidence in a court of law," and a case in America in 1856 where "one Ira Aldrich was disqualified as a witness in an Illinois case after he testified that he did not believe in a God that 'punishes people for perjury, either in this world or any other'." Martin's purpose is not to convert theists to atheists (at least not directly), but to "provide good reasons for being an atheist." In doing so he distinguishes between several types of atheists, including *positive atheists* who "disbelieve in god or gods" and *negative atheists* who "have no belief in a god or gods." Martin further classifies negative atheists into "*the broad sense of negative atheism,*" where there is "an absence of belief in any god or gods," and the "*narrow sense of negative atheism,* according to which an atheist is without a belief in a personal being who is omniscient, omnipotent, and completely good and who created heaven and earth." The reason for this latter classification is to distinguish between *deists, pantheists,* and *polytheists* who believe in an impersonal god or gods (small g) and *theists* who believe in a personal God who is all knowing, all powerful, all good, and created the universe. He does not stop there, further classifying *positive atheists:* "A positive atheist in the broad sense is a person who disbelieves that there is any god or gods and a positive atheist in the narrow sense is a person who disbelieves that there is a personal being who is omniscient, omnipotent, and completely good and who created heaven and earth."

How many angels can dance on the head of a pin? It depends on how you define *angel,* define *dance,* and define the size of the head of the pin. All these distinctions are useful for medieval theologians and modern academic philosophers seeking semantic precision and clarification for the many nuances of human thought, but I still find my own distinction between *atheism* (there is no God) and *nontheism* (no belief in God) as statements about personal beliefs to be adequate. How many deists, pantheists, or polytheists do you know? None I would guess. For our purposes at the beginning of a new millennium in the Western world, it is safe to assume that when we are discussing "God" we all know what we mean by this term: an all-powerful, all-knowing, all-good higher being who created the universe and us and grants ever-lasting life. If you believe this, you are a *theist*. If you do not believe this, you are a *nontheist*. If you believe God is unknowable through science or reason, you are an *agnostic*.

So much of this book deals with the history, psychology, sociology, anthropology, and evolutionary biology of religion that the Notes itself is the place to turn. But I especially want to mention a number of influential texts in this field that attempt to get at the answer to the question of why people believe in God and need religion. These include first and foremost David Wulff's *Psychology of Religion* and Ralph Hood, Bernard Spilka, Bruce Hunsberger, and Richard Gorsuch's *The Psychology of Religion*. On the anthropology of religion see Daniel Pals' *Seven Theories of Religion*, Brian Morris's *Anthropological Studies of Religion*, Arthur Lehmann and James Myers's *Magic, Witchcraft, and Religion*, Gerald Larue's *Ancient Myth and Modern Man*, Adolf Jensen's *Myth and Cult Among Primitive Peoples*, E. E. Evans-Pritchard's *Theories of Primitive Religion*, and Mircea Eliade's *The Sacred and the Profane* and *Myths, Dreams, and Mysteries*. On the sociology of religion one must begin with the founder's text, Max Weber's *The Sociology of Religion*; and two excellent reviews of the philosophy of religion are Basil Mitchell's *The Philosophy of Religion* and Michael Peterson, William Hasker, Bruce Reichenbach, and David Basinger's *Reason and Religious Belief*.

The body of literature associated with the God Question is Brobdingnagian. Visit any decent size library and look under God, theology, religion, philosophy, philosophy of religion, sociology of religion, anthropology of religion, morality, ethics, the Bible, biblical studies, biblical criticism, and other subjects and you will see what I mean. It is overwhelming. There are scholars who select one specialty and spend their entire careers in the narrow minutiae of that field. What I am aiming for here is clear communication about the God Question with virtually anyone interested in the subject, without sacrificing scholarly integrity.

To that end I also recommend the following edited volumes of readings that will provide a wide range of perspectives on the God Question: John Hick's *The Existence of God: From Plato to A. J. Ayer on the Question "Does God Exist?"* presents in depth all the major arguments for God's existence as well as the critiques of these arguments in a very balanced treatment; Peter Angeles's *Critiques of God:*

*A Major Statement of the Case Against Belief in God* has an obvious skeptical slant but presents essays by major thinkers such as Bertrand Russell, Antony Flew, and Sidney Hook; and Gordon Stein's *An Anthology of Atheism and Rationalism,* which is exactly what it says it is. Peter Angeles's *The Problem of God* provides an easy to read, succinct introduction to the debate, and Keith Parsons's *God and the Burden of Proof* provides responses to Alvin Plantinga's and Richard Swinburne's analytic defense of theism. Many of C. S. Lewis's books deal with these arguments from a Christian perspective, particularly *Mere Christianity, Miracles,* and *The Problem of Pain.* The most famous arguments for God's existence were laid down by St. Anselm in his eleventh-century book, *Proslogion,* and by St. Thomas Aquinas in his thirteenth-century book, *Summa Theologica,* the arguments with which I deal in Chapter 5.

Among modern authors I find Rabbi Harold Kushner's books, particularly *When Bad Things Happen to Good People* and *Who Needs God* to be highly readable, reasonable, and respectful of the reader's intelligence (although he is definitely a theist, most atheists and nontheists would find these works quite palatable). In the final chapter of *Who Needs God,* entitled "Why Is God So Hard to Find?," Kushner answers the question by admitting the human nature of religion: "Religion is first and foremost the community through which you learn to understand the world and grow to be human." If God is absent, it is "because we have stopped looking for Him." The implication of both statements (to me anyway) is that God exists in our minds and religion is the product of human culture.

In the arena of biblical studies, a subject not under my purview (with the exception of my analysis of the "Bible Code" in Chapter 5), the best place to begin is with the world-renowned biblical scholar Richard Elliott Friedman's *Who Wrote the Bible?* His book on *The Disappearance of God* is a fascinating read on the changing role of God from the early books of the Hebrew Bible to the later. Finally, his 1999 book, *The Hidden Book in the Bible,* puts forth a cutting-edge and potentially very controversial re-editing of the Old Testament in which Friedman pulls out of numerous books what he believes to be a continuous narrative, written by one author, that he calls the "first prose narrative." Reactions from the biblical scholarship community remain to be seen. There are a number of other excellent authors in this field as well. Burton Mack's *The Lost Gospel* and especially his *Who Wrote the New Testament?* are particularly good; the latter is particularly strong in giving cultural context to the construction of the New Testament. Randel Helms's *Gospel Fictions* and *Who Wrote the Gospels?* give a good perspective on these four books from a professor of literature who analyzes them as he would any important text in Western literature. Tim Callahan, the religion editor for *Skeptic* magazine, has written two splendid books that are at once comprehensive and well written: *Bible Prophecy: Failure or Fulfillment?* and *The Secret Origins of the Bible.* Callahan is especially good about cutting to the chase of an argument and exposing both its strengths and weaknesses. I have learned a lot in reading these works, especially about the antecedents to the Bible, most of which are shrouded in mystery and almost never discussed.

Finally, there is no excuse for not going to the primary source, and that is the Bible itself. To that end, my research has been greatly aided by several outstanding reference guides recognized by Bible scholars to be the very best places to begin a more thorough analysis of the good book (and available through any religious bookstore): *The Interpreter's Bible: A Commentary in Twelve Volumes, The Anchor Bible Dictionary, The Interpreter's Dictionary of the Bible, The Interpreter's One-Volume Commentary on the Bible, The New Oxford Annotated Bible with the Apocrypha, The Oxford Companion to the Bible, The Literary Guide to the Bible,* and *The Jerome Biblical Commentary.* For adding these to the Skeptics Society Research Library, and for keeping our analyses on religious and biblical matters on a professional level, I am grateful to Bruce Mazet.

# NOTES

## INTRODUCTION TO THE PAPERBACK EDITION

**PAGE xxiii.** For complete poll results see *Skeptic*, 8/1: 16.

**PAGE xxvi.** Barrow, J. D., and F. Tipler. 1988. *The Anthropic Cosmological Principle*. Oxford: Oxford University Press, p. vii.

**PAGE xxvii.** Dyson, F. 1979. *Disturbing the Universe*. New York: Harper and Row.

**PAGE xxvii.** Davies, P. 1999. *The Fifth Miracle*. New York: Simon and Schuster, p. 246.

**PAGE xxviii.** Meyer, S. C. 1999. "Word Games: DNA, Design, and Intelligence." *Touchstone*, 12 (4): 44–50.

**PAGE xxviii.** Hawking, S. 1996. "Quantum Cosmology." In Hawking, S., and R. Penrose. *The Nature of Space and Time*. Princeton, N.J.: Princeton University Press, pp. 89–90.

**PAGE xxix.** Smolin, L. 1997. *The Life of the Cosmos*. Oxford: Oxford University Press.

**PAGE xxix.** Gardner, J. N. 2000. "The Selfish Biocosm: Complexity as Cosmology." *Complexity*, 5 (3): 34–45.

**PAGE xxxii.** Gardner, M. 1983. *The Whys of a Philosophical Scrivener*. New York: William Morrow, pp. 209–211, 221, 223.

**PAGE xxxii.** Gardner, M. 1999. *The Whys of a Philosophical Scrivener*. (Paperback edition.) New York: St. Martin's Griffin, p. 384.

## CHAPTER 1: *Do You Believe in God? The Difference in Our Answers and the Difference It Makes*

**PAGES 3–4.** Lindsey, H. 1970. *The Late Great Planet Earth*. New York: Bantam, pp. 83, 86.

**PAGE 5.** Kushner, H. 1981. *When Bad Things Happen to Good People*. New York: Avon Books.

**PAGE 6.** Lindsey, H. 1970. *The Late Great Planet Earth*. New York: Bantam, pp. 40, 50, 55–56.

**PAGE 7.** Medawar, P. 1982. *Pluto's Republic: The Art of the Soluble and Induction and Intuition in Scientific Thought*. Oxford: Oxford University Press, p. 2.

**PAGE 7.** Huxley, T. H. 1894. *Collected Essays*, Vol. 5. New York: D. Appleton and Co., pp. 237–238.

**PAGES 7–8.** Elson, J. T. 1966. "Theology: Toward a Hidden God," *Time*, April 8: 85.

**PAGE 8.** Huxley, T. H. 1894. *Collected Essays*, Vol. 5. New York: D. Appleton and Co., p. 238.

**PAGE 9.** Shermer, M. 1997. "The Annotated Gardner: An Interview with Martin Gardner—Founder of the Modern Skeptical Movement," *Skeptic*, 4/1: 56–60.

**PAGE 10.** Symons, D. 1997. Personal correspondence. November 11.

**PAGES 10–11.** Sarich, V. 1997. Personal correspondence. November 11.

**PAGE 11.** Tillich quote from Edwards, P. (ed.). 1964. *The Existence of God.* New York: Macmillan, p. 2.

## CHAPTER 2: *Is God Dead? Why Nietzsche and* Time *Magazine Were Wrong*

**PAGE 17.** Nietzsche, F. 1954 (1883–1885). "Thus Spoke Zarathustra." In *The Portable Nietzsche,* (trans. and ed.) Walter Kaufmann. New York: Viking, p. 2.

**PAGE 17.** McGuire, B. 1965. "Eve of Destruction," ABC Dunhill Music.

**PAGE 17.** Elson, J. T. 1966. "Theology: Toward a Hidden God," *Time*, April 8: 82–87.

**PAGES 17–18.** Shute, N. 1957. *On the Beach.* New York: Ballantine Books, pp. 148–149.

**PAGE 19.** Lennon quote in Pareles, J., and P. Romanowski (eds.). 1983. *The Rolling Stone Encyclopedia of Rock and Roll.* New York: Rolling Stone, p. 34.

**PAGE 19.** Clarke, A. C. 1968. *2001: A Space Odyssey.* New York: Signet, pp. 34, 221.

**PAGE 20.** Elson, J. T. 1966. "Theology: Toward a Hidden God," *Time*, April 8: 82.

**PAGES 20–21.** Ibid.: 82–87.

**PAGE 21.** Gallup. G. H., Jr., and F. Newport. 1996. "Gallup Poll of American Religious Beliefs," *Wall Street Journal*, January 30.

**PAGE 22.** Barna, G. 1996. *Index of Leading Spiritual Indicators.* Dallas, Tex.: Word Publishing.

**PAGE 22.** Pew Research Center. 1997. "Poll Says 71% Believe in God," *San Francisco Chronicle.* AP Release. December 22.

**PAGE 22.** For a general discussion of the secularization thesis see Iannaccone, L. R. 1998. "Introduction to the Economics of Religion." *Journal of Economic Literature*, 36, September: 1465–1496.

**PAGE 23.** Greeley, A. 1997. "Pie in the Sky While You're Alive: Life after Death and Supply Side Religion." Paper delivered to the American Sociological Association annual meeting, Toronto, September, p. 2.
See also:

Finke, R. 1989. "Demographics of Religious Participation: An Ecological Approach, 1850–1971," *Journal for the Scientific Study of Religion*, 28: 45–58.

Finke, R. 1990. "Religious Deregulation: Origins and Consequences," *Journal of Church and State*, 32: 609–626.

Finke, R. 1992. "An Unsecular America." In *Religion and Modernization: Sociologists and Historians Debate the Secularization Thesis*. Oxford: Clarendon Press.

Chaves, M., and D. Cann. 1992. "Regulation, Pluralism, and Religious Market Structure: Explaining Religion's Vitality," *Rationality and Society*, 4(3): 272–290.

Stark, R., and L. R. Iannaccone. 1994. "A Supply-Side Reinterpretation of the 'Secularization' of Europe," *Journal for the Scientific Study of Religion*, 33:230–252.

**PAGE 23.** Smith, A. 1965 (1776). *An Inquiry into the Nature and Causes of the Wealth of Nations*. New York: Modern Library, pp. 3–4.

**PAGES 23–24.** Greeley, A. 1997. "Pie in the Sky While You're Alive: Life after Death and Supply Side Religion," pp. 16, 17.

**PAGE 25.** Finke, R., and R. Stark. 1992. *The Churching of America, 1776–1990: Winners and Losers in Our Religious Economy*. New Brunswick, N.J.: Rutgers University Press.

**PAGE 25.** Van Biema, D. 1997. "Does Heaven Exist?" *Time*, March 24: 73.

**PAGE 25.** Woodward, K. L. 1997. "Is God Listening?" *Newsweek*, March 31: 57–65.

**PAGE 25.** Koerner, B. I. 1997. "Is There Life after Death?" *U.S. News and World Report*, March 31: 59–66.

**PAGE 26.** Cheney, P. 1996. "Most Ontarians Believe in Miracles, Survey Finds," *The Toronto Star*, December 27: A10.

**PAGES 26–27.** Promise Keepers data cited in Stodghill, R. 1997. "God of Our Fathers," *Time*, October 6: 34–39.

**PAGE 27.** McCartney quote in Swomley, J. M. 1997. "Storm Troopers in the Culture War," *The Humanist*, September–October: 10–13.

**PAGE 27.** Television viewing data cited in Stein, J. 1997. "The God Squad," *Time*, September 22: 105–106.

**PAGE 27.** Reading data cited in Marquand, R. 1997. "Religious Reading Becoming More Popular," *Christian Science Monitor*, August 27.

**PAGE 28.** Friedman, R. E. 1995. *The Disappearance of God*. New York: Little, Brown, pp. 7, 284.

**PAGE 28.** Goodenough, U. 1998. *The Sacred Depths of Nature*. Oxford: Oxford University Press, p. xvii.

## CHAPTER 3: *The Belief Engine: How We Believe*

**PAGE 35.** Gallup, G. H., Jr., and F. Newport. 1991. "Belief in Paranormal Phenomena among Adult Americans," *Skeptical Inquirer*, 5(2): 137–147.

**PAGE 36.** Cosmides, L., and J. Tooby. 1994. "The Center for Evolutionary Psychology at the University of California, Santa Barbara" (Descriptive Brochure).
See also:

Barkow, J. H., L. Cosmides, and J. Tooby. 1992. *The Adapted Mind*. Oxford: Oxford University Press.

Miele, F. 1996. "The (Im)moral Animal," *Skeptic*, 4/1: 42–49.

**PAGE 36.** Pinker, S. 1997. *How the Mind Works*. New York: W. W. Norton, pp. 27-31.

**PAGE 37.** Noelle, D. C. 1998. Personal correspondence. March 19.
See also:

Karmiloff-Smith, A. 1995. *Beyond Modularity: A Developmental Perspective on Cognitive Science*. London: Bradford.

**PAGE 37.** Mithen, S. 1996. *The Prehistory of the Mind: The Cognitive Origins of Art, Religion, and Science*. London: Thames and Hudson, p. 163.

**PAGE 38.** For the relationship between magic and uncertainty see Vyse, S. A. 1997. *Believing in Magic: The Psychology of Superstition*. Oxford: Oxford University Press.

**PAGE 38.** For the relationship between worship and health see Schumaker, J. F. 1992. "Mental Health Consequences of Irreligion." In *Religion and Mental Health*, (ed.) J. F. Schumaker. Oxford: Oxford University Press.

**PAGE 38.** For the relationship between magic and power see Harris, M. 1974. *Cows, Pigs, Wars, and Witches: The Riddles of Culture*. New York: Vintage.

**PAGES 38–39.** Gould, S. J., and R. Lewontin. 1979. "The Spandrels of San Marco and the Panglossian Paradigm: A Critique of the Adaptionist Programme," *Proceedings of the Royal Society*, V. B205: 581–598.

**PAGES 39–41.** Evans-Pritchard, E. E. 1976 (1937). *Witchcraft, Oracles and Magic Among the Azande.* Oxford: Oxford University Press, pp. 178–179, 181.

**PAGE 41.** Oubré, A. 1996. "Plants, Property, and People," *Skeptic,* 4/2: 72–77.

**PAGE 41.** Chagnon, N. 1992. *Yanomamö,* 4th ed. New York: Harcourt, Brace, and Jovanovich, pp. 69–70, 105.

**PAGES 41–43.** Malinowski, B. 1954 (1925). *Magic, Science, and Religion.* New York: Doubleday, pp. 17, 29, 139–140.

**PAGE 44.** For medieval magical thinking examples see:

Seligman, K. 1948. *The History of Magic.* New York: Pantheon.

Thomas, K. 1971. *Religion and the Decline of Magic.* New York: Scribner's.

Grillot de Givry, E. 1973. *The Illustrated Anthology of Sorcery, Magic and Alchemy,* (trans.) J. Courtenay Locke. New York: Causeway Books.

Russell, J. B. 1980. *A History of Witchcraft: Sorcerers, Heretics and Pagans.* London: Thames and Hudson.

Surles, R. L. 1993. *Medical Numerology.* New York: Garland.

**PAGE 45.** Vyse, S. A. 1997. *Believing in Magic: The Psychology of Superstition.* Oxford: Oxford University Press, p. 105.

**PAGE 47.** Krakauer, J. 1997. *Into Thin Air.* New York: Villard, p. 128.

Vyse, S. A. 1997. *Believing in Magic: The Psychology of Superstition.* Oxford: Oxford University Press, pp. 84–85.

**PAGE 49.** See www.vanpraagh.com

**PAGES 49–50.** Van Praagh, J. 1997. *Talking to Heaven.* New York: Dutton.

**PAGE 50.** Witchel, A. 1998. "James Van Praagh Profile." *New York Times,* February 22.

## CHAPTER 4: *Why People Believe in God: An Empirical Study on a Deep Question*

**PAGE 61.** Humphrey, N. 1996. *Leaps of Faith: Science, Miracles, and the Search for Supernatural Consolation.* New York: Basic Books, pp. 153–155.

**PAGES 61–62.** Vyse, S. A. 1997. *Believing in Magic: The Psychology of Superstition.* Oxford: Oxford University Press, pp. 84–85.

**PAGE 62.** Kosko, B. 1993. *Fuzzy Thinking: The New Science of Fuzzy Logic.* New York: Hyperion, p. 278.

**PAGES 63–64.** Plomin, R. 1989. "Environment and Genes: Determinants of Behavior," *American Psychologist,* 44(2): 107.

**PAGE 64.** Waller, N. G., B. Kojetin, T. Bouchard, D. Lykken, and A. Tellegen. 1990. "Genetic and Environmental Influences on Religious Attitudes and Values: A Study of Twins Reared Apart and Together," *Psychological Science*, 1(2): 138–141.

**PAGE 64.** Segal, N. L. 1999. *Entwined Lives: Twins and What They Tell Us about Human Behavior.* New York: Dutton.

For a complete discussion of cultural and biological influences on religiosity see: Wulff, D. M. 1991. *Psychology of Religion: Classic and Contemporary Views.* New York: John Wiley and Sons.

Hood, R. W., B. Spilka, B. Hunsberger, and R. Gorsuch. 1996. *The Psychology of Religion: An Empirical Approach*, 2nd ed. New York: Guilford Press.

Parejko, K. 1998. "Selection for Credulity: A Biologist's View of the Millennium," *Skeptic*, 7/1: 37–39.

**PAGES 65–66.** Ramachandran, V. S., W. S. Hirstein, K. C. Armel, E. Tecoma, and V. Iragui. 1997. "The Neural Basis of Religious Experience." Paper delivered to the Annual Conference of the Society of Neuroscience. October. Abstract #519.1. Vol. 23, Society of Neuroscience.

**PAGE 65.** Hotz, R. L. 1997. "Brain Could Affect Religious Response, Researchers Report," *Los Angeles Times*, October 31: B1.

**PAGE 65.** Hotz, R. L. 1998. "Seeking the Biology of Spirituality," *Los Angeles Times*, April 26: A1, A32.

**PAGE 65.** Russell and Arbib quotes in Hotz, 1998.

**PAGE 66.** Ramachandran, V. S., and S. Blakeslee. 1998. *Phantoms in the Brain.* New York: Morrow.

**PAGE 66.** Persinger, M. A. 1987. *Neuropsychological Bases of God Beliefs.* New York: Praeger.

**PAGE 67.** Persinger, M. A. 1993. "Paranormal and Religious Beliefs May Be Mediated Differently by Subcortical and Cortical Phenomenological Processes of the Temporal (Limbic) Lobes," *Perceptual and Motor Skills*, 76: 247–251.

**PAGE 67.** Shermer, M. 1997. *Why People Believe Weird Things.* New York: W. H. Freeman and Company.

**PAGE 67.** For more background on Michael Persinger see Regush, N. 1995. "Brain Storms and Angels," *Equinox*, July–August: 63–75.

**PAGES 68–69.** Brierre de Boismont, A. J. F. 1859. *On Hallucinations: A History and Explanation of Apparitions, Visions, Dreams, Ecstasy, Magnetism, and Somnambulism*, (trans.) R. T. Hulme. London: Henry Renshaw, pp. 340, 346, 348, 378, 383.

**PAGE 69.** Persinger, M. A. 1987. *Neuropsychological Bases of God Beliefs.* New York: Praeger, p. 138.

**PAGE 69.** Noelle, D. C. 1998. Personal correspondence. March 31.

**PAGE 69.** For additional data on religiosity see Goldhaber, G. 1996. "Religious Belief in America: A New Poll," *Free Inquiry*, 16(3): 34–40.

**PAGES 69–70.** Dawkins, R. 1976. *The Selfish Gene.* Oxford: Oxford University Press, pp. 99, 192.

**PAGE 70.** Brodie, R. 1996. *Virus of the Mind: The New Science of the Meme.* Seattle, Wash.: Integral Press.

**PAGE 70.** Lynch, A. 1996. *Thought Contagion: How Belief Spreads through Society.* New York: Basic Books.

**PAGE 70.** Blackmore, S. 1997. "The Power of the Meme Meme: Religion as a Meme Suggests How a Science of Memetics Illuminates Human Evolution," *Skeptic*, 5/2: 46.

**PAGE 71.** Polichak, J. W. 1998. "Memes—What Are They Good For? A Critique of Memetic Approaches to Information Processing," *Skeptic*, 6/3: 46.

**PAGE 71.** Blackmore, S. 1999. *The Meme Machine.* Oxford: Oxford University Press.

**PAGE 72.** Leuba, J. H. 1916. *The Belief in God and Immortality: A Psychological, Anthropological and Statistical Study.* Boston: Sherman, French & Co.

**PAGES 72–73.** Larson, E. J., and L. Witham. 1997. "Scientists Are Still Keeping the Faith," *Nature*, 386: 435.

**PAGE 73.** Bergman, G. R. 1996. "Religious Beliefs of Scientists: A Survey of the Research," *Free Inquiry*, 16(3): 41–46.

**PAGE 73.** Leuba, J. H. 1934. "Religious Beliefs of American Scientists," *Harper's Magazine*, 169: 291–300.

**PAGES 73–74.** Bishop's data is reported in Huba, S. 1999. "Biblical Version of Creation OK by Americans," *The Detroit News*, April 6.

**PAGE 74.** Shermer, M. 1995. "Skeptics Society Survey," *Skeptic*, 3/4: 20.

**PAGE 74.** The Carnegie Commission study is reported in Stark, R., and L. R. Iannaccone. 1994. "A Supply-Side Reinterpretation of the 'Secularization' of Europe," *Journal for the Scientific Study of Religion*, 33: 230–252.

**PAGE 79.** U.S. Census Bureau data is reported in Day, J. C., and A. E. Curry. 1998. "Educational Attainment in the United States: March 1998 (Update)," *Current Population Reports.* Washington, D.C.: U.S. Department of Commerce.

**PAGE 80.** Brand, C. 1981. "Personality and Political Attitudes." In *Dimensions of Personality*, (ed.) R. Lynn. Oxford: Pergamon Press.

**PAGE 81.** Sulloway, F. 1996. *Born to Rebel: Birth Order, Family Dynamics, and Creative Lives.* New York: Pantheon, p. 269.

**PAGE 81.** U.S. Congressional data reported in Benson, P. L., and D. L. Williams. 1982. *Religion on Capitol Hill: Myths and Realities.* San Francisco: Harper & Row, p. 124.

**PAGE 81.** Wulff, D. M. 1991. *Psychology of Religion: Classic and Contemporary Views.* New York: John Wiley and Sons.

**PAGE 82.** For data on the five-factor model see McCrae, R. R., and P. T. Costa, Jr. 1987. "Validation of the Five Factor Model of Personality across Instruments and Observers," *Journal of Personality and Social Psychology*, 52: 81–90.
See also:

McCrae, R. R., and P. T. Costa, Jr. 1990. *Personality in Adulthood.* New York: Guilford Press.

**PAGE 82.** For the relationship between openness and birth order see Sulloway, F. 1996. *Born to Rebel: Birth Order, Family Dynamics, and Creative Lives.* New York: Pantheon.

**PAGE 82.** For data on the adjectives used in the inventory see Costa, P. T., Jr., and R. R. McCrae. 1992. *NEO PI–R Professional Manual: Revised NEO Personality Inventory (NEO PI–R) and NEO Five–Factor Inventory (NEO–FFI).* Odessa, Fla.: Psychological Assessment Resources, Inc.

**PAGE 85.** Attribution theory is discussed in Gilbert, D. T., B. W. Pelham, and D. S. Krull. 1988. "On Cognitive Busyness: When Person Perceivers Meet Persons Perceived," *Journal of Personality and Social Psychology*, 54: 733–739.

**PAGE 85.** Tavris, C., and C. Wade. 1997. *Psychology in Perspective*, 2nd ed. New York: Longman/Addison Wesley, p. 332.
See also:

Nisbett, R. E., and L. Ross. 1980. *Human Inference: Strategies and Shortcomings of Social Judgment.* Englewood Cliffs, N.J.: Prentice-Hall.

**PAGE 87.** Gallup poll data reported in Witham, L. 1997. "Many Scientists See God's Hand in Evolution," *The Washington Times*, April 11: A8.

**PAGE 87.** Levy, D. 1998. "Four Simple Facts Behind the Miracle of Life," *Parade Magazine*, June 21: 12.

**PAGE 88.** Gibbon, E. 1952 (1781). *The Decline and Fall of the Roman Empire.* Chicago: University of Chicago Press. Great Books of the Western World.

## CHAPTER 5: *O Ye of Little Faith: Proofs of God and What They Tell Us about Faith*

**PAGE 89.** Geivett, D. 1993. *Evil and the Evidence for God.* Philadelphia: Temple University Press.

**PAGE 89.** Geivett, D., and G. R. Habermas. 1997. *In Defense of Miracles.* Downers Grove, Ill.: InterVarsity Press.

**PAGE 90.** McDowell, J. 1972. *Evidence That Demands a Verdict.* San Bernardino, Calif.: Campus Crusade for Christ.

**PAGE 91.** Aquinas, T. 1952 (1273). *Summa Theologica.* Great Books of the Western World. R. M. Hutchins (ed. in chief). Chicago: University of Chicago Press.

**PAGE 92.** Guth, A. 1997. *The Inflationary Universe: The Quest for a New Theory of Cosmic Origins.* Reading, Mass.: Addison Wesley.

**PAGE 92.** Hawking, S. W. 1996. *The Cambridge Lectures.* New York: Dove Books.

**PAGE 92.** For Martin Gardner's opinions on the God question see Shermer, M. 1997. "The Annotated Gardner: An Interview with Martin Gardner." *Skeptic,* 5/2: 56–61.

**PAGE 93.** For Anselm quote see Hick, J. 1964. *The Existence of God: From Plato to A. J. Ayer on the Question "Does God Exist"?* New York: Collier Books.

**PAGE 95.** Lewis, C. S. 1947. *Miracles.* New York: Macmillan, p. 10.

**PAGE 96.** *"Pass It On": The Story of Bill Wilson and How the A.A. Message Reached the World.* 1984. New York: Alcoholics Anonymous. World Services, Inc.

**PAGE 97.** Gardner, M. 1983. *The Whys of a Philosophical Scrivener.* New York: William Morrow.

**PAGE 98.** For more on the role of personality in beliefs see Sulloway, F. 1996. *Born to Rebel: Birth Order, Family Dynamics, and Creative Lives.* New York: Pantheon.

**PAGE 99.** Ross, H. 1993. *The Creator and the Cosmos: How the Greatest Scientific Discoveries of the Century Reveal God.* Colorado Springs, Colo.: Navpress.

**PAGE 99.** Ross, H. 1994. *Creation and Time: A Biblical and Scientific Perspective on the Creation-Date Controversy.* Colorado Springs, Colo.: Navpress.

**PAGE 99.** Ross, H. 1996. *Beyond the Cosmos: What Recent Discoveries in Astronomy and Physics Reveal about the Nature of God.* Colorado Springs, Colo.: Navpress.

**PAGE 99.** For a refutation of Ross see Stenger, V. J. 1998. "The Functional Equivalent of God: A Refutation of the Cosmological Design Argument," *Skeptic*, 6/3: 89–91.

**PAGE 99.** Glynn, P. 1997. *God: The Evidence: The Reconciliation of Faith and Reason in a Postsecular World.* Rocklin, Calif.: Prima Publishing, pp. 165–166.

**PAGE 100.** For a refutation of Glynn see Shallit, J. 1998. "Designing the Designer. A Review of Patrick Glynn, *God: The Evidence*," *Skeptic*, 6/2: 80–82.

**PAGE 100.** Schroeder, G. L. 1997. *The Science of God: The Convergence of Scientific and Biblical Wisdom.* New York: Free Press, pp. 58, 61, 71.

**PAGE 101.** Raymo, C. 1998. *Skeptics and True Believers: The Exhilarating Connection between Science and Religion.* New York: Walker and Co., p. 8.

**PAGE 101.** Mazet, B. 1998. "A Case for God," *Skeptic*, 6/2: 50–55.

**PAGES 102–103.** White, M., and J. Gribbin. 1992. *Stephen Hawking: A Life in Science.* New York: Plume/Penguin, pp. 3, 166–167.

**PAGES 102–103.** Hawking, S. W. 1988. *A Brief History of Time.* New York: Bantam Books, pp. 140–141, 175.

**PAGE 103.** Davies, P. 1992. *The Mind of God: The Scientific Basis for a Rational World.* New York: Touchstone, p. 189.

**PAGE 103.** Barrow, J., and F. Tipler. 1986. *The Anthropic Cosmological Principle.* Oxford: Oxford University Press.

**PAGES 103–104.** Tipler, F. 1994. *The Physics of Immortality: Modern Cosmology, God and the Resurrection of the Dead.* New York: Doubleday, pp. 21–23, 677.

**PAGES 104–105.** Pannenberg, W. 1997. "Modern Cosmology: God and the Resurrection of the Dead." Lecture delivered at the Innsbruck Conference, June, p. 1.

**PAGES 105–106.** Tipler, F. 1997. Personal correspondence. October 17.

**PAGE 106.** Tipler, F. 1994. *The Physics of Immortality: Modern Cosmology, God and the Resurrection of the Dead.* New York: Doubleday, pp. 347, 356–357.

**PAGES 106–107.** Harmon N. (ed.). *The Interpreter's Bible.* Nashville, Tenn.: Abingdon Press, Vol. 1, pp. 874–875.

**PAGES 107–108.** Deutsch, D. 1997. The Fabric of Reality: *The Science of Parallel Universes—and Its Implications.* New York: Allen Lane/Penguin.

**PAGE 108.** Guth, A. 1997. *The Inflationary Universe: The Quest for a New Theory of Cosmic Origins.* Reading, Mass.: Addison Wesley, p. 276.

**PAGES 108–109.** Smolin, L. 1997. *The Life of the Cosmos.* Oxford: Oxford University Press, pp. 197–198, 201.

**PAGE 109.** For a history and analysis of the Louisiana case see Shermer, M. 1991. "*Science Defended, Science Defined: The Louisiana Creationism Case,*" *Science, Technology, and Human Values,* 16 (4): 517–539.

**PAGE 109.** Behe, M. 1996. *Darwin's Black Box: The Biochemical Challenge to Evolution.* New York: Free Press.

**PAGE 113.** Berta, A. 1994. "What Is a Whale?" *Science,* 263: 180–181.

**PAGES 113–114.** Behe, M. 1996. *Darwin's Black Box: The Biochemical Challenge to Evolution.* New York: Free Press, p. 39.

**PAGE 114.** Pennock, R. T. 1999. *Tower of Babel: The Evidence Against New Creationism.* Cambridge, Mass.: MIT Press.

**PAGE 115.** Johnson, P. 1991. *Darwin on Trial.* Downers Grove, Ill.: InterVarsity Press.

**PAGE 115.** Scott, E. C. 1998. "Two Kinds of Materialism: Keeping Them Separate Makes Faith and Science Compatible," *Free Inquiry,* Spring: 37–38.

**PAGE 115.** Dembski, W. A. 1998. *The Design Inference: Eliminating Chance through Small Probabilities.* Cambridge, England: Cambridge University Press.

For a complete refutation of Behe and the new intelligent design/irreducible complexity argument see Stenger, V. J. 1998. "The Functional Equivalent of God: A Refutation of the Cosmological Design Argument," *Skeptic,* 6/3: 89–91. See also:

Gilchrist, G. W. 1997. "The Elusive Scientific Basis of Intelligent Design Theory." *Report of the National Center for Science Education,* May–June: 14–15.

Coyne, J. A. 1996. "God in the Details: The Biochemical Challenge to Evolution," *Nature,* September 19: 227–228.

Miller, K. R. 1996. "Book Review of *Darwin's Black Box,*" *Creation/Evolution,* 16(2): 36–40.

See www.talkorigins.org for an ongoing discussion of intelligent design creationism.

**PAGE 116.** Dembski, W. A. (ed.). 1998. *Mere Creation: Science, Faith, and Intelligent Design.* Downers Grove, Ill.: InterVarsity Press.

**PAGE 117.** Drosnin, M. 1997. *The Bible Code.* New York: Simon and Schuster.

**PAGE 117.** Rips, E., D. Witztum, and Y. Rosenberg. 1994. "Equidistant Letter Sequences in the Book of Genesis." *Satistical Science,* 9 (4): 429–438.

**PAGE 118.** Drosnin, M. 1997. *The Bible Code*. New York: Simon and Schuster, p. 108.

**PAGE 119.** Hendel, R. S. 1997. "The Secret Code Hoax," *Bible Review*, August: 24.
See also:

Sternberg, S. 1997. "Snake Oil for Sale," *Bible Review*, August: 24–25.

**PAGE 119.** Woodward, K. L. 1997. "Is God Listening?" *Newsweek*, March 31: 57–65.

**PAGE 120.** Dart, J. 1997. "Does God's Hand Write in Code?" *Los Angeles Times*, June 10: A1, A24.

**PAGE 120.** Jeffrey, G. R., and Y. Rambsel. 1995. *The Signature of God*. Spring Arbor Publisher.

**PAGE 120.** McKay's findings are on his web page at cs.anu.edu.au./-bdm/dilugiss/statsci

**PAGE 120.** Friedman, R. E. 1987. *Who Wrote the Bible?* New York: Simon and Schuster.

**PAGE 120.** Friedman, R. E. 1998. *The Hidden Book in the Bible*. San Francisco: Harper.

**PAGE 121.** Hendel, R. S. 1997. "The Secret Code Hoax," *Bible Review*, August: 23.

**PAGE 121.** Larue, G. 1970. Old Testament Life and Literature. Boston: Allyn and Bacon.

**PAGE 121.** Drosnin, M. 1997. *The Bible Code*. New York: Simon and Schuster, pp. 58, 134, 142.

**PAGE 122.** Woodward, K. L. 1997. "Is God Listening?" *Newsweek*, March 31: 57–65.

## CHAPTER 6: *In a Mirror Dimly, Then Face to Face: Faith, Reason, and the Relationship of Religion and Science*

**PAGES 126–127.** Todd, D., and D. Menzel. 1954. *The Story of the Starry Universe*. New York: Popular Science Library, p. 381.

**PAGE 127.** Pope John Paul II. 1997. "Message to the Pontifical Academy of Sciences." Reprinted in *The Quarterly Review of Biology*, December, 72(4): 381–383.

**PAGE 130.** Draper, J. W. 1874. *History of the Conflict between Religion and Science*. New York: Appleton, p. ix.

**PAGE 130.** White, A. D. 1896. *A History of the Warfare of Science with Theology in Christendom*. New York: Appleton, p. viii.

**PAGE 130.** Larson, E. J. 1997. *Summer for the Gods: The Scopes Trial and America's Continuing Debate over Science and Religion*. New York: Basic Books.

**PAGES 130–131.** Dembski, W. A. (ed.). 1998. *Mere Creation: Science, Faith, and Intelligent Design*. Downers Grove, Ill.: InterVarsity Press, p. 14.

**PAGE 132.** Pannenberg, W. 1981. "Theological Questions to Scientists," *Zygon*, 16: 65–77.

**PAGE 132.** Tipler, F. 1994. *The Physics of Immortality: Modern Cosmology, God and the Resurrection of the Dead*. New York: Doubleday.

**PAGE 133.** Pope John Paul II. 1997. "Message to the Pontifical Academy of Sciences." Reprinted in *The Quarterly Review of Biology*, December, 72(4): 381–383.

**PAGE 133.** Gould, S. J. 1997. "Nonoverlapping Magisteria: Science and Religion Are Not in Conflict, for Their Teachings Occupy Distinctly Different Domains," *Natural History*, March: 16–20.

**PAGE 133.** Gould, S. J. 1999. *Rocks of Ages: Science and Religion in the Fullness of Life*. New York: Ballantine Books.

**PAGE 133.** Ruse, M. 1997. "John Paul II and Evolution," *The Quarterly Review of Biology*, December, 72(4): 394.

**PAGE 133.** Scott, E. C. 1997. "Creationists and the Pope's Statement," *The Quarterly Review of Biology*, December, 72(4): 406.

**PAGE 135.** Consolmagno quote in: Ortega, T. 1998. "High Priests of Astronomy," *Astronomy*, December: 61.

**PAGE 135–138.** Pope John Paul II. *Fides et Ratio*.

**PAGE 140.** Wallace, A. F. C. 1966. *Religion: An Anthropological View*. New York: Random House.

**PAGES 140–141.** Wilson, E. O. 1978. *On Human Nature*. Cambridge, Mass.: Harvard University Press, pp. 169–171.

**PAGE 141.** For data on church involvement after the L.A. riots see Melton, J. G. 1998. "Why People Need Religion," *Skeptics Society Annual Conference*. California Institute of Technology. May.

## CHAPTER 7: *The Storytelling Animal: Myth, Morality, and the Evolution of Religion*

**PAGE 143.** Gazzaniga, M. S. 1998. *The Mind's Past*. Berkeley: University of California Press, p. 27.

**PAGE 145.** For data in suport of an evolutionary analysis of diet see: Sapolsky, R. M. 1998. *Why Zebras Don't Get Ulcers: An Updated Guide to*

*Stress, Stress-Related Diseases, and Coping*. New York: W. H. Freeman and Company.
See also:

Widmaier, E. P. 1998. *Why Geese Don't Get Obese (And We Do): How Evolution's Strategies for Survival Affect Our Everyday Lives*. New York: W. H. Freeman and Company.

**PAGE 145.** Alcock, J. 1975. *Animal Behavior: An Evolutionary Approach*. Sunderland, Mass.: Sinauer Associates.

**PAGE 145.** Eibl-Eibesfeldt, I. 1970. *Ethology: The Biology of Behavior*. New York: Holt, Rinehart and Winston.

**PAGE 146.** Cosmides, L., and J. Tooby. 1992. "Cognitive Adaptations for Social Exchange." In *The Adapted Mind*, (eds.) J. H. Barkow, L. Cosmides, and J. Tooby. Oxford: Oxford University Press, p. 5.

**PAGE 146.** Wilson, E. O. 1975. *Sociobiology: The New Synthesis*. Cambridge, Mass.: Harvard University Press.

**PAGE 146.** Wilson, E. O. 1978. *On Human Nature*. Cambridge, Mass.: Harvard University Press.

**PAGES 146–148.** Wilson, E. O. 1998. *Consilience: The Unity of Knowledge*. New York: Knopf, pp. 127, 128, 149–150.

**PAGES 148–149.** Wason, P. 1966. "Reasoning." In *New Horizons in Psychology*, (ed.) B. M. Foss. Harmondsworth, England: Penguin.

**PAGE 149.** Cosmides, L., and J. Tooby. 1992. "Cognitive Adaptations for Social Exchange." In *The Adapted Mind*, (eds.) J. H. Barkow, L. Cosmides, and J. Tooby. Oxford: Oxford University Press, pp. 205, 221.

**PAGE 149.** Landau, M. 1984. "Human Evolution as Narrative," *American Scientist*, 72: 262–268.

**PAGE 150.** Brunvand, J. H. 1981. *The Vanishing Hitchhiker*. New York: W. W. Norton.

**PAGE 151.** Campbell, J. 1949. *The Hero with a Thousand Faces*. Princeton, N.J.: Princeton University Press, p. 382.

**PAGES 151–152.** Campbell, J. 1972. *Myths to Live* By. New York: Bantam Books, pp. 221–222.

**PAGES 154–157.** The mythic stories recounted and documented in Campbell, J. 1968. *The Masks of God: Creative Mythology*. New York: Viking, pp. 113–123.
See also:

Child, C. G. 1904. *Beowulf and the Finnesburg Fragment*. New York: Houghton Mifflin.

Lawrence, W. W. 1928. *Beowulf and the Epic Tradition.* Cambridge, Mass.: Harvard University Press.

**PAGE 155.** Konner, M. 1982. *The Tangled Wing: Biological Constants on the Human Spirit.* New York: Harper, pp. 5, 171.

**PAGE 156.** Wilson, E. O. 1998. Personal correspondence. July 7.

**PAGES 157–158.** For data and analysis on the evolution of language, see Johanson, D., and B. Edgar. 1996. *From Lucy to Language.* New York: Simon and Schuster, p. 106.
See also:

Pinker, S. 1994. *The Language Instinct.* New York: W. W. Norton.

Tattersall, I. 1995. *The Fossil Trail.* Oxford: Oxford University Press.

**PAGE 158.** Deacon, T. W. 1997. *The Symbolic Species: The Co-Evolution of Language and the Brain.* New York: W. W. Norton.
See also:

Cashdan, E. 1989. "Hunters and Gatherers: Economic Behavior in Bands." In *Economic Anthropology,* (ed.) S. Plattner. Stanford, Calif.: Stanford University Press.

Weissner, P. 1982. "Risk, Reciprocity and Social Influences on !Kung San Economics." In *Politics and History in Band Societies,* (eds.) F. Leacock and R. B. Lee. Cambridge, England: Cambridge University Press.

**PAGE 158.** Damas, D. 1972. "The Copper Eskimo." In *Hunters and Gatherers Today,* (ed.) M. G. Biccieri. Prospect Heights, Ill.: Waveland Press, p. 40.

**PAGE 158.** Isaac, G. L. 1978. "Food Sharing and Human Evolution: Archaeological Evidence from the Plio-Pleistocene of East Africa," *Journal of Anthropological Research,* 34: 311–325.

**PAGE 158.** Binford, L. 1981. *Bones: Ancient Men and Modern Myths.* New York: Academic Press.

**PAGE 158.** For a summary of the hunting debate see Cartmill, M. 1993. *A View to a Death in the Morning: Hunting and Nature Through History.* Cambridge, Mass.: Harvard University Press.

**PAGE 159.** Bettinger, R. L. 1991. *Hunter-Gatherers: Archaeological and Evolutionary Theory.* New York: Plenum Press, p. 158.

**PAGE 159.** Chagnon, N. 1992. *Yanomamö,* 4th ed. New York: Harcourt, Brace, and Jovanovich, pp. 80–86.

**PAGE 160.** Dunbar, R. 1996. *Grooming, Gossip and the Evolution of Language.* Cambridge, Mass.: Harvard University Press, pp. 61–79.

**PAGE 161.** Barkow, J. H., L. Cosmides, and J. Tooby. 1992. *The Adapted Mind.* Oxford: Oxford University Press, pp. 627–628.

**PAGE 164.** For data on families see Wilson, E. O. 1978. *On Human Nature.* Cambridge, Mass.: Harvard University Press.

**PAGE 164.** For discussion of inclusive fitness and reciprocal altruism see Alexander, R. D. 1979. *Darwinism and Human Affairs.* Seattle: University of Washington Press.
See also:

Miele, F. 1996. "The (Im)moral Animal," *Skeptic,* 4/1: 42–49.

**PAGE 165.** Sober, E., and D. S. Wilson. 1998. *Unto Others: The Evolution and Psychology of Unselfish Behavior.* Cambridge, Mass.: Harvard University Press, pp. 27, 92.

**PAGE 165.** Williams, G. C. 1966. *Adaptation and Natural Selection: A Critique of Some Current Evolutionary Thought.* Princeton, N.J.: Princeton University Press.

**PAGE 165.** Alexander, R. D. 1987. *The Biology of Moral Systems.* New York: Aldine De Gruyter.

**PAGE 165.** Ghiselin, M. 1974. *The Economy of Nature and the Evolution of Sex.* Berkeley: University of California Press, p. 247.

**PAGE 166.** Hamilton, W. D. 1975. "Innate Social Aptitudes of Man: An Approach from Evolutionary Genetics." *In Biosocial Anthropology,* (ed.) R. Fox. New York: John Wiley and Sons, pp. 135–136.

**PAGE 166.** For fuzzy logic see Kosko, B. 1993. *Fuzzy Thinking: The New Science of Fuzzy Logic.* New York: Harper.

**PAGE 166.** Axelrod, R. 1984. *The Evolution of Co-operation.* New York: Penguin.

**PAGE 167.** Smith, J. M. 1982. *Evolution and the Theory of Games.* Oxford: Oxford University Press, p. 204.
See also:

von Neuman, J., and O. Morgenstern. 1944. *Theory of Games and Economic Behavior.* Princeton, N.J.: Princeton University Press.

**PAGE 167.** Irons, W. 1996. "In Our Own Self Image: The Evolution of Morality, Deception, and Religion," *Skeptic,* 4/2: 50–61.

**PAGE 169.** Tylor, E. B. 1871. *Primitive Culture: Researches into the Development of Mythology, Philosophy, Religion, Language, Art, and Custom.* London: John Murray.

**PAGE 169.** Frazer, J. G. 1924. *The Golden Bough: A Study in Magic and Religion.* New York: Macmillan.

**PAGE 169.** Freud, S. 1927. *The Future of an Illusion,* (trans.) J. Strachey. New York: W. W. Norton.

**PAGE 169.** Durkheim, E. 1912. *Elementary Forms of the Religious Life,* (trans.) J. W. Swain. New York: Collier Books.

**PAGE 169.** Marx, K. 1978 (1869). *The Marx–Engels Reader,* (ed.) R. C. Tucker. New York: W. W. Norton.

**PAGE 169.** Eliade, M. 1957. *The Sacred and the Profane: The Nature of Religion,* (trans.) W. R. Trask. New York: Harcourt Brace.

**PAGE 169.** Evans-Pritchard, E. E. 1965. *Theories of Primitive Religion.* Oxford: Clarendon Press.

**PAGE 169.** Geertz, C. 1966. "Religion as a Cultural System." In *Anthropological Approaches to the Study of Religion,* (ed.) M. Banton. London: Tavistock Press.

**PAGE 169.** Pals, D. L. 1996. *Seven Theories of Religion.* Oxford: Oxford University Press, p. 269.
See also:

Wulff, D. M. 1991. *Psychology of Religion*: Classic and Contemporary Views. New York: John Wiley and Sons.

Hood, R. W., B. Spilka, B. Hunsberger, and R. Gorsuch. 1996. *The Psychology of Religion: An Empirical Approach,* 2nd. ed. New York: Guilford Press.

## CHAPTER 8: *God and the Ghost Dance: The Eternal Return of the Messiah Myth*

**PAGE 176.** Harrison quote in Milligan, E. A. 1976. *Dakota Twilight.* Hicksville, N.Y.: Exposition Press, p. 121.

**PAGES 176–178.** Mooney, J. 1896. *The Ghost Dance Religion and the Sioux Outbreak of 1890,* (ed.) A. F. C. Wallace. Chicago: University of Chicago Press, pp. 771–772, 780–781.

**PAGES 179.** Mooney, J. 1896. *The Ghost Dance Religion and the Sioux Outbreak of 1890,* (ed.) A. F. C. Wallace. Chicago: University of Chicago Press, p. 798.

**PAGES 180–181.** Harrison and Miles quote in Utley, R. M. 1963. *The Last Days of the Sioux Nation.* New Haven, Conn.: Yale University Press, pp. 105, 111, 127.

**PAGE 182.** For more information on Wounded Knee and the Ghost Dance see Miller, D. H. 1959. *Ghost Dance.* Lincoln: University of Nebraska Press.
See also:

Utley, R. M. 1993. *The Lance and the Shield: The Life and Times of Sitting Bull.* New York: Macmillan.

Utley, R. M. 1973. *Frontier Regulars: The United States Army and the Indians, 1866–1890*. New York: Macmillan.

Marshall, S. L. A. 1972. *Crimsoned Prairie: The Indian Wars on the Great Plains*. New York: Scribner's.

Brown, D. 1970. *Bury My Heart at Wounded Knee*. New York: Bantam.

**PAGES 182–183.** Mooney, J. 1896. *The Ghost-Dance Religion and the Sioux Outbreak of 1890*, (ed.) A. F. C. Wallace. Chicago: University of Chicago Press, p. 1.

**PAGES 183–185.** La Barre, W. 1970. *The Ghost Dance: Origins of Religion*. Garden City, N.Y.: Doubleday, pp. 233–234, 238–239.

**PAGE 185.** For examples and a discussion of cargo cults see Harris, M. 1974. *Cows, Pigs, Wars, and Witches: The Riddles of Culture*. New York: Vintage, p. 133.

**PAGE 186.** Worsley, P. 1958. *The Trumpet Shall Sound: A Study of Cargo Cults in Melansia*. New York: Shocken, p. 226.

**PAGE 186.** Helms, R. 1988. *Gospel Fictions*. Buffalo: Prometheus Books, p. 10.

**PAGE 187.** Mack, B. L. 1995. *Who Wrote the New Testament?: The Making of the Christian Myth*. New York: HarperCollins, pp. 43, 226.

## CHAPTER 9: *The Fire That Will Cleanse: Millennial Meanings and the End of the World*

**PAGE 192.** For analysis of the Bahai's sect see Balch, R. W., J. Domitrovich, B. L. Mahnke, and V. Morrison. 1997. "Fifteen Years of Failed Prophecy." In *Millennium, Messiahs, and Mayhem*, (eds.) T. Robbins and S. J. Palmer. New York: Routledge.

**PAGE 197.** Reagan quote in Abanes, R. 1998. *End Times Visions*. New York: Four Walls Eight Windows.

**PAGE 197.** Wojcik, D. 1997. *The End of the World as We Know It: Faith, Fatalism, and Apocalypse in America*. New York: New York University Press, pp. 209–210.

**PAGE 197.** Lindsey, H. 1984. *There's a New World Coming*. Irvine, Calif.: Harvest House.

**PAGE 198.** Augustine. 1952. *The City of God*. Great Books of the Western World. Chicago: University of Chicago Press, 22:30.

**PAGE 198.** For a discussion of Columbus' prophetic beliefs see Walls, P. M. 1985. "Prophecy and Discovery: On the Spiritual Origins of Christopher Columbus' 'Enterprise of the Indies,'" *American Historical Review*, 900(1): 73–102.

**PAGE 199.** Gould, S. J. 1997. *Questioning the Millennium. A Rationalist's Guide to a Precisely Arbitrary Countdown.* New York: Harmony Books/ Random House.

**PAGE 200.** Michelet, J. 1844. *History of France.* Whittaker and Co., p. 143.

**PAGES 200–201.** Mackay, C. 1841. *Extraordinary Popular Delusions and the Madness of Crowds.* New York: Crown, p. 258.

**PAGE 201.** Schwartz, H. 1990. *Century's End: A Cultural History of the Fin de Siècle from the 990s through the 1990s.* New York: Doubleday.

**PAGE 201.** Stearns, P. N. 1996. *Millennium III, Century XXI.* New York: Westview Press, p. 26.

**PAGES 201–202.** Landes, R. 1997. "The Apocalyptic Year 1000." In *The Year 2000: Essays on the End,* (eds.) C. B. Strozier and M. Flynn. New York: New York University Press, pp. 15, 17.

**PAGE 201.** O'Leary, S. D. 1994. *Arguing the Apocalypse: A Theory of Millennial Rhetoric.* Oxford: Oxford University Press.

**PAGE 202.** Balch, R. W., J. Domitrovich, B. L. Mahnke, and V. Morrison. 1997. "Fifteen Years of Failed Prophecy." In *Millennium, Messiahs, and Mayhem,* (eds.) T. Robbins and S. J. Palmer. New York: Routledge.

**PAGE 203.** O'Leary, S. D. 1994. *Arguing the Apocalypse: A Theory of Millennial Rhetoric.* Oxford: Oxford University Press, pp. 93–110.

**PAGE 203.** Penton, M. J. 1997 (1985). *Apocalypse Delayed: The Story of Jehovah's Witnesses,* 2nd ed. Toronto: University of Toronto Press, p. 100.

**PAGE 203.** Watchtower Bible and Tract Society. 1966. *Life Everlasting in Freedom of the Sons of God.* Brooklyn, N.Y., p. 29.

**PAGE 205.** Story about God's Salvation Church in Shermer, M. 1998. "The End of the World . . . Again," *Skeptic,* 6/1: 12–13.

**PAGE 205.** Jones tape transcript in Dwyer, S. 1989. "A Revolutionary Suicide: Jonestown, Guyana." In *Rapid Eye.* London: Creation Books, pp. 224–229.

**PAGE 205.** Applewhite quotes in Shermer, M. 1997. "What's the Harm in Believing in UFOs and Pseudoscience? Heaven's Gate Cult Mass Suicide Answers the Question," *Skeptic,* 5/1: 10–11.

**PAGE 205.** Abanes, R. 1998. *End Times Visions.* New York: Four Walls Eight Windows.

**PAGE 206.** Sheler, J. L. 1997. "Dark Prophecies," *U.S. News and World Report,* December 15: 63.

**Page 206.** Leslie, J. 1996. *The End of the World: The Science and Ethics of Human Extinction.* New York: Routledge.

**PAGE 207.** Thompson, D. 1996. *The End of Time: Faith and Fear in the Shadow of the Millennium.* Hanover, N.H.: University Press of New England, p. 325.

**PAGE 209.** Buttrick, G. A., and N. B. Harmon (eds.). 1994. *The Interpreter's Bible: A Commentary in Twelve Volumes.* Nashville, Tenn.: Abingdon Press, Vol. 5, p. 755.

**PAGE 210.** Taussig, M. T. 1980. *The Devil and Commodity Fetishism in South America.* Chapel Hill: University of North Carolina Press.

**PAGE 210.** Fukuyama, F. 1992. *The End of History and the Last Man.* New York: Free Press.

**PAGE 210.** Rand, A. 1957. *Atlas Shrugged.* New York: Random House, p. 1159.

**PAGES 210–211.** Eisler, R. 1987. *The Chalice and the Blade: Our History, Our Future.* San Francisco: Harper & Row, pp. xvi, 201.

**PAGE 212.** For pre-biblical ethical/law codes see Cohn, N. 1993. *Cosmos, Chaos, and the World to Come: The Ancient Roots of Apocalyptic Faith.* New Haven, Conn.: Yale University Press.

For data on belief in the Second Coming see Sheler, J. L. 1997. "Park Prophecies." *U.S. News and World Report*, December 15: 63–71.

## CHAPTER 10: *Glorious Contingency: Gould's Dangerous Idea and the Search for Meaning in an Age of Science*

**PAGE 215.** Weinberg, S. 1977. *The First Three Minutes.* New York: Basic Books.

**PAGE 216.** Gould, S. J. 1989. *Wonderful Life: The Burgess Shale and the Nature of History.* New York: W. W. Norton.

**PAGES 216–217.** Fontana, W., and L. Buss. 1994. "What Would Be Conserved If 'The Tape Were Played Twice'?" In *Complexity*, (eds.) G. Cowan, D. Pines, and D. Meltzer. Reading, Mass.: Addison Wesley.

**PAGE 217.** Lorenz, E. 1979. "Predictability: Does the Flap of a Butterfly's Wings in Brazil Set Off a Tornado in Texas?" Address at the AAAS annual meeting, Washington, D.C., December 29.

**PAGE 217.** Kauffman quote in: Kauffman, S. A. 1993. *The Origins of Order: Self-Organization and Selection in Evolution.* Oxford: Oxford University Press. See also:

Kauffman, S. A. 1995. *At Home in the Universe.* Oxford: Oxford University Press, p. 13.

**PAGE 217.** Cohen, J., and I. Stewart. 1991. "Chaos, Contingency, and Convergence," *Nonlinear Science Today*, 1(2): 9–13.

**PAGE 217.** Kelly, K. 1994. *Out of Control.* Reading, Mass.: Addison Wesley, p. 410.

**PAGES 217–218.** McRae, M. W. 1993. "Stephen Jay Gould and the Contingent Nature of History," *Clio*, 22(3): 239–250.

**PAGE 218.** Gould, S. J. 1989. *Wonderful Life: The Burgess Shale and the Nature of History*. New York: W. W. Norton, p. 283.

**PAGES 218–219.** Dennett, D. C. 1995. *Darwin's Dangerous Idea: Evolution and the Meanings of Life*. New York: Simon and Schuster, pp. 306, 307.

**PAGE 219.** Gould, S. J. 1989. *Wonderful Life: The Burgess Shale and the Nature of History*. New York: W. W. Norton, p. 289.

**PAGES 219–220.** Gould, S. J. 1978. "The Panda's Thumb." Reprinted in *The Panda's Thumb*. 1980. New York: W. W. Norton, pp. 19–26.

**PAGE 220.** Gould, S. J. 1987. "The Panda's Thumb of Technology," *Natural History*, 1: 22.

**PAGE 220.** For the history of the typewriter and the Qwerty keyboards see Dvorak, A. 1936. *Typewriting Behavior*. New York: American Book Company. See also:

Masi, F. T. (ed.). 1985. *The Typewriter Legend*. Secaucus, N.J.

Cassingham, R. C. 1986. *The Dvorak Keyboard*. Arcata, Calif.: Freelance Communications.

David, P. 1986. "Understanding the Economics of QWERTY: The Necessity of History." In *Economic History and The Modern Economist*, (ed.) W. N. Parker. New York: Basil Blackwell.

Romano, F. J. 1986. *Machine Writing and Typesetting*. Salem, NH: Gama.

Hoke, D. R. 1990. *Ingenious Yankees: The Rise of the American System of Manufacturers in the Private Sector*. New York: Columbia University Press.

**PAGE 221.** McRae, M. W. 1993. "Stephen Jay Gould and the Contingent Nature of History," *Clio*, 22(3): 244.

**PAGE 221.** Gould, S. J. 1987. "The Panda's Thumb of Technology," *Natural History*, 1: 22.

**PAGE 221.** Gould, S. J. 1989. *Wonderful Life: The Burgess Shale and the Nature of History*. New York: W. W. Norton, p. 289.

**PAGES 221–222.** Dennett, D. C. 1995. *Darwin's Dangerous Idea: Evolution and the Meanings of Life*. New York: Simon and Schuster, p. 308.

**PAGE 222.** Shermer, M. 1996. "An Urchin in a Haystack: An Interview with Stephen Jay Gould," *Skeptic*, 4/1: 88.

**PAGE 222.** Dennett, D. C. 1995. *Darwin's Dangerous Idea: Evolution and the Meanings of Life*. New York: Simon and Schuster, pp. 75, 76.

**PAGE 223.** Marx, K. 1852. "The Eighteenth Brumaire of Louis Bonaparte." In *The Marx–Engels Reader*, 2nd ed., (ed.) R. C. Tucker. New York: W. W. Norton, p. 594.

**PAGE 223.** Hempel, C. G. 1942. "The Function of General Laws in History." In *Theories of History*, (ed.) P. Gardiner. New York: Free Press, p. 346.

**PAGES 223–224.** Gould, S. J. 1989. "The Horn of Triton," *Natural History*, 12: 18–24.

**PAGE 227.** For numerous historical examples of the model of Contingent-Necessity see Shermer, M. 1995. "Cycles and Curves" *Skeptic*, 3/3: 58–61.

Shermer, M. 1993. "The Chaos of History: On a Chaotic Model That Represents the Role of Contingency and Necessity in Historical Sequences," *Nonlinear Science Today*, 2(4): 1–13.

Shermer, M. 1997. "The Crooked Timber of History," *Complexity*, 2(6): 23–29.

**PAGE 227.** Prigogine, I., and I. Stengers. 1984. *Order Out of Chaos*. New York: Bantam, p. 169.

**PAGES 228–229.** Gould, S. J. 1993. "Fungal Forgery," *Natural History*, 9: 12–21.

**PAGE 229.** Wallace, A. R. 1903. *Man's Place in the Universe: A Study of the Results of Scientific Research in Relation to the Unity or Plurality of Worlds*. New York: McClure Phillips and Co., p. 73.

**PAGE 231.** Dyson, F. 1988. *Infinite in All Directions*. New York: Harper & Row.

**PAGE 231.** Tipler, F. 1994. *The Physics of Immortality: Modern Cosmology, God and the Resurrection of the Dead*. New York: Doubleday.

**PAGE 231.** Ruse, M. 1996. *Monad to Man: The Concept of Progress in Evoluntionary Biology*. Cambridge, Mass.: Harvard University Press, pp. 131–132.

**PAGE 231.** Gould, S. J. 1996. *Full House: The Spread of Excellence from Plato to Darwin*. New York: Harmony Books, p. 33.

**PAGE 232.** Gould, S. J. 1996. *Full House: The Spread of Excellence from Plato to Darwin*. New York: Harmony Books, p. 132.

**PAGE 233.** Gould, S. J. 1996. *Full House: The Spread of Excellence from Plato to Darwin*. New York: Harmony Books, pp. 169–172, 173.

**PAGES 233–234.** Dennett, D. C. 1995. *Darwin's Dangerous Idea: Evolution and the Meanings of Life*. New York: Simon and Schuster, p. 300.

**PAGE 234.** Gould, S. J. 1989. *Wonderful Life: The Burgess Shale and the Nature of History*. New York: W. W. Norton, p. 284.

## AFTERWORD TO THE SECOND EDITION: *God on the Brain*

**PAGES 239–240.** Newberg, A., E. D'Aquili, and V. Rause. 2001. *Why God Won't Go Away*. New York: Ballantine Books.

**PAGES 241–43.** Blanke, O., S. Ortigue, T. Landis, and M. Seeck. 2002, "Neuropsychology: Stimulating Illusory Own-Body Perceptions." *Nature*, 419, September 19: 269–270.

For a popular account of the research see Verrengia, J. 2002. "Misfiring Brain May Cause 'Out-of-Body Experiences' among patients." AP wire story. September 19.

For a general discussion of brain-generated psychological states and experiences, see Damasio, A. 2000. *The Feeling of What Happens: Body, Emotions and the Making of Consciousness*. London: Vintage.

**PAGE 243.** Lommel, P. V., R. V. Wees, V. Meyers, and I. Elfferich. 2001. "Near-Death Experience in Survivors of Cardiac Arrest: A Prospective Study in the Netherlands." *Lancet*, 358 (9298): 2039.

**PAGES 243–244.** Brugger, P. 2002. Paper presented at the annual conference of the Federation of European Neuroscience Societies in Paris. Reported in *New Scientist*. July.

**PAGES 244.** Scripps Howard News Service and the E. W. Scripps School of Journalism at Ohio University sponsored the survey. The telephone poll was conducted October 21 through November 1 among 1,127 adults living in all 50 states and the District of Columbia. Households were selected randomly by computer. Journalism professor Tom Hodges and survey manager Robert Owen supervised the interviewing at the Scripps Survey Research Center. The survey has a 4 percent margin of error.

**PAGE 245–246.** Gallup News Service. 2001. "Americans' Belief in Psychic Paranormal Phenomena is up Over Last Decade." June 8. For the full report go to www.gallup.com/poll/releases/pr010608.asp

**PAGES 246–248.** National Science Foundation. 2002 Science Indicators Biennial Report. The section on pseudoscience, "Science Fiction and Pseudoscience," is in Chapter 7. www.nsf.gov/sbe/srs/seind02/c7/c7h.htm

**PAGE 249.** Walker, W. R., S. J. Hoekstra, and R. J. Vogl. 2002. "Science Education is no Guarantee of Skepticism." *Skeptic*, 9/3: 24–27.

**PAGES 249–250.** Brooks, D. J. 2001. "Substantial Numbers of Americans Continue to Doubt Evolution as Explanation for Origin of Humans." Gallup News Service. March 5.

**PAGE 250.** Shermer, M. 2002. "The Gradual Illumination of the Mind." *Scientific American*, February: 32.

**PAGES 252–253.** Barrett, D. B., G. T. Kurian, and T. M. Johnson (eds.). 2001. *World Christian Encyclopedia: A Comparative Survey of Churches and Religions in the Modern World.* 2 vols. 2nd ed. New York: Oxford University Press.

**PAGES 253–254.** Gaustad, E. S., and P. L. Barlow (eds.). 2001. *New Historical Atlas of Religion in America.* New York: Oxford University Press.

**PAGE 254.** Snowdon, D. 2001. *Aging with Grace: What the Nun Study Teaches Us about Leading Longer, Healthier, and More Meaningful Lives.* New York: Bantam Books.

**PAGES 254–255.** Pargament, K. 2002. *Archives of Internal Medicine.* August.

**PAGE 255.** Barnes, M. H. 2000. *Stages of Thought: The Co-Evolution of Religious Thought and Science.* New York: Oxford University Press.

**PAGES 255–256.** Smith, H. 2001. *Why Religion Matters: The Fate of the Human Spirit in an Age of Disbelief.* San Francisco: Harper San Francisco.

**PAGES 256–257.** Boyer, P. 2001. *Religion Explained: The Evolutionary Origins of Religious Thought.* New York: Basic Books.

**PAGE 257.** Brasher, B. 2001. *Give Me That Online Religion.* New York: Jossey-Bass/Wiley.

**PAGES 257–259.** Flanagan, O. 2002. *The Problem of the Soul: Two Visions of Mind and How to Reconcile Them.* New York: Basic Books.

**PAGE 259.** Larson, E. J., and L. Witham. 1997. "Scientists Are Still Keeping the Faith." *Nature*, 386: 435.

**PAGES 259–262.** Sherrill, M. 2000. *The Buddha from Brooklyn.* New York: Random House.

## APPENDIX I: *What Does It Mean to Study Religion Scientifically? Or, How Social Scientists "Do" Science*

**PAGES 263–264.** Sulloway, F. 1996. *Born to Rebel: Birth Order, Family Dynamics, and Creative Lives.* New York: Pantheon.

**PAGE 264.** For correlation data on twins and religiousity see Waller, N. G., B. Kojetin, T. Bouchard, D. Lykken, and A. Tellegen. 1990. "Genetic and Environmental Influences on Religious Attitudes and Values: A Study of Twins Reared Apart and Together," *Psychological Science*, 1(2): 140.

**PAGE 264.** Jensen, A. R. 1971. "Note on Why Genetic Correlations Are Not Squared," *Psychological Bulletin*, 75(3): 223.

**PAGES 264–265.** Sulloway, F. 1996. *Born to Rebel: Birth Order, Family Dynamics, and Creative Lives.* New York: Pantheon, pp. 371–372.

**PAGE 265.** Pargament, K. 1997. *The Psychology of Religion and Coping.* New York: Guilford.

**PAGE 265.** For more on the power and importance of correlations see Rosenthal, R., and R. L. Rosnow. 1984. *Essentials of Behavioral Research: Methods and Data Analysis.* New York: McGraw-Hill.

**PAGES 265–266.** For a general survey of the field see Wulff, D. M. 1991. *Psychology of Religion: Classic and Contemporary Views.* New York: John Wiley and Sons. See also:

Hood, R. W., B. Spilka, B. Hunsberger, and R. Gorsuch. 1996. *The Psychology of Religion: An Empirical Approach*, 2nd ed. New York: Guilford Press.

Pals, D. L. 1996. *Seven Theories of Religion.* Oxford: Oxford University Press.

### *A Bibliographic Essay on Theism, Atheism, and Why People Believe in God*

**PAGE 280.** Smith, G. H. 1989. *Atheism: The Case Against God.* Buffalo, N.Y.: Prometheus Books, pp. 3, 7.

**PAGE 281.** Taylor, A. E. 1947. *Does God Exist?* New York: Macmillan, p. 158.

**PAGE 281.** Martin, M. 1990. *Atheism: A Philosophical Justification.* Philadelphia: Temple University Press, pp. 4, 5, 24, 464–465.

**PAGE 282.** Wulff, D. M. 1991. *Psychology of Religion: Classic and Contemporary Views.* New York: John Wiley and Sons.

**PAGE 282.** Hood, R. W., B. Spilka, B. Hunsberger, and R. Gorsuch. 1996. *The Psychology of Religion: An Empirical Approach*, 2nd ed. New York: Guilford Press.

**PAGE 282.** Pals, D. L. 1996. *Seven Theories of Religion.* Oxford: Oxford University Press.

**PAGE 282.** Morris, B. 1987. *Anthropological Studies of Religion.* Cambridge, England: Cambridge University Press.

**PAGE 282.** Lehmann, A. C., and J. E. Myers (eds.). 1993. *Magic, Witchcraft, and Religion: An Anthropological Study of the Supernatural*, 3rd ed. Mountain View, Calif.: Mayfield.

**PAGE 282.** Larue, G. 1975. *Ancient Myth and Modern Man.* New York: Prentice-Hall.

**PAGE 282.** Jensen, A. E. 1963. *Myth and Cult Among Primitive Peoples*, (trans.) M. T. Choldin and W. Weissleder. Chicago: University of Chicago Press.

**PAGE 282.** Evans-Pritchard, E. E. 1965. *Theories of Primitive Religion.* Oxford: Clarendon Press.

**PAGE 282.** Eliade, M. 1957. *The Sacred and the Profane: The Nature of Religion*, (trans.) W. R. Trask. New York: Harcourt Brace.

**PAGE 282.** Eliade, M. 1967. *Myths, Dreams, and Mysteries: The Encounter Between Contemporary Faiths and Archaic Realities.* New York: Harper Torchbooks.

**PAGE 282.** Weber, M. 1924. *The Sociology of Religion*, (trans.) E. Fischoff. Boston: Beacon Press.

**PAGE 282.** Mitchell, B. (ed.). 1971. *The Philosophy of Religion.* Oxford: Oxford University Press.

**PAGE 282.** Peterson, M., W. Hasker, B. Reichenbach, and D. Basinger (eds.). 1991. *Reasons and Religious Belief: An Introduction to the Philosophy of Religion.* Oxford: Oxford University Press.

**PAGES 282–283.** Hick, J. 1964. *The Existence of God: From Plato to A. J. Ayer on the Question "Does God Exist?"* New York: Collier Books.

**PAGE 283.** Angeles, P. A. 1976. *Critiques of God: A Major Statement of the Case against Belief in God.* Buffalo, N.Y.: Prometheus Books.

**PAGE 283.** Angeles, P. A. 1980. *The Problem of God: A Short Introduction.* Buffalo, N.Y.: Prometheus Books.

**PAGE 283.** Stein, G. (ed.). 1980. *An Anthology of Atheism and Rationalism.* Buffalo, N.Y.: Prometheus Books.

**PAGE 283.** Parsons, K. M. 1989. *God and the Burden of Proof: Plantinga, Swinburne, and the Analytic Defense of Theism.* Buffalo: Prometheus Books.

**PAGE 283.** Lewis, C. S. 1943. *Mere Christianity.* New York: Macmillan.

**PAGE 283.** Lewis, C. S. 1947. *Miracles.* New York: Macmillan.

**PAGE 283.** Lewis, C. S. 1962. *The Problem of Pain.* New York: Macmillan.

**PAGE 283.** Aquinas, T. 1952 (1273). *Summa Theologica.* Great Books of the Western World. R. M. Hutchins (ed. in chief). Chicago: Encyclopaedia Britannica.

**PAGE 283.** Kushner, H. 1981. *When Bad Things Happen to Good People.* New York: Avon Books.

**PAGE 283.** Kushner, H. 1989. *Who Needs God?* New York: Pocket Books/Simon and Schuster, pp. 194, 201.

**PAGE 283.** Friedman, R. E. 1987. *Who Wrote the Bible?* New York: Simon and Schuster.

**PAGE 283.** Friedman, R. E. 1995. *The Disappearance of God.* New York: Simon and Schuster.

**PAGE 283.** Friedman, R. E. 1998. *The Hidden Book in the Bible.* San Francisco: Harper San Francisco.

**PAGE 283.** Mack, B. L. 1993. *The Lost Gospel: The Book of Q and Christian Origins.* San Francisco: Harper San Francisco.

**PAGE 283.** Mack, B. L. 1995. *Who Wrote the New Testament? The Making of the Christian Myth.* New York: HarperCollins.

**PAGE 283.** Helms, R. 1988. *Gospel Fictions.* Buffalo: Prometheus Books.

**PAGE 284.** Helms, R. 1997. *Who Wrote the Gospels?* Altadena, Calif.: Millennium Press.

**PAGE 284.** Callahan, T. 1997. *Bible Prophecy: Failure or Fulfillment?* Altadena, Calif.: Millennium Press.

**PAGE 284.** Callahan, T. 2000. *The Secret Origins of the Bible.* Altadena, Calif.: Millennium Press.

**PAGE 284.** Buttrick, G. A., and N. B. Harmon (eds.). 1994. *The Interpreter's Bible: A Commentary in Twelve Volumes.* Nashville, Tenn.: Abingdon Press.

**PAGE 284.** Freedman, D. L. (ed.). 1992. *The Anchor Bible Dictionary.* New York: Doubleday.

**PAGE 284.** Laynon, C. (ed.). 1971. *The Interpreter's One-Volume Commentary on the Bible.* Nashville, Tenn.: Abingdon Press.

**PAGE 284.** Metzger, B., and R. Murphy. 1991. *The New Oxford Annotated Bible with the Apocrypha.* Oxford: Oxford University Press.

**PAGE 284.** Metzger, B., and M. Coogan (eds.). 1993. *The Oxford Companion to the Bible.* Oxford: Oxford University Press.

**PAGE 284.** Alter, R., and F. Kermode (eds.). 1987. *The Literary Guide to the Bible.* Cambridge: Harvard University Press.

**PAGE 284.** Brown, R., J. Fitzmyer, and R. Murphy (eds.). 1968. *The Jerome Biblical Commentary.* New York: Prentice Hall.

# CREDITS

Grateful acknowledgment is made for permission to reproduce the following:

**PAGE 18.** *Time* Magazine Cover, April 8, 1966: Is God Dead? Courtesy of Time/Life.

**PAGES 18–19.** Excerpt from "The Hollow Men" in *Collected Poems 1909–1962* by T.S. Eliot, copyright 1936 by Harcourt, Inc. Copyright © 1964, 1963 by T. S. Eliot, reprinted by permission of the publisher.

**PAGE 35.** The Virgin Mary appears in Clearwater, Florida. Courtesy of Gary Posner, Tampa Bay Skeptics.

**PAGE 40.** Earliest Evidence of Hominid Magical Thinking. Courtesy of Ralph Solecki.

**PAGE 42.** Enhanced Magical Thinking by the Yanomamö. Courtesy of Napoleon Chagnon.

**PAGE 114.** *Ambulocetus natans,* a Transitional Fossil Between Land and Marine Mammals. Courtesy of *Science.*

**PAGE 131.** The Scopes Trial. Photograph by the author.

**PAGE 132.** The Institute for Creation Research. Photograph by the author.

**PAGE 139.** Martin Luther King Jr.'s Ebenezer Baptist Church and Monument to His Moral Courage and Nobility of Spirit. Photograph by the author.

**PAGE 163.** Bio-Cultural Pyramid. Designed by author. Rendered by Pat Linse.

**PAGE 192.** W. B. Yeats poem excerpt from *Collected Poems.* Courtesy of Simon and Schuster.

**PAGE 194.** *The Opening of the Fifth and Sixth Seals, the Distribution of White Garments Among the Martyrs and the Fall of Stars* (1498) by Albrecht Dürer. Woodcut from The Revelation of Saint John (Rev. VI, 9–15). Courtesy of Giraudon/Art Resource, New York.

**PAGE 194.** *The Last Judgment* (1558) by Pieter Brueghel the Elder. Pen and ink. Graphische Sammlung Albertina, Russia. Courtesy of Foto Marburg/Art Resource, New York.

**PAGE 195.** Detail from *The Last Judgment* (1536–1541) by Michelangelo Buonarroti. Sistine Chapel, Vatican Place, Vatican State. Courtesy of Alinari/Art Resource, New York.

**PAGE 238.** Machu Picchu and Chartres Cathedral. Photographs by the author.

Data from Chapter 4, rendered into graphs by Pat Linse from Frank Sulloway's original graphs, and presented in Appendix II:

**PAGE 242.** Three-Dimensional Surface Reconstruction of the Right Hemisphere of the Brain from Magnetic-Resonance Imaging. Courtesy of *Nature*.

**PAGE 246.** Changing Belief in Paranormal Phenomena. Courtesy of Gallup News Service.

**PAGE 250.** Why People Do Not Accept Evolution.

**PAGE 275.** Scientists Belief in God and Immortality, 1916 v. 1996.

**PAGE 275.** Skeptics Belief in God.

**PAGE 276.** General Belief in God.

**PAGE 276.** Religious Upbringing and Belief in God.

**PAGE 277.** Religious Upbringing and Religiosity.

**PAGE 277.** Education and Religiosity.

**PAGE 278.** Age and Belief in God.

**PAGE 278.** Age and Religiosity.

**PAGE 279.** Parental Conflict and Belief in God.

**PAGE 279.** Parental Conflict and Religiosity.

# INDEX


Abanes, Richard, 205
Abbot, Edwin, 12–15
*Adaptation and Natural Selection* (Williams), 165
*Adapted Mind, The* (Barkow, Tooby, and Cosmides), 146
African Americans, 174
  Ghost Dance of 1990s, 183
  and religion, 139–140
afterlife
  belief in, 22–24
  and near-death experience, 25–26
  and talking with the dead, 48–50
age
  and belief in God, 274, 278
  and religiosity, 80, 264, 270, 271, 274, 278
Age of Science, 45, 151
agnosticism, agnostics, 7-9, 75, 282
agreeableness and religiosity, 82
Aish Ha Torah's Discovery Seminars, 117
alchemy, 45
Alcock, John, 145
Aldrich, Ira, 281
Alexander, Richard, 165
alien abductions, 67
Allen, Steve, xiii
Altai people, 184
alternative medicine, 247
altruism, 160–161, 164–167, 169
*Ambulocetus natans*, 112, 114
American Booksellers Association, 27
American Enterprise Institute, 99
*American Men and Women of Science*, 72, 73
amygdala, 66, 67
*Anchor Bible Dictionary, The*, 284
*Ancient Myth and Modern Man* (Larue), 282
Angeles, Peter, 283
angels, 189
  belief in, 21, 22, 26, 244–245
*Animal Behavior* (Alcock), 145
animal myths, 155
*Annals of the Old Testament, Deduced from the First Origin of the World* (Ussher), 198
Anselm, Saint, 93, 283
*Anthology of Atheism and Rationalism, An* (Stein), 283
*Anthropic Cosmological Principle, The* (Tipler), 103
Anthropic Principle, 99, 105
*Anthropological Studies of Religion* (Morris), 283
Antichrist, 207
anxiety, 38, 42, 44
apocalypse, apocalypticism, 193, 196–198
*Apocalypse Now* (film), 193
Apollonius of Tyana, 186
apologetics, 90–91
Applewhite, Marshall, 189, 205
Aquinas, St. Thomas, 89, 91–94, 283
Arbib, Michael, 65
Archilochus, 144
arguments from ignorance, 115
Aristotle, 224
Armageddon, 196
Arnold, Matthew, 2
art of the insoluble, 7–10
Asimov, Isaac, 132
astrology, 35, 44, 140
astronomers
  and belief in God, 72
astronomy, 45
*Atheism* (Martin), 281
*Atheism* (Smith), 280–281
atheism, atheists, 8, 9, 20, 24, 74, 75, 280–282
Atlantis, 35
*Atlas Shrugged* (Rand), 210

attribution biases
  in religous beliefs, 85–86

attribution theory, 85
Augustine, Saint, 96, 137, 198
*Australopithecines,* 36, 230–231
*Australopithecus africanus,* 157, 230
Axelrod, Robert, 166–167
Azande, 39–41

baby-boomer generation, 17
Baha'i sect, 202
Bahai Under the Provisions of the Covenant, 192
Barkow, Jerome, 146
"Barn, The" (La Crescenta, Calif.), 3
Barna, George, 22
Barnes, Michael, 255
baseball
  as a system of change, 231–232
  superstitions, 45
Basinger, David, 282
"Battle Hymn of the Republic," 168
B.C.–A.D. chronological system, 199, 200
beautiful people myth, 207–208, 210
Beauvoir, Simone de, 20
behavioral geneticists
  and religiosity, 64
Behe, Michael, 109–110, 113, 116
Belief Engine, 37–58
belief formation, 38
belief in God, *see* God, belief in
Bentley, Richard, 281
Beowulf, 156–157
Bergman, Gerald, 73
Berlin Wall, 225
Bettinger, Robert, 159
*Beyond the Cosmos* (Ross), 99
biases in attribution, 85–86
Bible, 122, 207, 284
  authorship of, 120–121
  Heaven on Earth metaphors, 208–209
  science facts revealed in, 132
  study of, 283–284
*Bible Code, The* (Drosnin), 27, 116–122
*Bible Prophecy* (Callahan), 284
biblical literalism, 74
Big Foot, Chief, 181
Binford, Lewis, 158
bioaltruism, 164
Bio-Cultural Pyramid, 162–164
biologists and belief in God, 72
*Biology of Moral Systems, The* (Alexander), 165
*Biophilia* (Wilson), 164
birth order and openness to experience, 247
Bishop, George, 73–74
black religious experience, 139–140
Black Coyote, 182
Blackmore, Susan, 70, 71
Blanke, Olaf, 241–243
body and soul, 129
Boggs, Wade, 45
*Book of Prophecies* (Columbus), 198
book publishing, 27
*Born to Rebel* (Sulloway), 263–264
Bouchard, Thomas, 64
Boyer, Pascal, 256–257
brain
  chemistry of, 243–244
  circuits of, 37
  evolution of, 36–37, 158
  God module in, 65–69
  and language, 158
  modular view of, 36–37
  and number of relationships, 160
  and religiosity, 65–69
  and storytelling, 143
Branch Davidians, 205–206
Brand, Chris, 80
Brasher, Brenda, 257
Brett, George, 232
*Brief History of Time, A* (Hawking), 30, 102
Brierre de Boismont, Alexandre, 67–68
Broca's area, 36
Brodie, Richard, 70
Brown, Denise, 48
Browning, Robert, 88
Brueghel, Pieter, the Elder, 192, 195
Brugger, Peter, 243
Brunvand, Jan Harold, 150
Bryan, William Jennings, 123
Buckley, William F., 109
*Buddha from Brooklyn, The* (Sherrill), 259–262
Burgess Shale, 217, 219
Buss, Leo, 216–217
butterfly effect, 217, 227

Callahan, Tim, 284
Calvin, John, 96
Campbell, Joseph, 151, 157
capitalism, 210
Capra, Frank, 235
Carew, Rod, 232
Cargo Cult Ghost Dance, 104–106
Carlyle, Thomas, 32
Carnegie Commission study of scientists and religiosity (1969), 74
Casti, John, 104
Catholicism, Catholics
  belief in afterlife, 23
  belief in evolution, 127–129
  belief in God, 76
  recruitment and conversion, 71
*Century's End* (Schwartz), 201
Chagnon, Napoleon, 41, 42, 159
chain letters, 47
*Chalice and the Blade, The* (Eisler), 210–211
chance, 218, 265
chaos, chaos theory, 225–227
Chartres Cathedral, 237, 238
Chaucer, Geoffrey, 161
chemistry, 45
Chen, Heng-ming, 204–205
Chiang, Chin-Hung, 205
Chot Chelpan, 184
Christian atheists, 20
Christ's Soon Return, 188
church attendance
  and belief in God, 76
  and openness to experience, 271
  and religiosity, 79–80, 270
church membership rates, 25
Church of England, 160
Church Universal and Triumphant, 197–198
Churchill, Winston, 168
*Churching of America, 1776–1990, The*, 25
*City of God, The* (Augustine), 198
Clarke, Arthur C., 19
Clearwater, Florida
  Virgin Mary appearance, 34, 35
Clough, Arthur H., 171
cognitive fluidity, 37
Cohen, Jack, 217, 218
cold readings, 33, 52
Coles, Michael, 138–140
Columbus, Christopher, 198
conditional apocalypticism, 197, 198
conflicting-worlds model of science and religion, 130, 138
conscience, 165–166
conscientiousness
  and religiosity, 82, 270
conservatives
  and belief in God, 80–82
*Consilience* (Wilson), 146
consilience of inductions, 144
Consolmagno, Guy, 135
Constantine, 96
contingency, 216–222, 230–233
  and freedom, 233–235
  and *Homo sapiens*, 228–231
  and necessity, 223–227
  and problem of emphasis, 218, 221–222
  and problem of meaning, 218–221
contingent-necessity, 224–228, 230, 234, 236
convention
  and observance of religion, 83
convergence of evidence, 144
*Conversations with God* (Walsh), 27
cooperation, 164–167, 169
  and human evolution, 159
Copernicus, 127
Copper Eskimo, 158
Coppola, Francis Ford, 193
correlation coefficient (*r*), 264
correlations, 265
Cosmides, Leda, 36, 37, 146, 149
cosmogonies, 100
cosmogony myths, 133–134
cosmological arguments for God's existence, 99, 101–109
cosmology, 132, 134, 141
  *see also* New Cosmology
*Cosmos* (TV series), 29
courage, moral, 138–141
cranes, 222
  *see also* skyhooks
creation, 198–199
creationism, 73
  in public schools, 109, 248
  *see also* New Creationism

creationists
  example of human eye evolution, 110–114
  on science, 130–131, 133, 249–252
*Creator, and the Cosmos, Creation and Time, The* (Ross), 99
Credo Quia Consolans argument (for God's existence), 97–98
Crichton, Michael, 45
*Critiques of God* (Angeles), 283
Cro-Magnons, 36
cryonics, 257
cryptozoology, 153
cult leaders, 54
cults, fringe, 81
cultural evolution, 146–148
culture and genes, 147
cyberspirituality, 257

Damas, David, 158
danger and magic, 43
D'Aquili, Eugene, 239
Darwin, Charles, 59, 142, 263
*Darwin on Trial* (Johnson), 115
Darwinian literalists, 165
Darwinian Revolution, 263, 264
*Darwin's Black Box* (Behe), 109
Davies, Paul, 103
Dawkins, Richard, 29, 69–70, 162, 165
Deacon, Terrence, 158
dead, talking to, 48–58
death, 55–58
  of God, 16–31
  and science, 60–61
*Decline and Fall of the Roman Empire, The* (Gibbon), 88
deists, 281
Dembski, William, 115, 130–131
Dempsey, Father Robert, 128
Dennett, Daniel, 217–219, 221–222, 233–234
Descartes, René, 95
Design/Teleological argument (for God's existence), 94
Desmond, Adrian, 263
destruction-redemption myth, 196, 202, 205
determinism, 234
Deutsch, David, 107–108
devil, 21
Diamond, Jared, 153
Diana, Princess, 142–143, 161
Dickerson, Willard, 27
Diprima, Dominque, 171, 174
*Disappearance of God, The* (Friedman), 28, 283
Disch, Thomas, 151
dispositional attribution, 85
distracted atheists, 20
divine intervention, 26
*Does God Exist?* (Taylor), 281
domain-general processors, 37, 38
domain-specific processors, 37, 38
Doors, The, 193
Dow Jones Industrial Average, 191–192
Downs, Hugh, 55
dragon myths, 154
Draper, John William, 130
dreams and storytelling, 143
Drosnin, Michael, 27, 116–122
Dunbar, Robin, 159, 160
Dürer, Albrecht, 192, 195
Durkheim, Émile, 23, 169
Dyson, Freeman, 231

Earth
  age of, xiv
ebene powder, 42
Ebenezer Baptist Church, 139
Ecker, Don, 171
economic theories of religion, 23–24
Edson, Hiram, 203
education
  and belief in God, 83–84, 272
  and religiosity, 79, 80, 264, 270, 274, 277
Eibl-Eibesfeldt, Irenäus, 145
*Eighteenth Brumaire, The* (Marx), 223
Eisler, Riane, 210–211
Eliade, Mircea, 169, 282
Eliot, T. S., 18–19
Elson, John T., 17, 20
emotions and beliefs, 83–88
"End, The" (song), 193
*End of History, The* (Fukuyama), 210
*End of the World as We Know It, The* (Wojcik), 197

end-times, 192–193, 197–199
*Entwined Lives* (Segal), 64
environment
and genes, 63–64
and magical thinking, 39–43
and religiosity, 64–65
environmont of ovolutionary adaptation (EEA), 36
environmentalists, 210
epigenetic rules, 146–148
for mythmaking, 153, 154, 156
oqualitarian partnership model, 211
eschatology, 193
eschatology myths, 196
*Essay on Man* (Pope), 50, 208
ethical behavior, 163, 164
ethology, 145–146
*Ethology* (Eibl-Eibesfeldt), 145
Evans-Pritchard, Edward E., 39–40, 169, 282
"Eve of Destruction" (song), 17
Evidence Amendment Act (1869), 281
*Evidence That Demands a Verdict* (McDowell), 90
Evil, Problem of, 5
evils, historical, 5
evolution, 233
of brain, 36
and contingent-necessity, 230
of the eye, 110–113
fossil record, 113, 114
human, 145, 229
evolution theory, 144–146
acceptance by Pope John Paul II, 127–129
and belief in God, 73–74, 248–252
and religion, 133
evolutionary biology, 141, 145–145
evolutionary history, 144
Evolutionary Level Above Humans, The, 205
evolutionary psychology, 36–37
Evolutionary Stable Strategy, 167
exclusive fitness, 164
*Existence of God, The* (Hick), 282–283
existentialist atheists, 20
extinction, mass, 219
*Extraordinary Popular Delusions and the Madness of Crowds* (Mackay), 200–201
extrasensory perception (ESP), 35, 245
extraversion and religiosity, 82
Extropians, 59–61
Extropy Institute, 59
eye
evolution of, 110–113
Eysenck, Hans, 63

faith, 3, 15, 83, 89–91, 103, 123
in God, 7–8
and God's existence, 99
and miracles, 95
negation of, 117
and reason, 129, 131, 133, 135–138
and science, 123, 130
faith, leap of, 97, 136
falsehoods, 38, 40–41
*see also* thinking errors and hits
falsifiability problem, 119
family, 163–164
Farrakhan, Louis, 172–173
feedback loop, 226, 227
feminism, 210
Festinger, Leon, 202
Fideism argument (for God's existence), 97–98
fideism, fideists, 8–9
*Fides et Ratio of the Supreme Pontiff to the Bishops of the Catholic Church on the Relationship Between Faith and Reason* (John Paul II), 133, 135–138
*Firing Line* (TV program), 109
First Cause argument (for God's existence), 92
*First Three Minutes, The* (Weinberg), 215
First Vatican Council, 136
Five Factor Model (of personality), 82
Flanagan, Owen, 257–259
*Flatland* (Abbot), 12–15
Flew, Antony, 283
Fludd, Robert, 46
FM-2030, 60
Fontana, Walter, 216–217
fossil record, 113, 114

Four Horsemen of the Apocalypse, 192
Franz, Frederick, 203
Franz Ferdinand, Archduke, 226
Frazer, James, 169
Free Will, Problem of, 5
freedom and contingency, 233–235
Freud, Sigmund, 169
Friedman, Richard Elliott, 28, 107, 120, 283
Fukuyama, Francis, 210
*Full House* (Gould), 231
Fuller, R. Buckminster, 1
"Function of General Laws in History, The" (Hempel), 223
fundamentalism, Christian, 24, 130, 140–141
fundamentalism, paranormal, 61
fuzzy logic, science of, 60
*Fuzzy Thinking* (Kosko), 62

Galileo, 127, 130
Gallup poll of belief in supernatural (1991), 35
gambling as a form of magical thinking, 50–51
Gardner, Martin, xiii, 8–9, 92, 97
Gazzaniga, Michael, 143
Geertz, Clifford, 169
Geivett, Dr. Doug, 89
gender
  and belief, 83, 84
  and belief in God, 271, 272
  and religiosity, 79, 264, 270
Genesis myth, 231
General Social Surveys (1973–1994), 24
gene-culture coevolution, 146
genes
  and culture, 147
  and environment, 63–64
  and religiosity, 63–65, 75
  and religious attitudes, 64
Genesis and science, 100
genetic programming
  of belief in God, 63–65
Ghiselin, Michael, 165
Ghost Dance, 173–186
ghosts, 35
Gibbon, Edward, 88
Gibson, Charles, 48
Gingrich, Newt, 208
Glaber, Radulfus, 202
global intelligence, 37
glorious contingency, 231
Glynn, Patrick, 99
God, 169–170
  death of, 16–31
  defining, 10
  described by believers, 22
  disbelief in, 9
  existence of, 8–9, 89–123, 132
  hidden face of, 28
  insolubility of, 10
  as a meme, 70
  nonbelief in, 87–88
  pattern of, 63
  philosophical argument for existence of, 91–98
  and religion, 168–170
  scientific arguments for existence of, 98–122
  seeking pattern of, 61–63
  social indicators of, 24–28
  synonyms for, 10
  understanding existence or nonexistence of, 15
God, belief in, xvii–xix, 2–15, 22, 58–88
  and age, 274, 278
  as art of the insoluble, 8
  and church attendance, 76
  correlations, data, and statistics on, 267–279
  and education, 83–84
  evolution of, 144–145
  and gender, 271, 272
  and genetic programming, 63–65
  and parental conflict, 274, 279
  polls, 21, 22
  reasons for, 85–88, 168, 271, 272
  and religious upbringing, 274–277, 279
  by scientists, 72–74, 275
  and skeptics, 268–269, 275
  variables that shape, 250
  why people believe, 59–88
God, kingdom of, 187
*God and the Burden of Proof* (Parsons), 283
God Experience, 68–69

"God is dead." (Nietzsche), 16
God module (in brain), 65–69
God of the Gaps argument, 95, 115, 116
God Question, xiii–xviii, 8, 10–11, 63, 91, 282–283
*God: The Evidence* (Glynn), 99
*God's Descending in Clouds (Flying Saucers) on Earth to Save People* (Chen), 204
God's Salvation Church, 204
Goodenough, Ursula, 28
gossip, 161
Gorsuch, Richard, 282
*Gospel Fictions* (Helms), 283–284
Gould, Stephen Jay, 29, 38, 69, 133, 199, 214, 216–224, 231–234
Graham, Billy, 27
Great Disappointment, 202–203
Greeley, Andrew, 22–23
Grendel, 156–157
Grenier, Jean, 154–155
Gribbin, John, 102
Grodin, Charles, 48
*Grooming, Gossip and the Evolution of Language* (Dunbar), 159
group selection, 165
Guth, Alan, 92, 108

Halley's comet, 201
hallucinations, 67–68
hallucinogenic drugs, 42
Hamilton, William, 165
Hammurabi code, 212
Hanes, Leah, 53
hard-core atheists, 24
Hardison, Richard, 6–7, 97
Harris, Marvin, 184, 185
Harris, Sidney, 95
Harrison, President Benjamin, 176, 180
Hasker, William, 258
Hawking, Stephen, 29–30, 92, 105, 106, 135
  on God, 101–103
heaven, 21, 22, 25, 209
Heaven on Earth, 207–209
Heaven's Gate, 189, 202, 205
heavens, secular, 210–211
Hebrew Bible, *see* Old Testament
hell, 21
Helms, Randel, 186, 283–284
Heltzer, Ruth, 61
Hempel, Carl, 223
Hendel, Ronald, 119, 121
heritability
  of religious tendencies, 64–65
*Hero with a Thousand Faces, The* (Campbell), 151
Herod, King, 186–187, 199
Hick, John, 282–283
*Hidden Book in the Bible, The* (Friedman), 120–121, 283
historical sequences, 224–227
history
  general laws in, 223
  and social class, 263–264
*History of the Conflict Between Religion and Science* (Draper), 130
*History of the Warfare of Science with Theology, A* (White), 130
Hitchens, Christopher, 201
Hobbes, Thomas, 44
Holocaust, 5
home base hypothesis, 158
*Homo habilis,* 157
*Homo sapiens,* 36, 219, 228–231
*Homo symbolicus,* 158
Hood, Ralph, 282
Hook, Sidney, 283
hope, 89
hot readings, 53
Hotz, Robert Lee, 65
how questions, 144–145
*How to Argue and Win Every Time* (Spence), 148
Howe, Julia Ward, 168
Hubble Telescope, 237
human behavior, 145
human rights, 168
*Humani Generis* (Pius XII), 127–128
Humphrey, Nicholas, 61
Hunsberger, Bruce, 282
hunter-gatherer communities, 158–159
Hutterites, 159
Huxley, Thomas Henry, 7, 8, 55
Huxley–Wilberforce evolution debate (1850), 130
Hyde, Charles L., 179

"Imagine" (song), 19
imitation, *see* meme
immortality, 72, 275
inclusive fitness, 161, 164
*Index of Leading Spiritual Indicators* (1996), 22
indirect altruism, 164
*Infinite in All Directions* (Dyson), 231
*Inflationary Universe, The* (Guth), 92, 108
insoluble, art of the, 7–10
insolubility of God, 8–10, 15
Institute for Creation Research, 132–133
intellectual attribution bias, 85–86
intellectual reasons for belief in God, 85–88
Intelligent Design Theory, 249–252
interpreter brain mechanism, 143
*Interpreter's Bible, The,* 107, 284
*Interpreter's Dictionary of the Bible, The,* 284
*Interpreter's One-Volume Commentary on the Bible, The,* 284
*Into Thin Air* (Krakauer), 47
Irons, William, 167
irreducible complexity (of life), 109–113
Isaac, Glynn, 158
*It's a Wonderful Life* (film), 235–236

Jagodzinski, Wolfgang, 24
jaguar myths, 41
James, William, 97, 239
Jeffrey, Grant, 120
Jehovah's Witnesses, 202–204
Jensen, Adolf, 282
Jensen, Arthur, 264
Jensen, Dr. Leland, 192, 202
*Jerome Biblical Commentary, The,* 284
Jesus Christ, 196, 197, 199, 206, 209
  and messiah myth, 186–189
  *see also* Second Coming
Jews
  belief in afterlife, 23
  belief in God, 76
Johanson, Donald, 29, 158
John Paul II, Pope, 133, 135–138
  and acceptance of evolution theory, 127–129
John the Divine, Saint, 195, 196
Johnson, Phillip, 109, 115
Jones, Jim, 205
Jonestown disaster, 205
Judaism
  and belief in God, 76
  and new members, 70–71
  and political liberalism, 80–81
Judgment Day depictions, 195

Kanizsa-square illusion, 62
Kauffman, Stuart, 217
Kelly, Kevin, 217
Kicking Bear, 179, 181
kin altruism, 164
King, Dr. Martin Luther, Jr., 139
kingdom of God, 187
Koenig, Harold, 255
Komp, Diane, 25–26
Konner, Melvin, 155
Koppel, Ted, 173
Koresh, David, 205
Kosko, Bart, 60, 62
Kostner, Kevin, 193
Krakauer, Jon, 47
!Kung San people, 155
Kushner, Rabbi Harold, 5, 283

La Barre, Weston, 183–185
Landau, Misia, 149
Landes, Richard, 201–202
language, 146
  evolution of, 157–159
Larson, Edward, 72–74
Larue, Gerald, 121, 282
*Last Judgment* (Brueghel painting), 195
*Last Judgment* (Michelangelo painting), 195
*Late Great Planet Earth, The* (Lindsey), 3–4, 6, 193, 212
leap of faith, 97, 136
*Leaps of Faith* (Humphrey), 61
Lehmann, Arthur, 282
Lennon, John, 19
Leuba, James, 72, 73
Levy, David, 87
Lewis, C. S., 95, 283
Lewontin, Richard, 38

liberalism, political
and belief in God, 80, 82
and religiosity, 271
life after death, *see* afterlife
*Life Everlasting in Freedom of the Sons of God* (Watchtower Society), 203
*Life of the Cosmos, The* (Smolin), 108
Lindsey, Hal, 3–4, 6, 193, 197, 212
Linse, Pat, 267
*Literary Guide to the Bible, The,* 284
Lorenz, Edward, 217
Los Angeles riots (1992), 141
*Los Angeles Times* poll on belief in God (1991), 22
*Lost Gospel, The* (Mack), 283
Luke, 188
Luther, Martin, 68, 96, 106, 198
Lynch, Aaron, 70

McCartney, Bill, 27
McDowell, Josh, 90
McGuire, Barry, 17
Machu Picchu, 237, 238
Mack, Burton, 187, 283
McKay, Brendan, 119–120
Mackay, Charles, 200–201
MacLaine, Shirley, 102
McLaughlin, James, 180, 181
Macquarrie, Dr. John, 8
McRae, Murdo William, 217, 221
Madonna, 142–143
*Magic, Witchcraft, and Religion* (Lehmann and Myers), 282
magical thinking, 35, 38, 43, 47
Azande people, 39–40
and environment, 39–43
and medieval mind, 44–45
Neanderthal man, 40
Sherpas, 47
Yanomamö, 41
Malinowski, Bronislaw, 41–43
*Man's Place in the Universe* (Wallace), 229
Maori Ghost Dance, 183–184
Martin, Michael, 257
Marx, Karl, 169, 223, 263
Marxism, 210, 234
*Master of the Universe* (TV program), 102
materialism, *see* philosophical naturalism
mathematicians
and belief in God, 72
Maven, Max, 53
Mazet, Bruce, 101
Medawar, Sir Peter, 7
medicines
magical use of, 39–40
Melanesians
Cargo Cult Ghost Dance, 185
meme, 69–72
and religion, 70–71
*Meme Machine, The* (Blackmore), 71
mental modules, 36, 38
mentalists, 33
*Mere Christianity* (Lewis), 283
messiah myth, 174, 182, 190, 196
and Cargo Cult Ghost Dance, 185
Jesus as, 186–189
Messiah Cults, 186
methodological naturalism, *see* philosophical naturalism
Michelangelo, 192, 195
Middle Ages
Belief Engine, 43–46
Miles, General Nelson, 180, 181
militant atheists, 74
millennial meanings, 191–213
millennium, 193–200
and apocalypse, 196
attraction of, 206–207
Miller, Dr. David, 89
Miller, William, 202–203
Millerites, 202–203
*Mind of God, The* (Davies, 103
*Mind's Past, The* (Gazzaniga), 143
Minnesota twins study, 64
Minsky, Marvin, 60
miracles, 21–22, 26, 95
*Miracles* (Lewis), 283
Miracles argument (for God's existence), 95
Mitchell, Basil, 282
Mithen, Steven, 37
monotheism, 10
monsters and beasts myths, 153–154
Mooney, James, 176, 182–183

Moore, Gordon, 61
Moore, James, 263
Moore's law, 61
Moral argument (for God's existence), 98
moral courage, 138–141
morality, 156–161
  evolution of, 162–168
  and evolution of religion, 142–170
More, Max, 60
Morgan, Thomas J., 176
Mormons, 177
Morris, Brian, 282
Morrison, Jim, 193
Morrow, Tom, 60
multiple regression analysis, 264–265
multiverse model, 108
Mundus Intellectualis (Fludd), 46
Murphy, Nancey, 65
Mutual Assured Destruction (MAD), 17
Myers, James, 282
mysterian mystery, 92
"Mystery of the Human Head" (Fludd), 46
Mystical Experience argument (for God's existence), 96–97
*Myth and Cult Among Primitive Peoples* (Jensen), 282
mythmaking, 143, 150–161, 208
mythology, 151
myths, 149, 151, 154–155
  and evolution of religion, 142–170
  functions of, 151–152
  and survival, 41
  types of, 152–153
*Myths, Dreams, and Mysteries* (Eliade), 282
*Myths to Live By* (Campbell), 151

Nahmanides, 100
nanotechnology, 61
narratology, *see* storytelling
Nation of Islam, 172, 173
National Academy of Sciences, 73
National Opinion Research Center survey for belief in afterlife, 24
Native Americans, 174, 176
  Ghost Dance, 173–186
natural sciences and history, 223
natural selection, 38, 110–113, 145, 221, 222, 234
naturalism, philosophical, 115, 131
nature and nurture
  and religiosity, 64
*Nature* magazine survey of scientists and belief in God, 72
Neanderthal man, 40, 157, 230
near-death experiences, 25–26, 67, 241, 243
necessity, 230
  and contingency, 223–227
  *see also* contingent-necessity
negative atheism and atheists, 281
neocortex, 160
Nero, 207
Nettles, Bonnie Lu, 189
neural nets, 62
neuroticism and religiosity, 82
New Age spiritual movements, 81
New Cosmology, 98–109
New Creationism, 109–116, 249–252
*New Oxford Annotated Bible with the Apocrypha, The,* 284
New Testament, 283–284
New York Yankees, 199
*New Yorker* magazine, 193
Newburg, Andrew, 239
*Newsweek* magazine, 25
Newton, Isaac, 45
Nietzsche, Friedrich, 16–17, 20
*Nightline* (TV program), 173
no-boundary universe, 135
nobility of spirit, 138–141
Noelle, David, 37, 69
nonbelief in God, 87–88, 269
  *see also* agnosticism, agnostics; atheism, atheists
nontheism, nontheists, 9, 75, 282
North, Jay, 21
Nostradamus, 207
Novak, Michael, 20
nuclear bombs, 199–200
Nun Study, 254

Old Testament, 28, 121, 283
O'Leary, Stephen, 201
Omega Point Theory, 104–108
*On Hallucinations* (Brierre de Boismont), 67–68

*On Human Nature* (Wilson), 140–141
*On the Beach* (Shute), 17–19
ontological argument (for God's existence), 93–94
*Opening of the Fifth and Sixth Seals, . . . (Dürer painting),* 195
openness (to experience)
and religiosity, 82–84, 270–272
oppression-redemption myth, *see* messiah myth
orientation association area, 239–240, 241
origin myths, 149
*Other Side, The* (TV program), 49, 53
Oubré, Alondra, 41
out-of-body experiences, 241–242
*Oxford Companion to the Bible, The,* 284
Pacific islanders
Cargo Cult Ghost Dance, 184–185
Paine, Thomas, 125
Paiute Indians, 175
Paley, William, 87
Pals, Daniel, 169, 282
Panda Principle, 220–221
"Panda's Thumb, The" (Gould), 219–220
Pannenberg, Wolfhart, 104–106, 132
pantheists, 281
paranormal fundamentalism, 61
Parcells, Bill, 45
parental conflict
and belief in God, 274, 279
and religiosity, 79–80, 264, 270, 274, 279
parents and religiosity, 264, 270
Parents Television Council survey, 27
Pargament, Kenneth, 265
Parsons, Keith, 283
Pascal, Blaise, 95–96
Pascal's Wager argument (for God's existence), 95–96
path dependency, 220
pattern-seeking, 34–43, 61–63, 119, 143, 148–150
and millennium, 206, 207
Paul, 188
pea instanton, 105
"Peace of God" movements, 201
Peirce, Charles, 97
Pennock, Robert, 114
Perfection/Ontological argument (for God's existence), 93–94
Persinger, Michael, 66–67
personality
and beliefs, 97–98
and religiosity, 81–84
Peter, 187
Peterson, Michael, 282
petitionary prayer, 26
Pew Research Center survey on belief in God (1997), 22
philosophical naturalism, 115, 131
*Philosophy of Religion, The* (Mitchell), 282
*Physics of Immortality, The* (Tipler), 103–105, 231
physicists
and belief in God, 72, 73
piety, 82, 83
Pine Ridge Reservation (S.D.), 180
Pinker, Steven, 36, 37
Pittman, Sandy, 47
Pius XII, Pope, 127–128
Plantinga, Alvin, 283
plants, magical, 41
Plomin, Robert, 63–64
plutonium, 200
Polichak, James, 71
political beliefs and religiosity, 80
polytheists, 257
Pope, Alexander, 50, 208
positive atheists, 281
Possibility and Necessity argument (for God's existence), 92
*Postman, The* (film), 193
postmillennial Christians, 197
postmillennial secularists, 197
practical atheists, 20
pragmatism, 97
prayer, petitionary, 26
praying, 25
premillennial Christians, 197
premillennial secularists, 197
Pre-Millennial Syndrome (PMS), 201
Preskill, John, 29
Prigogine, Ilya, 227
primates, 230

Prime Mover argument (for God's existence), 7, 91–92
Princeton Survey Research Associates survey on prayer, 25
Prisoner's Dilemma, 166–167
probability (*p*) and correlations, 265
problem of emphasis
  and contingency, 218, 221–222
Problem of Evil, 5
Problem of Free Will, 5
*Problem of God, The* (Angeles), 283
problem of meaning
  and contingency, 218–221
*Problem of Pain, The* (Lewis), 283
Promise Keepers, 26–27
prophecy failure, 200–206
Prophet, Elizabeth Claire, 198
*Proslogion* (St. Anselm), 283
Protestant Reformation, 81
Protestants' belief in afterlife, 23
psychic experience, 35
*Psychology of Religion* (Hood et al.), 282
*Psychology of Religion* (Wulff), 282
*Psychology of Religion and Coping, The* (Pargament), 265

*Questioning the Millennium* (Gould), 199
QWERTY Principle, 220–221

radical contingency, 233
Ramachandran, Dr. Vilayanur, 65–66, 69
Rambsel, Yacov, 120
Rand, Ayn, 210
Rapoport, Anatol, 167
rationalizations
  and failed prophecy, 200, 203
Raymo, Chet, 101
readings (by mentalists), 33, 52–53
Reagan, Ronald, 197
reason
  and faith, 129, 131, 133, 135–138
  and God's existence, 99
*Reason and Religious Belief* (Peterson et al.), 282
Reasons to Believe, 99
reciprocal altruism, 160–161, 164, 166, 167, 169
Reichenbach, Bruce, 282
reinforcement, 51
religion, 162–168
  anthropology of, 282
  as art of the insoluble, 7
  biocultural theory of, 169
  decline of, and rise of science, 22–23
  economic theories of, 23–24
  evolution of, 142–170
  and God, 168–170
  Middle Ages, 44
  models of, 129–135, 138
  philosophy of, 282
  purposes of, 143–145
  and science, 22–23, 123, 126–141
  scientific study of, 263–266
  secular, 29, 61
  secularization of, 22–24
  sociology of, 282
  supply-side, 23–24
  theory of, 169
  as a virus, 70
religiosity
  and age, 80, 264, 270, 271, 274, 278
  and agreeableness, 82
  and brain, 65–69
  and church attendance, 79–80, 246
  and conscientiousness, 82, 270
  and education, 79, 80, 264, 270, 274, 277
  and environment, 64–65
  and gender, 79, 264, 270
  and genes, 63–65, 75
  interest in science and, 270, 271
  and openness to experience, 82–84, 270–272
  and parental conflict, 79–80, 264, 270, 274, 279
  and parents, 264, 270
  and personality, 81–84
  and political beliefs, 80
  and political liberalism, 271
  predictors of, 79
  and skeptics, 76
  and tender-mindedness, 271
  of twins, 63–64, 264

and upbringing, 264, 270, 271, 274, 277
variables that shape, 274
religious apocalyptic scenarios, 193
religious attribution bias, 85–88
religious belief
undermining of, 83
religious books, 27
religious competition, 22–24
religious conviction, 267
religious doubt, 271
religious naturalism, 28
religious right, 80
religious tendencies
heritability of, 64–65
renewal myth, 196
residue problem, 116
revelation
truth through, 136–137
Rips, Eliyahu, 117, 122
Ritter, Bill, 55
Roman empire, 193
Roosevelt, Franklin, 168
Rosenberg, Milt, 32, 33
Rosenberg, Yoav, 117
Ross, Hugh, 99
Ruse, Michael, 133, 231
Russell, Bertrand, 283

*Sacred and the Profane, The* (Eliade), 282
*Sacred Depths of Nature, The* (Goodenough), 28
sacred science, 29–31
Sagan, Carl, 29, 214–215
Saint-Jouin de Marnes, Carlulaire, 197
same-worlds model of science and religion, 131–134, 138
Sarich, Vince, 10–11
Satan, 68, 196, 197
Schlessinger, Dr. Laura, xiii, xv–xvi
Schroeder, Gerald, 100
Schultz, Dwight, 171
Schwartz, Hillel, 201
science
as art of the soluble, 7
and Extropians, 60
and faith, 123, 130
and Genesis, 100
interest in, and religiosity, 270, 271
and patterns, 34
and religion, 123, 126–141
and religion, models of, 129–135, 138
and religious intensity, 80
rise of, and decline of religion, 22–23
as a type of myth, 153
Science, Age of, 45, 151
science, sacred, 29–31
*Science of God, The* (Schroeder), 100
scientism, 29, 60, 61, 115, 138
scientists
belief in God and immortality, 72–75, 258–259, 275
Scopes "Monkey Trial," 109, 123, 130, 131
Scott, Eugenie, 115, 133
Second Coming, 203, 206, 210
*Secret Origins of the Bible, The* (Callahan), 284
secular apocalyptic scenarios, 193–194
secular millennialists, 197, 206
secular religion, 29, 61
secular religion of progress, 231
secularization of religion, 22–24
Segal, Nancy, 64
*Selfish Gene, The* (Dawkins), 69–70
self-deception, 54
separate-worlds model of science and religion, 133–135, 141
*Seven Promises,* 26
*Seven Theories of Religion* (Pals), 282
sexes, equality between the, 211
shapeshifting, 155
Sherpas, 47
Sherrill, Martha, 259–262
Shroud of Turin, xiv
shunning, 159
Shute, Nevel, 17
Siberian Ghost Dance, 184
*Signature of God, The* (Jeffrey and Rambsel), 120
Simpson, Nicole Brown, 48
Sitting Bull, 176, 180, 181
situational attribution, 85
*Skeptic* magazine, xiii–xiv, xv, xxi, 63, 141

skeptics
on belief and nonbelief in God, 74–78, 268–269, 275
Skeptics Society, xiii–xiv, xv, xxi, 63, 74, 75
*Skeptics and True Believers* (Raymo), 101
Skeptics Survey of belief in God, 75–78
Skinner, B. F., 51
skyhooks, 222
Smith, Adam, 23
Smith, George, 280–281
Smith, John Maynard, 167
Smolin, Lee, 108–109
snakes
fear and fascination of, 147
Snowden, David, 254
Sober, Elliott, 165
social class, 263–264
social environment, 146
social scientists, 263–266
*Sociobiology* (Wilson), 146
socioeconomic status (SES), 263
*Sociology of Religion, The* (Weber), 282
soft-core atheists, 24
soluble, art of the, 7
Something There, 61, 63
soul, 106, 129
Soviet Union, 225
spandrels, 38–39, 69
species altruism, 164
speech
evolution of, 157
Spence, Gerry, 148
Spilka, Bernard, 282
Spinoza, Baruch, xxi
spirit, nobility of, 138–141
spirituality
biology of, 65
*Star Trek* (TV program), 151
*Statistical Science* journal, 117
Stearns, Peter, 201
Stein, Gordon, 283
*Stephen Hawking's Universe* (TV series), 29
Stewart, Ian, 217, 218
Stewart, Jimmy, 235
storytelling, 142–170, 207, 208
*see also* mythmaking
*Street Science* (radio show), 171
Sulloway, Frank, 75, 76, 78, 81, 82, 263–265, 267
Sulloway–Shermer survey of belief in God, 76–85
*Summa Theologica* (Aquinas), 283
supernatural, 35, 115
supernatural consolation, 61
superstition, 41–45, 47
supply-side religion, 23–24
survival
and Belief Engine, 38
and myths and superstitions, 41
Swinburne, Richard, 283
Symons, Donald, 10
synapses, 62

*Talking to Heaven* (Van Praagh), 27, 49
taste aversion, 146
Tavibo, 175
Tavris, Carol, 85
Taylor, A. E., 281
television programs
and religious or spiritual themes, 27
temporal lobe epilepsy (TLE), 66
temporal lobe transients, 66–69
tender-mindedness
and religiosity, 271
theism, theists, 8, 9, 75, 281
theology and cosmology, 132
*Theories of Primitive Religion* (Evans-Pritchard), 282
*There's a New World Coming* (Lindsey), 197
thinking errors and hits, 38–41, 45, 148, 153, 206
Thomists, 7
Thompson, Damian, 207
Thorne, Kip, 29, 30, 104
thought contagion, 70
*Thus Spoke Zarathustra* (Nietzsche), 16–17
Tillich, Paul, 11
*Time* magazine
cover stories on religion, 25

"Is God Dead?" cover story, 7, 17–21
*Time/CNN* end of the world poll, 206, 211–212
Tipler, Frank, 103–108, 132, 231
Tit for Tat program, 167
Tooby, John, 37, 146, 149
Torah, 117, 120–121
trigger effect, 227, 236
triskaidekaphobia, 47
Trobriand Islanders, 41–43
*Trumpet Shall Sound, The* (Worsley), 186
truth
believing, 38
rejecting, 38, 40
through revelation, 136–137
*see also* thinking errors and hits
*Truth Cannot Contradict Truth* (John Paul II), 127–128, 133
*20/20* (TV program), 54–55, 56, 57, 102
twins and religiosity, 63–64, 264
*2001: A Space Odyssey* (Clarke), 19
Tylor, Edward, 169

UFOs, 172–173, 204, 205
Unamuno, Miguel, 97
unconditional apocalypticism, 197
Unitarians and belief in God, 76
United States Energy Department, 199–200
universal spirit, 22
universe, 105, 135
*Unsolved Mysteries* (TV program), 53
*Unto Others* (Sober and Wilson), 165
upbringing
and belief in God, 274–277
and religiosity, 264, 270, 271, 274, 277
urban legend, 150
*U.S. News and World Report*
cover stories on religion, 25
Second Coming poll, 206
Ussher, Archbishop James, 198–199, 202–203

Van Praagh, James, 27, 48–57
*Vanishing Hitchhiker, The* (Brunvand), 150
Variable Ratio Schedule of reinforcement, 51
Velikovsky, Immanuel, 141
Virgin Mary
appearances of, 34, 35
Vyse, Stuart, 45, 47, 61

Wade, Carole, 85
Walford, Roy, 60
Wallace, Alfred Russel, 229
Wallace, Anthony, 140
Walsh, Neale Donald, 27
Walters, Barbara, 55
warm readings, 52
Wason, Peter, 148–149
Watchtower Society, 203
*Waterworld* (film), 193
*Wealth of Nations, The* (Smith), 23
Weber, Max, 282
Weinberg, Steven, 215
werewolf myths, 154–155
*When Bad Things Happen to Good People* (Kushner), 5, 283
Whewell, William, 144
White, Andrew Dickson, 130
White, Michael, 102
*Who Needs God* (Kushner), 283
*Who Wrote the Bible?* (Friedman), 120, 283
*Who Wrote the Gospels?* (Helms), 284
*Who Wrote the New Testament?* (Mack), 283
*Why People Believe Weird Things* (Shermer), 67
why questions, 144–145
*Whys of a Philosophical Scrivener, The* (Gardner), 97
Williams, George, 165
Williams, Ted, 231–232
Wilson, Bill, 96
Wilson, David Sloane, 164, 165
Wilson, Edward O., 29, 140–141, 146–148, 156
Wilson, Jack, *see* Wovoka
*Witchcraft* (Evans-Pritchard), 39

Witchel, Alex, 48
Witham, Larry, 72–74
Witztum, Doron, 117
Wodziwob, 175
Wojcik, Daniel, 197
*Wonderful Life* (Gould), 216–219, 221, 222
world, end of the, 191–213
Worsley, Peter, 186, 187
Wounded Knee, 181–182
Wovoka (Jack Wilson), 175–179
Wulff, David, 81, 282

Xhosa tribe, 183

Yanomamö, 41, 42, 159
year A.D. 1000, 200–202
Yeats, William Butler, 192
Yellow Bird, 182

# About the Author

Michael Shermer, Ph.D., is the founding publisher of *Skeptic* magazine (www.skeptic.com), the director of the Skeptics Society, the host of Skeptics Lecture Series at Caltech, and a contributing editor of and monthly columnist for *Scientific American*. He is the author of *Why People Believe Weird Things*, *Denying History*, and *The Borderlands of Science*. He lives in Southern California.